# Sustainability in Denim

AF736963

## The Textile Institute Book Series

Incorporated by Royal Charter in 1925, The Textile Institute was established as the professional body for the textile industry to provide support to businesses, practitioners and academics involved with textiles and to provide routes to professional qualifications through which Institute Members can demonstrate their professional competence. The Institute's aim is to encourage learning, recognise achievement, reward excellence and disseminate information about the textiles, clothing and footwear industries and the associated science, design and technology; it has a global reach with individual and corporate members in over 80 countries.

The Textile Institute Book Series supersedes the former 'Woodhead Publishing Series in Textiles', and represents a collaboration between The Textile Institute and Elsevier aimed at ensuring that Institute Members and the textile industry continue to have access to high calibre titles on textile science and technology.

Books published in The Textile Institute Book Series are offered on the Elsevier web site at: www.elsevier.com/books-and-journals and are available to Textile Institute Members at a substantial discount. Textile Institute books still in print are also available directly from the Institute's web site at: www.textileinstitute.org

To place an order, or if you are interested in writing a book for this series, please contact Matthew Deans, Senior Publisher: m.deans@elsevier.com

## Recently Published and Upcoming Titles in The Textile Institute Book Series:

Handbook of Technical Textiles, Volume 1, 2nd Edition, A. Richard Horrocks and Subhash C. Anand, 9781782424581
Handbook of Technical Textiles, Volume 2, 2nd Edition, A. Richard Horrocks and Subhash C. Anand, 9781782424659
Geotextiles, Robert Koerner, 9780081002216
Advances in Braiding Technology, Yordan Kyosev, 9780081009260
Antimicrobial Textiles, Gang Sun, 9780081005767
Active Coatings for Smart Textiles, Jinlian Hu, 9780081002636
Advances in Women's Intimate Apparel Technology, Winnie Yu, 9781782423690
Smart Textiles and Their Applications, Vladan Koncar, 9780081005743
Advances in Technical Nonwovens, George Kellie, 9780081005750
Activated Carbon Fiber and Textiles, Jonathan Chen, 9780081006603
Performance Testing of Textiles, Lijing Wang, 9780081005705
Colour Design, Janet Best, 9780081012703
Forensic Textile Science, Debra Carr, 9780081018729
Principles of Textile Finishing, Asim Kumar Roy Choudhury, 9780081006467
High-Performance Apparel, John McLoughlin and Tasneem Sabir, 9780081009048

The Textile Institute Book Series

# Sustainability in Denim

*Edited by*

***Subramanian Senthilkannan Muthu***

The Textile Institute

ELSEVIER

WP
WOODHEAD
PUBLISHING
An imprint of Elsevier

Woodhead Publishing is an imprint of Elsevier
The Officers' Mess Business Centre, Royston Road, Duxford, CB22 4QH, United Kingdom
50 Hampshire Street, 5th Floor, Cambridge, MA 02139, United States
The Boulevard, Langford Lane, Kidlington, OX5 1GB, United Kingdom

Copyright © 2017 Elsevier Ltd. All rights reserved.

No part of this publication may be reproduced or transmitted in any form or by any means, electronic or mechanical, including photocopying, recording, or any information storage and retrieval system, without permission in writing from the publisher. Details on how to seek permission, further information about the Publisher's permissions policies and our arrangements with organizations such as the Copyright Clearance Center and the Copyright Licensing Agency, can be found at our website: www.elsevier.com/permissions.

This book and the individual contributions contained in it are protected under copyright by the Publisher (other than as may be noted herein).

**Notices**
Knowledge and best practice in this field are constantly changing. As new research and experience broaden our understanding, changes in research methods, professional practices, or medical treatment may become necessary.

Practitioners and researchers must always rely on their own experience and knowledge in evaluating and using any information, methods, compounds, or experiments described herein. In using such information or methods they should be mindful of their own safety and the safety of others, including parties for whom they have a professional responsibility.

To the fullest extent of the law, neither the Publisher nor the authors, contributors, or editors, assume any liability for any injury and/or damage to persons or property as a matter of products liability, negligence or otherwise, or from any use or operation of any methods, products, instructions, or ideas contained in the material herein.

**Library of Congress Cataloging-in-Publication Data**
A catalog record for this book is available from the Library of Congress

**British Library Cataloguing-in-Publication Data**
A catalogue record for this book is available from the British Library

ISBN: 978-0-08-102043-2 (print)
ISBN: 978-0-08-102044-9 (online)

For information on all Woodhead Publishing publications
visit our website at https://www.elsevier.com/books-and-journals

ELSEVIER
Book Aid International
Working together to grow libraries in developing countries
www.elsevier.com • www.bookaid.org

*Publisher:* Matthew Deans
*Acquisition Editor:* David Jackson
*Editorial Project Manager:* Andrae Akeh
*Production Project Manager:* Poulouse Joseph
*Designer:* Victoria Pearson

Typeset by TNQ Books and Journals

*This book is dedicated to:*

*The lotus feet of my beloved Lord Pazhaniandavar.*

*My beloved late father.*

*My beloved mother.*

*My beloved wife Karpagam and daughters Anu and Karthika.*

*My beloved brother.*

*Last but not least,*
*To everyone working in the global textile supply chain*
*who make it sustainable.*

# Contents

# List of contributors

**M.J. Abreu** University of Minho, Guimarães, Portugal

**K. Amutha** Bharathiar University, Coimbatore, India

**S.G. Annapoorani** Bharathiar University, Coimbatore, India

**S. Basak** ICAR-Central Institute for Research on Cotton Technology, Mumbai, India

**I. Braga** Federal University of Piauí, Teresina, Brazil

**V. Buscio** Universitat Politècnica de Catalunya-BarcelonaTech (UPC), Terrassa, Spain

**K.N. Chatterjee** The Technological Institute of Textile & Sciences, Bhiwani, India

**S.K. Chattopadhyay** ICAR-Central Institute for Research on Cotton Technology, Mumbai, India

**A.K.R. Choudhury** Retired Professor, Government College of Engineering and Textile Technology, Serampore, India

**C. Gutiérrez-Bouzán** Universitat Politècnica de Catalunya-BarcelonaTech (UPC), Terrassa, Spain

**T. Karthik** PSG College of Technology, Coimbatore, India

**J. Militky** Technical University of Liberec, Liberec, Czech Republic

**R. Murugan** PSG College of Technology, Coimbatore, India

**H. Pal** BPS Mahila Vishwavidyalaya, Sonipat, Haryana, India

**A.P. Periyasamy** Technical University of Liberec, Liberec, Czech Republic

**R. Rathinamoorthy** PSG College of Technology, Coimbatore, India

**K.K. Samanta** ICAR-National Institute of Research on Jute and Allied Fibre Technology, Kolkata, India

**D. Sharma** Amity University Haryana, Gurgaon, India

**J. Wiener** Technical University of Liberec, Liberec, Czech Republic

# Preface

I am sure that I do not have to begin this preface with an introduction to denim because it is an icon and one of the most familiar products within the textile industry that attracts all age groups. The idea and interest in working on such a book, entirely devoted to the environmental impacts and sustainability aspects of denim, have been with me since 2005, when I was working in the manufacturing sector of denim. During that period, the concept of environmental assessment and tools such as life-cycle assessment and an interest in sustainability were not as developed and familiar as they are now. I have been witnessing the environmental damage created by this denim industry since 2005, and at the same time I am aware that many manufacturers that produce denim are keen to alleviate these impacts.

There are a lot of attempts being taken globally to work on quantifying and reducing the impacts created by a denim product over its entire life cycle. To make denim more sustainable, encompassing the use of sustainable raw materials for the production of denim, sustainable processes/manufacturing methods which have comparatively fewer impacts, and the use of sustainable chemicals in the entire production chain of denim, knowledge of the sources of impacts and understanding of the measures to mitigate them are essential. When we look at the entire life cycle of a denim product, the effort to make denim sustainable should come from every key player involved: i.e., raw material manufacturers (including chemicals, auxiliaries, and packaging), manufacturers (spinning, weaving, processing, and garment mills), distributors and, importantly, consumers also have a major role in terms of using, caring, maintaining and disposing of a denim product in a sustainable way.

With this idea and framework in mind, the chapters in this book are planned in such a way that readers will get the complete package of sustainability aspects related to denim from one source. Complete dissemination of information to readers covering basic aspects to the advanced sustainability aspects of denim is attempted through 12 informative chapters written by academic scholars.

I take this wonderful opportunity to thank all of the contributors of this book for their earnest efforts and hard work to bring out this book successfully. I am sure that the readers of this book will be highly benefited.

**Dr Subramanian Senthilkannan Muthu**

# Introduction to denim

# 1

*S.G. Annapoorani*
Bharathiar University, Coimbatore, India

## 1.1 Introduction

Denim, a cotton woven fabric, may be the most considered article of style today. Denim is cut and sewn in an assortment of clothing types fit for all ages, seasons and events. This is what has made denim a profitable fashion item, and the increase in the quantities of clients buying it is truly obvious. Blue jeans are presumably what one considers when denim is concerned; however the use of denim has really gone pass the standard pattern. Also, with the development of weave structure and comfort in denim, it is now the most favoured classification of fabric employed in the style business. Also, as expected, 10% of aggregate cotton generation in the world is meshed into denim.

One noteworthy feature of denim is the sum and assortment of off-structure impacts that can be composed on it. There are many denim wash innovations that confer variable shading results. The most fascinating portion of denim fabric is presumably its inconsistency. Whereas most producers work hard to keep fabrics free from auxiliary and configuration mistakes, denim is purposely made somewhat disgusting and faddish. Truth be told, a denim jeans producer spends more to scratch out and tear up a couple of jeans than what is spent to sew it. The utility of denim is also a major explanation for its value. Although it was developed for cowboys, who used it as work wear, now sewn denim is even being used as attractive tights for young ladies.

Denim has achieved so much fame that if we glance around, we will doubtlessly see some individual wearing denim in our midst. Currently, more than simply supplementing a tough style, the denim has ended up as appropriate for any event. Denim is being worn regardless of demographic contrasts. Denim material is identical to familiar jeans and is a tough twill material that has the well-known slanting ribbing. There are around 20 denim producers in India alone, taking into account local and foreign markets. Manufacturing amenities are quickly getting up to speed in Bangladesh, Pakistan and Vietnam. Today denim is available in shades of blue, dark and cocoa; each has diverse impacts produced by washing.

Denims are stylish wear for everyone. Today's denim fashion comes in an assortment of hues, styles and textures. Without a doubt, fashion designers' attire is comparatively more costly, yet most of the time it is justified regardless of the expense.

It is a great opportunity to leave outdated jeans behind and experiment with the most recent styles that have arrived. Now and again one needs to make a special effort to locate the perfect denim wear, whether it is jeans, denim coats, shirts or even shorts. Designer denim will knock some people's heads off each time it is worn to demonstrate the venture one has made.

**Sustainability in Denim. http://dx.doi.org/10.1016/B978-0-08-102043-2.00001-0**
Copyright © 2017 Elsevier Ltd. All rights reserved.

There are some solid reasons why some people want to spend more to purchase famous designer denim than the typical stuff accessible to the market. The foremost reason is the quality. The fabrics, sewing methods and patterns are special to such an extent that one can discover the best quality with the designer brands. They use the best fabrics and most are cut and sewn independently, in contrast with the normal denim organizations, which may use shabby fabrics and line in huge parcels. Besides, designer jeans will last more than ordinary jeans and will not tear or become dull immediately even when they are worn consistently.

One primary explanation for wearing fashioner denim is style remainders. It is obvious that designers' items showcase the most recent patterns on the market, and wearing one of them will put the individual in an attractive light. Fashion designers show new patterns each season to overhaul one's design sensibilities.

The designer brands guarantee that their apparels give the best fit to their clients. Because top-notch fabrics are used, the jeans fits better over a time frame, as opposed to turning out to be tight or loose. Thus, numerous individuals select designer denim. In comparison, the standard jeans company delivers jeans that are less costly but may not be consistent as to size, fit and fabric. In any case, there is a consistency in the denims of the fashioner brands even after years.

Various new trends are launched for every season by designer brands and are craze among men, ladies and kids. Shaded denims have turned out to be exceptionally prevalent. It is stunning to see the quantity of à la mode and appealing items the planner denim brands have come up for genuine denim lovers.

New and innovative patterns, fine shapes and various hues are the most recent choices to fulfill each client. Some of the new patterns in denims for ladies are designed denims, denim T-shirt dresses, jeans with zipper pockets, square-shaped jean vests, jeans with molded joints, dyed print denims and so forth. Furthermore, energetic hued leggings for ladies are in vogue. Tribal and ikat prints are the pace setters. New lightweight denims give a milder vibe to the skin. High midriff and lower leg length jeans are generally appreciated. Body-embracing denim will illuminate the glitz remainder. In addition, Hudson brings ribbed fabric denims that will give a smooth feel. One can browse these examples, styles, cuts and hues and bring home one's most loved creator brands.

Denims for children offer a variety of garments. Thin jeans, obliterated jeans, sweetheart jeans, shaded jeans and leggings are the most sizzling patterns for children. In addition, biker-styled denim is incredible for both young men and young ladies. Different patterns such as various identification and washed patches, and bare-shade denim are especially in style.

Today's children are adding taste to their look with brilliant jeans. They are accessible in hot pinks, reds and blues and an assortment of brilliant shades. For young ladies, other than jeans, skirts, waistcoats and leggings are most sought after; for kids, denim coats, dungarees, and shirts are the top of the season. For young ladies, leggings are the best alternative, and are accessible with ease and style.

A denim coat represents solid manliness and masculinity for men. It can be well arranged with any garment. They give a tough and easygoing look and are the best to keep the chill away. Denim coats have been well known since the 1980s and each fashion brand has presented its own designer version.

The trend in denim for men is to have more enumerating, bleached impacts, appliqués and so on that can withstand numerous new washes. Mainstream designer brands such as Valentino, Louis Vuitton and YMC have come up with exemplary denim coats for men, whereas Acne and Richard Nicoll have gone for strange constructions such as a round denim coat with an edited round neck and a stone-washed coat offering significance to minor details.

'Denim' has been synonymous with 'jeans' since time immemorial. It is unrealistic to separate one word from the other. Denim jeans are only the most commonly worn form of denim apparel.

The word 'denim' refers to a colour, not refer the type of fabric. The shading indigo is blended with a large group of different hues, for example, blue, dark, white or dark, to give a variety of hues.

At first, denim jeans were just used as easygoing wear; however denim jeans are currently entering the meeting rooms of vast organizations as 'easygoing formals'. Casual-formals are particularly intended for individuals to whom solace and style mean the same thing.

In the past denim apparel was considered to be weekend parlour wears; however individuals now wear them for work as well as gatherings. Stroll into the ritziest of gatherings, and more than half of people there may be found in denim. A straightforward fabric has snatched the consideration of everybody and has had an enormous effect. It is a worthy bit of apparel to a house spouse and also to a design-keen model on the runway. The denim business has been recognized as a gainful decision right from the retailer offering in the mass business sector to the elite-strength shop proprietor. At first it was considered to be a self-contradictory item. Although jeans have been acknowledged as essential in everyone's closet, the substantial sticker price was not considered worth it. Be that as it may, later on, when more producers entered the business and the opposition warmed up, denim array was sold with a moderate sticker price to suit the necessities of each client. Denim jeans are especially vital to youngsters of today. They are an image of youth, dynamic quality and utility. Thus, denim jeans can be found at pretty much every retail location and markdown shop available.

This chapter deals with the origins of denim and its history, manufacture and market Status.

## 1.2 What is denim?

Denim is a solid cotton warp confronted material in which the weft goes under two or more warp strings. This twill weaving produces a corner-to-corner ribbing that distinguishes it from cotton duck.

The most well-known denim is indigo denim, in which the twist string is coloured whereas the weft string is left white. As a consequence of the twist-confronted twill weaving, one side of the material is ruled by the blue twist strings and the opposite side is overwhelmed by the white weft strings. This causes Levi's to be white within. The indigo colouring process, in which the centre of the twist strings stays white, makes denim's mark-blurring qualities.

Denim is a strong cotton twill material ordinarily used to make jeans, overalls, and other clothes. To make denim fabric, the weft (flat strings) goes under two or more twists (vertical strings). This procedure makes the slanting ribbing of denim that distinguishes it from cotton duck, another twill fabric.

Denim is the name of the fabric that is used to make jeans. It is a fabric made of cotton twill that is 100% cotton and extremely agreeable. This fabric is employed everywhere throughout the world to make jeans, coats, shirts, totes, sacks and numerous different garments for men and ladies of any age. Nonetheless, 'jeans' are the clothing that have given this fabric a personality. This fabric is woven with twist yarn that is blue cotton and a filling yarn that is white cotton to make twill that has corner-to-corner running parallel lines. Denim is so highly used to create ready-made pieces of clothing and extras that today it is found in all stores offering ready-made wear.

The word 'denim' originated from is a constriction or subordinate of the French expression *serge de Nîmes*. Denim was generally shaded blue with indigo colour to make blue 'jeans'; however 'jean' then meant an alternate, lighter cotton material. The contemporary use of 'jean' originates from the French word for Genoa, Italy (*Gênes*), from which the principal denim trousers were made.

Denim is a kind of cotton material known for its usefulness in Levi's and other garments. It requires a solid twill weave with a trademark inclined ribbing. Initially employed for labourers' garments, denim is now universal and has even entered the universe of high form. Nearly everyone has more than one piece of clothing made from this fabric in the storage room nowadays.

Today design is deficient without denim. Denim comes in all structures, looks and washes to coordinate with each form of dress. It is hard to believe that the same denim was initially used in apparel for jeans and overalls worn by mine workers on the West Coast of the United States. Various innovative components have been added to make denim the design symbol that it is today, incorporating incomprehensible changes in **spinning**, **weaving**, **finishing**, etc.

### *1.2.1 Types of denim*

Although the first denim was a 100% cotton serge material, you can now get it in an assortment of materials, including mixes that give you the same magnificent look of 100% cotton denim with some extraordinary extra components. Denim's one of a kind look originates from the rich indigo blue in some shade woven together with white strings to give the depth that individuals partner with denim. Today, some denims no longer use indigo, but instead different hues with the white restricting strings, delivering denim in a rainbow of shades.

Types of denim are broadly categorized as:

1. dry
2. selvage
3. stretch
4. poly
5. ramie cotton

1. **Dry denim**
   Dry or crude denim, rather than washed denim, is a denim fabric that is not washed subsequent to being coloured amid its creation. Most denim is washed subsequent to being created into a piece of attire with the specific end goal of making it gentler and taking out any shrinkage which could make it not fit after the owner washes it. Notwithstanding being washed, nondry denim is at times misleadingly 'troubled' to accomplish a ragged look. A significant part of the claim for dry denim is that, with time, the fabric will blur in the way of industrial facility troubled denim. With dry denim, notwithstanding, such blurring is influenced by the body of the individual who wears the jeans and the exercise in his or her everyday life. This makes what numerous jeans-lovers feel to be a more unique, one of a kind look than predistressed denim. To encourage the common troubling procedure, a few wearers of dry denim will frequently refrain from washing their jeans for over 6 months; however it is no needed for blurring. Prevalently found in premium denim lines, dry denim speaks to a small specialty in the general overall market.
2. **Selvage denim**
   Selvage denim (also called selvedge denim) is a sort of denim which shapes a perfect regular edge that does not disentangle. It is generally introduced in the unwashed or crude state. Normally, the selvage edges will be situated along the outside crease of the jeans, making it obvious when sleeves are worn. Although selvage denim is not totally synonymous with unwashed denim, the nearness of selvage ordinarily infers that the denim used is of a higher quality. 'Selvage' originates from the expression 'selfedge' and signifies that denim is made on old-style transport looms. These looms weave fabric with one continuous cross thread (the weft) that is passed back and forth all the way down the length of the bolt. As the weft circles once again into the edge of the denim it makes this 'selfedge' or selvage. Selvage is alluring because the edge cannot shred like lower-quality denims that have separate wefts which leave an open edge that must be sewn. Transport approaching is an additional tedious weaving process that produces denim of a more tightly weave bringing about a heavier weight fabric that endures. Transport looms weave a smaller bit of fabric, and in this manner a more extended bit of fabric is required to make a couple of jeans (around 3 yards). To amplify the yield, customary jean creators use the fabric the full distance to the selvage edge. At the point where the sleeve is turned up, the two selvage edges, where the denim is sewn together, can be seen. The selvage edge is typically sewn with hued string: green, white, chestnut, yellow and (red is the most widely recognized). Fabric factories used these hues to separate fabrics.
3. **Stretch denim**
   Stretch denim is more often than not around 98% cotton and 2% Spandex for a touch of that easygoing stretch we love as a whole. This blend gives wonderful ease of movement and at the same time some support for those 'trouble spots' such as around the hips or thighs. Stretch denim jeans are one of the quickest developing segments for ladies' jeans manufacturers.
4. **Poly denim**
   The poly blend is for individuals who love the look of denim yet favor polyester mixes that wash and dry rapidly and are of lighter weight and somewhat dressier. As a rule these speak to a somewhat more established business sector, but on the other hand are discovering support for jeans suits and so on when the look is intended to be 'dressy but easy-going'.
5. **Ramie cotton denim**
   Ramie cotton denim is found in an assortment of blends, with a wide value difference. Ramie is a plant fibre more often included because it diminishes wrinkling and adds a luxurious brilliance to the fabric. It is not as solid as cotton, however, so it must be mixed with this more grounded material with the goal of standing up as a denim material.

## 1.3 History of denim

The history of jeans goes back even before the time of Levi Strauss, which has a famous relationship with it. Denim is a fabric used to make jeans. Although denim is all the more prevalently connected with America, India has the credit for its initial history. At first jeans were requested for Italian mariners from the French city of Nîmes. The fabric was coloured from a blue colour made of the indigo plant that was developed in India.

Jeans has an even more intriguing association with India. Over time, the denim clothes Italian mariners changed into boiler suits to secure their garments. This general clothing was produced using coarse and undyed calico fabric. This fabric is essentially connected with Calicut in India: thus the name 'calico'. These overalls were for the most part fabricated in the Dongri range in Mumbai, and henceforth gained the name 'dungaree'.

The fabric *serge de Nîmes, made in the French city of Nîmes,* has paved the way for denim. It has been associated with America since the late 18th century, when it was given a blue hue with indigo colour to make blue 'jeans', a kind of cotton jeans. Jeans originate from 'Qualities', a name given by the French to Genoa and the general population from Genoa where the cotton trousers were made. In the 1800s, in the season of the Gold Rush, American gold excavators required garments that were solid, kept going longer and did not tear effortlessly. Levi Strauss, a specialist, and Jacob Davis, a tailor, supplied excavators with denim jeans that were produced using sturdy material and fortified with bolts at the spots where jeans tended to tear, which extended the lifetime of the jeans. This signaled the start of the legend of jeans; the brand Levi Strauss is still immensely popular.

Denim was initially used for garments worn by specialists as a result of its high sturdiness. It turned became famous in the 1930s when Hollywood began making cowboy motion pictures in which the performers wore jeans. With the start of the World War II, the creation of jeans dropped but the world still required denim when American fighters began wearing jeans when they were on leave. When the war finished, different companies that made denim began operating, including Wrangler and Lee. Youngsters began wearing denim in the 1950s as a sign of defiance. This style was also motivated by Hollywood: for example, by Marlon Brando in his 1953 film *The Wild One*, and by James Dean's motion picture *Rebel Without a Cause* in 1955. Some public places such as schools and theatres banned jeans because of what they symbolized. Denim crossed from counterculture to design in the 1960s and 1970s when producers began to make diverse styles with jeans.

Unique denim was coloured using the plant *Indigofera tinctoria*. Currently denim is coloured with engineered indigo. Denim is regularly coloured with indigo and dried many times over to get more grounded shading that will not blur rapidly. After the denim is made into a garment it is frequently washed to make it milder and to lessen or prevent shrinkage. Denim that is not washed is called dry or crude denim. Denim blurs over time and achieves a ragged outlook which is regularly coveted as a design point of interest. Some denim is misleadingly distressed, so a ragged outlook can be obtained even before wear. Denim that is not made of unadulterated cotton but has a similar flexible quality (frequently elastin) is stretch denim. Other than indigo colouring, denim can be hued with sulphur colouring, which is utilized to colour denim as a part of hues other than indigo.

Denim is more than just a cotton fabric; it inspires the hearts of students of history, planners, adolescents, movie stars, columnists and authors.

Legend and truth are interlaced when researchers examine the origin of the name 'denim' itself. Most reference books say that denim is an English derivation of the French *serge de Nîmes*, a serge fabric from the town of Nîmes in France. Be that as it may, a few researchers have started to scrutinize this convention.

*Serge de Nîmes* was known in England before the end of the 17th century. The question then arises: Is it true that this is fabric was imported from France or is it an English fabric bearing the same name? According to Ms Gorguet-Ballesteros, fabrics which were named for a specific geographic area were regularly made somewhere else as well; the name was employed to loan a specific cachet to the fabric when it was offered as available for purchase. Accordingly *serge de Nîmes* bought in England was likely also made in England and not in Nîmes, France.

There remains the topic of how 'denim' is famously thought to be derived from *serge de Nîmes*. *Serge de Nîmes* was made of silk and fleece; however denim has always been made of cotton. What we have once more, I believe, is a connection between fabrics that exists only in name; however both fabrics are a twill weave. Is the genuine root of the word 'denim' *serge de nim*, which means a fabric that looked like the part-fleece fabric called *nim*? Was *serge de Nîmes* all the more outstanding, and is it safe to say that this was word was mistranslated when it crossed the English Channel? On the other hand, did British traders choose to give a zippy French name to an English fabric to give it more cachet? It is reasonable to expect that we will never truly know. To confound things considerably more, there also existed at this time a fabric known as 'jean'. Research on this material demonstrates that it was a fustian (a cotton, cloth and/or fleece mix) and that the fustian of Genoa, Italy, was called 'jean'. Here we see proof of a fabric being named from the very beginning. It was evidently prominent, and was imported into England in vast amounts during the 16th century. Before the end of this period, jean was being created in Lancashire. By the 18th century, jean material was made totally of cotton, and was used to make men's attire, esteemed particularly for its property of solidity even after numerous washings. Denim's ubiquity was also on the ascent. It was more grounded and costly than jean; however the two fabrics were fundamentally the same except for one noteworthy thing: denim was made of one shaded string and one white string whereas jean was woven of two strings of the same shading. Moving over the Atlantic, we discover American material factories beginning on a small scale in the late 18th century, for the most part as an approach to begin autonomy from foreign producers.

From their earliest beginnings, cotton fabrics were an essential segment of such product offerings. The industrial resources in the state of Massachusetts facilitated weaving both denim and jeans. President George Washington visited this factory in 1789 and saw a demonstration of the apparatus, which wove denim, which had both twist and fill made of cotton.

One of the initially printed references to 'denim' in the United States was in this same year: a Rhode Island daily paper covered the neighbourhood generation of denim (among different fabrics). The book, *The Weavers Draft Book and Clothiers Assistant*, published in 1792, contains specialized representations of weaving strategies for an assortment of denims.

In 1864, an East Coast wholesale house publicized that it conveyed 10 different types of denim, including 'New Creek Blues' and 'Madison River Browns'. *Webster's Dictionary* of that year contained 'denim', alluding to it as 'a coarse cotton penetrating utilized for overalls, and so on'.

Research demonstrates that jeans and denim were two altogether different fabrics in 19th-century America. They also varied in they way they were used. In 1849, a New York apparel producer publicized topcoats, vests or short coats in chestnut, olive, dark, white and blue jeans. Fine trousers were offered in blue jeans; overalls and trousers made for work were offered in blue and fancy denim. Other American notices show working men wearing apparel that delineates this distinction in use amongst jeans and denim. Mechanics and painters wore overalls made of blue denim; working men (counting those not occupied with difficult work) wore more custom-made trousers made of jeans.

Denim thus appears to have been saved for work garments, when both solidness and comfort were required. Jeans were a work-wear fabric without the additional advantages of denim. In *Staple Cotton Fabrics* by John Hoye, distributed in 1942, jeans are recorded as cotton serge with a twist and woof of the same shading, used for overalls, work and game shirts, specialists' and medical attendants' regalia and as linings for boots and shoes. Hoye says, 'The most critical fabric of the work-dress gathering is denim. Denims are solid and serviceable; they are especially solid in the twist course, where the fabric is subjected to more noteworthy wear than the filling'.

The tale of jeans starts in the mid-19th century at the same time in two European cities: from Nîmes in France, where the fabric itself was made, the so-called *serge de Nîmes*, which prompted the name 'denim'; and from Genoa in Italy, from where the colour comes, French *bleude gênes*, which prompted the expression 'blue jeans'. The fabric and shading met up to become the jeans, denim.

In these two urban areas the solid and safe fabric began being used for the creation of work wear, particularly for mariners and dockworkers. This blue fabric, which eventually supplanted traditional chestnut and beige work wear, set up itself numerous kilometres far from Europe, in the United States. A youthful migrant from Germany called Levi Strauss began promoting the new article of clothing with bolted copper catches, which impressively fortified pockets. Jeans turned out to be increasingly famous in the 1920s and 1930s and became equally popular as a typical work article of clothing.

Little by little, an industry was set up and it came back to Europe as some recently established brands (for example, Morris Cooper Overall, later called Lee Cooper). As in the United States, the main European jeans were planned for the work market. In the 1950s, jeans entered another stage in the United States as an article of style. They made progress because of their appearance in Hollywood movies, where popular actors such as James Dean, Marlon Brando, and Marilyn Monroe wore them.

Along these lines, jean rapidly turned into an object of numerous youngsters' longing. This new design pattern came to Europe after World War II, not without challenges, which were to be expected because of the financial and trade issues of the time.

In the 1960s, brands such as American Levi's and Wrangler set up their first outlets in Europe in light of consistently increasing demand. Later, others such as Lee began operations. Until then colouring and manufacture of jeans was an exceedingly relentless and not extremely beneficial undertaking. However, in the 1960s new colouring frameworks were presented and the creation of the denim fabric expanded and was streamlined. Gradually, the principal European jeans brands showed up: Carrera in Italy, Lois in Spain and Chipie in France.

Although they were widely available, jeans were not sufficiently acknowledged in the style business. Gradually, by the start of the 1970s new brands showed up. They put emphasis on sales strategies that were completely different from those established so far and provided denim fashion with a definite impulse. It was combined in the 1980s with the formation of brands such as Goldie, Diesel, and Replay.

Today denim is a piece of clothing as well as a way of life. Brands clarify their commitment to the protection of the environment. The use of natural cotton and compound items with a low effect on the environment is getting to be normal, as well as modern procedures which minimize the use of both water and buildup produced by the procedure, from string assembly to the last part of making the piece of clothing. This is an approach which must be taken for the advantage of future generations.

Denim will continue 'reinventing' itself every day in each brand, each new collection, creating new fashion trends. It will probably stay on the theme of duty to the earth, similarly as common assets are worried and also the vocation of dyestuffs and substance items used as a part of the generation of the last jeans piece of clothing. Agreeable, unsafe, bold, unwinding, marvelous, alluring, forceful, savvy, easygoing, amusing, changing, lively, in vogue, down to earth, inventive … widespread: Every one of these descriptive words characterizes denim as a style in design as well as something more: **a lifestyle**.

## 1.4 Jeans, the cultural icon

### *1.4.1 Evolution of jeans*

Jeans have been and are an imperative item of garments. Denim articles of clothing appeared in the 18th century, when there was plentiful generation of cotton. In that period, it became significant because of its solidness, and because it was not easily torn, which benefited physical workers. In the 19th century denim was advanced by Levi Strauss, an outstanding name in the jeans world. Denim attire was made by them for diggers. The 20th century saw jeans as a craze of popular society. Western rancher motion pictures depicted individuals wearing Levi's. The 1950s saw the fabric turning into a frenzy of the high school populace. Producers used the circumstance to think of imaginative ideas such as new hues, weaving and fix works. In the 1960s and 1970s, bellbottoms and hip hugger designs became well known.

In the 1980s denims began drawing the consideration of style fashion. New and charming examples were made and were conveyed to the limelight by mainstream models in ads. Puffy jeans, puffy shoulder braces and puffy hair turned into a style. In the 1990s jeans became a standout amongst the most looked-for article of clothing. To skim the cream of the denim market, producers then thought of different imaginative ways to use denim fabric, such as poodle skirts, tops, purses and so forth. During that time, denim deals soared. Currently denim is in every home around the world.

### 1.4.2 Denim as a fashion icon

From the place of normal clothing, denim has ascended to become a style symbol and is being shown off by design models. It has turned into an image of cutting-edge society. The transformation of jeans from a product to a design happened amid the 1990s, when jeans were processed into various different things, such as coats, shorts, skirts and so on, separate from typical jeans. Prior various models such as flower child bellbottoms and skinny legs, were famous. Current patterns are about an assortment and perspectives of denim as having intense styles to keep pace with the most optimized way of life.

### 1.4.3 Resurrection of denim as a fashion brand

Any style magazine will have photos of models wearing a couple of most recent designers' jeans. Today's generation give up spending money on other accessories only to save them for buying branded designer denim. What is the explanation for this enlivened resurgence? The exposure given by famous people is one. Celebrities are spotted wearing designer denim wears even for events where formal attire is normally expected. Gwyneth Paltrow, a famous actress, was seen wearing a couple of pairs of jeans to New York fashion shows. Later, the blue trendy turned out to be popular to the point that it was even called Gwyneth Jeans. Famous stars such as Jennifer Aniston, Jennifer Lopez, Kate Hudson, Britney Spears, and Brittany Murphy spotted wearing jeans clothing enhance the fame of denim.

Denim makers also have an impact on this area by continually redesigning the business sector with new items. For example Levi Strauss did away with its trademark image of jeans and Lee Cooper is thinking about doing the same thing. Arvind Mill has gotten rid of its Flying Machine, and Raymond, its GAS, an Italian denim brand. Lee Cooper presented the Beatles Collection, in collaboration with British music. The trademarked jeans business sector was assessed to have a normal yearly development of 30%–40%.

### 1.4.4 Denim trends

A current pattern is crude denim, jeans which have not been washed or treated with chemicals. They are virgin fabric that comes straight from the weaving machine. These denims are hard at first and later grow and conform to the wearer's body, becoming distinctive to the client's shape. Boot legs and boot cuts are trendy and remain the most loved of men. The straight-legged zip fly is more favored than the catch fly. Boot-legged denims that touch the floor are men's favorite accoutrement. Skinny jeans are closely fit and give a great look to the wearer. Baggy-hole jeans are favored by both men and ladies. Selvage jeans originate from extraordinary weavers who use more string to make heavier fabric that does not easily become torn.

There are lightweight jeans that accompany thin legs and that scrunch up. Back pocket zippers, catch folds, and beautiful joins attract some individuals. Chime-legged jeans joined with a water wash are likewise in the business sector. Lightweight jeans with a multidirectional weave and nitty-gritty back pockets give a distinctive look to the wearer. Thin jeans in indigo and dark shades, and denim tops and coats are most sought after by ladies. Washed denims give an easygoing look and attractive persona that ladies wish to have. Denims are accessible in rainbow hues.

Aside from ordinary jeans, different things such as skirts, coats, tops, wallets, shoes, purses and so forth have also made their way into the style world as design additions. Denim packs are in vogue; they are tough and are prominent among the family of in-vogue individuals. Prewashed denim caps with buckled ribbon, hook and loop tape along with logos and motifs embroidered are in vogue.

## 1.5 Denim: manufacture

The name 'denim' originated from the city of *Nîmes* in France, where *serge de Nîmes* was made. Denim is produced using a vat colour, indigo, which is connected to cotton fabric in an inexactly held structure in layers. The manufacturing process of denim is similar to that of grey fabric up to the process of weaving, with the main difference that in the case of denim fabric, it is dyed at the stage of sizing, whereas in the case of grey fabric, the decision regarding the dyeing stage depends on the finished product. The details of each process are given in the flowchart.

**Flowchart of Denim Production**.

Spinning
↓
Bale warping
↓
Rope dyeing
↓
Long chain beaming
↓
Sizing
↓
Weaving
↓
Finishing
↓
Inspection & folding
↓
Packing
↓
Dispatch

### 1.5.1 Spinning

The beginning phase of denim production is opening and blending. Opening starts with baled cotton fibre being isolated into little tufts. A mix of cotton strands is made on every opening line. These parcels are chosen using US Department of Agriculture High-Volume Instrument information, and PCCA's Portable Computer And Communications Association one of a kind PC mixing programming produces the ideal yarn quality.

Cotton is conveyed via air suction from the opening and blending lines through extra cleaning and mixing machines, to the cards. The real functions of carding are to expel remote matter and short filaments, shape the cotton into a web, and change the web into a rope-like structure known as a bit.

The drawing procedure delivers a solitary, uniform fragment from six card bits. The extra mixing, paralleling of filaments, and cleaning in this procedure deliver a bit for open end and ring spinning. For ring spinning, the bit must go through an extra procedure called roving.

Cotton fibres are framed into a yarn by diffusive activity in open-end spinning. Singular fibres are laid down in the groove of a fast-spinning rotor and twisted into yarn. After the cotton strands are spun into yarn, the yarn is twisted into an expansive bundle.

#### 1.5.1.1 Open-end spinning

Open-end spinning machines have robots on every side which naturally piece up (repair broken closures). On an alternate track, they have another robot that naturally doffs (expels full bundles) and begins up another bundle. The size and nature of every yarn end are checked by the Barco Profile System to guarantee consistency.

#### 1.5.1.2 Ring spinning

In ring spinning, the spinning frames obtain roving by means of a travel framework from the wandering machine. Yarn is shaped from cotton filaments that are turned together subsequent to being drafted by going between three steel rolls and three elastic rolls. The yarn is then wrapped on a bobbin as it twists on a shaft using a voyager. The relationship between move speeds, explorer paces and shaft speeds controls the measure of turn in the yarn. Close-down levels and creation data are assembled by the Uster Ring Expert System. The turning outlines naturally doff bobbins loaded with yarn and send them to bundle winding.

Denim produced using this kind of yarn has yarn character and surface intrigue that cannot be accomplished with conventional open-end yarn.

### 1.5.2 Fabric weaving

#### 1.5.2.1 Grey yarn on cones

Ordinarily yarns obtained for weaving in cone structures are either from ring turning or from open-end turning in single or twofold overlap as required. For weaving, the yarn is classified into:

- warp
- weft

Normally, to weave, yarn used as twist ought to be adequately solid to withstand stretch and strains applied amid weaving operations. Thus they have count strength product and are further measured to build quality. The weft yarn is used straightforward on weaving machines and, if required, is also rewound to upgrade its execution in weaving.

#### *1.5.2.2 Warping on sectional/direct warping*

The twist yarn needs to be made into a sheet structure for the weaving machines. At twisting, the individual cones are put into the creel (the quantity of cones relies on fabric development) and yarn from individual cones is pulled together into a sheet structure, twisted on a barrel called distorting shafts (for direct distorting) or on weaving bars (for sectional warping). Ordinarily if the twist sheet has various shaded yarns it is prepared on a sectional distorting machine.

#### *1.5.2.3 Sizing of yarn in set/beam to beam position*

The object of sizing is to enhance the quality of the yarn by synthetically restricting the filaments with each other, enhancing its grating resistance limit by artificially covering the surface of yarn/strands. Furthermore, the number of strings in the wrapper's shaft sheet is less than against the number of strings required in the entire width of fabric. Subsequently sheets are duplicated by drawing yarns together from numerous twist shafts; one sheet is also made on the measuring machine. Approximately 8%–12% size material on twist string is regularly connected. This change in quality and frictional resistance that is normal for twist yarn is crucial because during weaving, yarn experiences serious strain, push and frictional operations.

#### *1.5.2.4 Drawing-in*

Weaving is fundamentally the interlacement of two sets (i.e., twists and weft) of strings in a desired arrangement. To achieve this interlacement, the twisted yarn sheet is bifurcated and opened as two layers per sheet and the weft string is embedded into the two twist sheets. This operation is called shedding. To perform shedding the twist yarn should go through the heald eyes of the heald shafts. This operation is known as **drawing-in**.

#### *1.5.2.5 Beam gaiting or knotting on a loom*

The drawn weavers beams are settled on weaving machines, strings are tied and head shafts are coupled. This operation is called beam gaiting. On the off chance that undrawn twist strings are hitched directly to the strings of completed bars, it is called knotting. These operations are crucial because ordinarily a weavers pillar can convey just a certain length of twist sheet on it, and so when it is woven, the entire length is becomes the fabric by the weaving machine. A further twist length is required to be sustained, which should be possible by tying or gaiting different bars on the weaving machine.

#### 1.5.2.6 Weaving

Weaving is interlacing two sets of yarn and making fabric. One set is called the warp, which is in the sheet frame; the other one is called the weft, which is embedded between two layers of warp sheet by a method for an appropriate carrier (i.e., shuttle, projectile, rapier, air current, water current and flow, and so on). The distinctive types of technologies accessible for weaving machines are:

- conventional shuttle weaving system by ordinary looms or automatic looms
- shuttleless weaving system by air jet/water jet/rapier/projectile

A shuttle loom is technology with significantly less generation by virtue of its slow speed and excessive wear and tear of machinery. The shuttle loom is now out of date. Denim is woven through a shuttleless weaving system using 96 Tsodakoma Corporation's ZAX-e type air jet looms, rapier looms or projectile looms.

##### Air jet looms

These kinds of weaving machines are an improvement in weaving technology in which weft inclusion is finished with the assistance of compressed air. A high weft addition rate up to 1800 metres for each moment is accomplished. Compared with rapier and projectile looms, these looms are less flexible but are efficient and are used as a part of mass textile manufacturing such as for denim.

### 1.5.3 Finishing

#### 1.5.3.1 Grey fabric

Woven fabric or grey fabric, as it is famously called, is wound on a cloth roll and is taken from weaving machines at specific intervals and checked on inspection machines for conceivable outcomes of any weaving deficiency. If such defects are seen anywhere in the fabric during review, certain remediations are taken at weaving, warping, sizing, and so forth. This is a quality control exercise.

#### 1.5.3.2 Denim fabric

Denim fabrics woven of 100% cotton would be exceptionally solid and strong. Generally blue denim is twist-confronted cotton fabric with 3 × 1 twill development with the twist coloured in a strong shading and the weft left uncoloured. The look and nature of the denim fabric should be enhance subsequent to colouring, the procedure of which varies from plant to plant. Typically the procedure of colouring manages the innovation of denim assembly.

Colouring of denim fabric happens at the sizing stage. There are well-known techniques for colouring denim fabric:

- rope dyeing
- sheet dyeing

An organization can use any of the strategies. In nations such as India, sheet dyeing is normally used to assemble its denim fabrics, which has the following advantages over rope dyeing:

- The innovation requires less capital venture.
- The innovation is demonstrable.
- The expense of generation is lower.
- The process time is lower.
- Sheet dyeing machines are easy to work with.

The main problem with sheet colouring is that there is an issue with choosing the selvedge shade variety.

#### 1.5.3.3 Sheet dyeing process

This procedure disposes of a few in-between procedures of rope dyeing. The yarn sheet is washed with chemicals such as caustic and washing soda, and after squeezing the excess water out, the yarn sheet is allowed to pass through dyeing troughs one time for oxidation and the development of dye on yarn. After colouring, the coloured yarn is washed again with water two to three times and finally is pressed before it goes through six drying chambers. The coloured yarn enters a starching gadget and the measuring is finished. After measuring, the estimated twist shaft enters weaving. After weaving, the woven denim fabrics undergoes different completing procedures composed of brushing, searing, washing, impregnation for dressing and drying. Brushing and singeing should eliminate impurities and help to even out the surface of the fabric. Dressing controls the hand and inflexibility of the fabric whereas compressive contracting manages its dimensional dependability.

Indeed, denim fabric without indigo dyeing is not called reliable denim. At first, when denim fabric entered the style market, denim makers used natural indigo dye, which was excessive and gave a characteristic finish. Although synthetic indigo dye has progressively replaced natural indigo dye, some manufacturers still incline toward the natural dye and charge more because they market them as have used a 'natural dye'.

#### 1.5.3.4 Making up

Weaving of fabrics on such multiple sizes is not efficient; subsequently a standard width of fabrics is then forwarded to making up. Fabrics are cut into the necessary width according to estimates required for the machine. Denim fabric and grey fabric are checked for different sorts of imperfections:

- weaving defects
- uneven dyeing
- bleaching and dyeing defects
- oil stains
- patches

The last item is classified as a quality problem. The items then determined to be acceptable are isolated and sent to the bundling division while deficient ones are sent for rectification. After assessment, the sets are wrapped with polythene covers and sent for dispatch according to the purchaser's needs.

### Dispatch

Rolls and sets framed and pressed according to the purchaser's prerequisites are then sent for final dispatches.

## 1.6 Chemistry of denim

Denim jeans are an exceptionally prevalent type of clothing that has been in style for a long time. Denim is a solid twill fabric made out of cotton and dyed blue with indigo colour. Denim is made by a weaving procedure. In the weaving procedure, the 'warp' is the term for the thread or yarn that runs the long way. The 'weft' is the string or yarn that is drawn up through the warp. At the point when weaving denim, the warp is coloured and the weft is left white. Accordingly, jeans are blue on the outside and lighter or white within. This procedure creates denim's exceptional blurring attributes and isolates denim from different materials.

**Cotton is composed of**:

- cellulose $(C_6H_{10}O_5)_n$ 91.00%
- water $(H_2O)$ 7.85%
- protoplasm, pectins (HMDB03402) 0.55%
- waxes, fatty substances 0.40%
- mineral salts 0.20%

**Indigo dye's qualities are**:

- chemical formula: $C_{16}H_{10}N_2O_2$
- dark blue crystalline powder
- density: 1.199 g/cm$^3$
- melting point of 734–738°F
- decomposes at boiling point

**Main chemicals, compounds and components used in denim are**:

- cotton [cellulose $(C_6H_{10}O_5)_n$]
- water $(H_2O)$,
- protoplasm
- pectins (HMDB03402)

The delicate and fluffy material called cotton develops from seeds of the class Gossypium in the group Malvaceae. The cotton seeds are developed in a 'boll', which is a seed-bearing container. Cotton is primarily developed in America, Africa, India, Mexico and Australia. Cotton is practically unadulterated cellulose. Although cotton is to some degree tolerant to dry spells, it needs daylight, a warm atmosphere, water, and dirt with supplements to develop. Cotton is frequently spun into yarn and used to make material. Well-known material items include terrycloth towels, denim and shirts.

### 1.6.1 Indigo dye = $C_{16}H_{10}N_2O_2$

Indigo's first real focal point of creation was in India. Indigo was esteemed and predominantly used as a dye. Indigo dye is one of the most established dye materials. Throughout the years, the generation of indigo dye itself has changed enormously. Today, indigo dyes are simpler and less costly to make. Indigo can be created normally or artificially. Traditionally an indigo dye, which is a natural compound, was removed from plants. Today, most indigo dyes are made artificially. For the most part, indigo dyes are used to dye cotton, which is then used to make denim. The normal measure of indigo used as a part of a couple of jeans is 3–12 g.

#### 1.6.1.1 Chemistry's role

The cotton in denim is natural. Although cotton develops normally, it has turned out to be genetically adjusted. Genetically changed cotton withstands pesticides by introducing *Bacillus thuringiensis* (Bt) to cotton. Bt is a normally created compound that is unsafe to some irritating bugs.

The cellulose in cotton is a polymer of glucose and is developed by unique plants. Glucose atoms can join to each other from by various sorts of bonds. In cotton, glucose atoms are created so that the polymer is the most expanded structure conceivable. Every glucose unit has three OH bunches (hydroxyls) that can bond hydrogen to adjoining chains. The strong bond of chain, called intermolecular forces, makes cotton tough so that it can be strung into textiles for clothing.

Indigo dye can be made naturally and artificially. Most characteristic indigo dye originates from a variety of Indigofera plants. There are a few stages to making characteristic indigo because the dye itself does not exist in nature. At first a concoction called indican is extracted from the natural plant leaves. The indican is put through a progression of tanks. The upper tank contains the aging vessel where the crisply sliced plants are put through aging. In this procedure, indimulsin is added to hydrolyse or separate the indican into indoxyl and glucose. Carbon dioxide is radiated and the liquid in the tank transforms into dim yellow shading. After around 14 h, the remaining fluid is depleted into the second tank. In this tank, an indoxyl-rich blend is mixed with paddles to blend with the surrounding air. Air is added to the indoxyl-rich blend, permitting the indoxyl blend to oxidize and shape into indigotin. The result of this progression settles to the base of the tank. After evacuating the top layer of fluid, the dye that settles at the base of the tank is decanted into the third and last tank. To stop the maturation procedure, the shade is warmed. The last blend of the third tank is separated to evacuate pollution and is dried, creating a thick glue that can be used as dye.

Indigo dye can be artificially created as well. The various procedures for making the manufactured dye include consolidating a progression of responses to substances under controlled conditions. These reactants experience numerous responses that format indigo particles. Numerous results are created during the response. These techniques are all done inside a stainless-steel or glass vessel. These vessels have coats which permit steam or icy water to stream around the bundle as the response takes place. The multifaceted nature of this item often constrains numerous makers of indigo to make engineered indigo in huge amounts. Chemicals used as part of creating engineered indigo shade and throughout the assembly procedure incorporate aniline, sulphur, sodium hydroxide, hydrosulphate and formaldehyde. For indigo to append itself to yarn fully, it must become water solvent. The indigo dye should be blended with a decreasing operator, sodium hydrogen sulphite, turning the dye yellow in shading. Although the dye is yellow, it can connect to the cotton yarn. The yarn will seem yellow at first; yet the dye will noticeably respond to the oxygen around it. After its introduction to oxygen, the dyed yarn will soon seem indigo blue.

## 1.7 Market for denim and today's status

The denim industry is the most segmented in the apparel business. Because of the stiff competition with its substitute products such as cargos, khaki trousers, and cotton combat, denim sales witnessed a slight drop during the recent past. The business sector likewise met with a downturn in 2005 because of an overabundance of clothing from developing nations. This downturn in the business sector prompted the advancement of new fabrics and patterns in the denim business in the developing world. In 2005–6, the industry rose again with a newly refreshed clothing design, back on track. Couture architects worldwide are attempting different strategies with the fabric to enhance its worth. Their imagination and longing to capture the denim business sector is found in the form of patches, rhinestones, appliqués and sparkles in denim attire. Smart fabrics are also used inventively. Smart fabrics are also inventively use chemical finishes including dry foam finish, flat mercerized, tinted, pigment blotch in denim apparel.

World manufacturing of denim fabric achieved 2.7 billion metres in 2006 and established companies such as Levi Strauss & Co. are confronting rivalry from Asian makers. The Asian denim market represents around 50% of the world denim output and a large proportion of it is produced in China and India. Part of the developing market has occurred in countries such as Bangladesh, Indonesia, Pakistan and Turkey. It has been calculated that the worldwide demand for denim will increase by 5%–6% whereas the supply will increase by 8%; thus it will be a buyer's market.

The use of denim items in the world has expanded in past years. The jeans market sector was worth US$49 billion in 2004 and rose to US$55 billion in 2005. Americans are true devotees of jeans. In 2005, the aggregate use of jeans in America was roughly US$3.9 billion, 7.96% of the world's total use. The estimated amount of American expenditure of denim was expected to fall by 1.16% in 2006.

The worldwide denim business sector is conjectured to reach an extraordinary $64.1 billion soon. Within this scenario, the Indian denim industry has demonstrated nonstop development as well. Currently, the country has a denim manufacturing capacity of around 1.2 billion meters per annum operating at a utilization level of about 80–85%. However, despite the great numbers, the Indian denim fabrication industry contributes only five for each penny to the worldwide business sector, generally mirroring the general execution of the nation's materials industry. Nevertheless, denim is the main segment in the Indian materials industry that can experience development, in contrast with others.

### 1.7.1 Indian market for denims

India is being looked upon by various worldwide denim producers as a rising source of local goods owing to its high accessibility of cotton, high quality and a vast pool of talented workers. In addition, the costs of fabricating denim jeans expenses are significantly lower in India compared with other developed and developing countries.

On the local front, the denim wear business sector is being driven by various components, such as good demographics, expanding dispensable livelihoods, the westernization of the work society and an ascent in denim jeans as casual business wear. Because of continuing globalization, youthful India is also inclining toward denim as a component of daily wear. Each of these components bodes well for development.

As far as demographics are concerned, the median age of Indian shoppers is 26 years; most fall into the 15- to 59-year age group. By 2020, India will include another 140 million individuals in the expanding age group. This population has more aspirations, is more aware, has a higher spending power, and is expected to consume across a greater number of categories than the previous generation. As indicated in an evaluation report, the quantity of family units with a yearly wage of $7000 or more will increase from around 30 million to about 100 million by 2020. There will be 400 million people from the middle pay level to a high level of pay by then. Taken together, these components fuel the per capita use of denim.

Currently the Indian market is facing an oversupply situation, following the capacity expansion post-2011 owing to a lower incubation period after the demand for denim picked up as a result of a recovery in the country's economy. The household denim industry has seen this cycle twice. The first was a significant downturn somewhere around 1998 and 2001, which was an aftereffect of the high limit established in China, trailed by a time of fast development popular in 2002–7. To meet expanding demands, denim fabric makers developed the current limit. Nevertheless, after 2008, residential and outsider interest in denim fabric in India declined because of the worldwide financial crash, which influenced the usage rates of denim fabric makers. With continuous recuperation in the worldwide economy and declining cost-intensity of China, interest in denim fabric regained after 2011, which interested players to again expand limits.

Nonetheless, numerous assembling units cannot use their abilities completely, because of oversupply in the business sector, which negatively affects the working edge of these players. Limit expansion has generally become the biggest problem for the business sector, and factories with the smallest business are gravely influenced. Large and middle-size plants in the business have lesser effects from this circumstance.

Meanwhile, denim production capacity has grown by 500 million metres, from 700 million metres in 2010 to 1.2 billion metres in 2015. It is projected to reach 1.5 billion metres by 2020, growing at a CAGR of 4.6%. The share of denim export is projected to increase from 35% in 2015 to 45% in 2020. The two major denim producers, China and the United States, have cut down production of denim, as a result of rising labour and energy costs.

This will prompt critical development in interest in Indian denim. India has been a noteworthy worldwide supplier of denim fabric, yet the nation is behind other contending countries regarding denim apparel such as jeans. The vast majority of worldwide brands outsource denim clothing work to Indians but a lot of it is meant for residential markets and is not sent out.

In addition, the yearly consumption of denim jeans in India is just 453 million pairs. The normal per capita use is a tiny 0.3 pairs, which is much less than in China and the United States, where the normal per capita figures are 1.0 and 8.0 each. However, the per capita consumption is expected grow to 1.0 owing to the development of extra cash, good demographics, and expanding infiltration into provincial markets. The sky is the limit along these lines.

Another source of problems for development is the way in which the Indian denim business sector is skewed toward paying so little in the large cities. Records show roughly 49% share of the country's denim market. Although the business sectors of other urban zones in India contribute on high-volume terms, their consolidated offer to their worth in the business sector was just 51 for every penny in 2015. Indeed, even today, the greater part of interest from small urban groups and country ranges are satisfied by unbranded nearby denim producers. This is ripe ground for marked producers to connect and grow their proximity in those business sectors. Once infiltration of marked denim escalates and mindfulness about its constructive characteristics (for example, quality, plan and fit) increases in these up-and-coming urban groups and country ranges, individuals will pay for marked items, prompting increases in the esteem of the general business sector.

## 1.8 Sustainable perspectives on environmental, social, and economic aspects of denim

Sustainability has turned into a unique selling proposition or unique selling point for some denim jeans producers. It takes around 1800 gallons of water to develop enough cotton to deliver one pair of Levi's, not even mentioning the water needed to wash them properly. Numerous makers and brands are eager to tell potential clients what they are doing to make their generation of denim more supportable, in view of the earth's fragile climate. New manufacturing plants are being built with sun-powered boards on the rooftop to warm water for bathrooms, outside dividers are canvassed in vines to hold in the brilliant warm sunshine and keep the building cool, and movement sensors are used to kill the lights afterward. Wastewater from hand sinks is being employed to flush toilets and normal ventilation through windows on inverse finishes of the building and through the rooftop is preferred for aeration and cooling. Global brands tell us on their hang tags about the names of the endeavors they are making to produce their jeans in a more supportable way.

## 1.9 Challenges and opportunities for the worldwide denim market

A decent pair of jeans is a central component of a modern wardrobe which is both elegant and strong. In denim production generally the effluents which are discharged from the procedures do not harm the environment because they are considered to be eco-friendly. Nevertheless denim organizations are constantly attempting to lessen these environmental results with a sustainable approach.

Jeans, a worldwide fashion, consistently offers roughly 6 billion pairs far and wide. It was foreseen that the profits from jeans would have crossed $65 billion by 2015. Despite predicting such a prosperous situation in the denim business, denim organizations want to change their techniques for assembling jeans to handle environmental issues.

With such stunning varieties to offer, clearly the ways in which jeans fabricated have a major impact on the health of factory specialists and the earth. Workers are in danger of tuberculosis or silicosis during sandblasting procedures, to give an immaculate 'worn' look to jeans. These lethal techniques also generate destructive gasses which affect the earth.

The worldwide denim business faces an extensive burden. The worldwide subsidence of a years ago has left an enduring effect on shopper spending. Costly is out; shabby is in. Credit-energized spending has offered an approach to tight spending plans as purchasers keep paying off their debt.

Therefore, the purchase of jeans in the United States, for instance, has declined whereas athleisure wear has increased significantly. There was a time when couple of pairs of jeans for a youthful buyer turned into a prized ownership, just as smartphones and other gadgets compete with jeans for consumers' attention. Beyond a doubt, the extraordinarily high end of the jeans showcase stays strong; however the mass market for denim around the globe has slowed down.

Another variable has influenced the denim showcase as well: manufactured fibres. With the end goal of bringing down the cost of mass-market jeans, numerous producers have supplanted cotton with polyester. Overcapacity has promoted rock-bottom value levels. It has created the impression that prices will continue to fall for the next few years. Over that time, overabundance will be shed around the globe. However, although that limit will decrease in total over the world, a few nations will venture into the development of new factories.

Initially fabricating jeans using ordinary techniques may not appear to be a major issue regarding sustainability; however, if the situation of the division and the techniques for jeans era are considered it, will have a great impact. One can envision the effect of the denim industry on the earth if every pair of jeans on the planet were delivered using eco-accommodating strategies.

Luckily, numerous denim organizations are endeavoring to find greener strategies and are attempting to discover new procedures for creating jeans as a piece of their business techniques to ensure the earth's sustainability. They have comprehended the significance and the need to assemble a supportable business.

Also, denim organizations comprehend the practices completed by providers. Although the inventory network is spread across various nations, when they trade across moral and ecological models, the brand name is harmed.

## 1.10 Sustainable denim

Conventional jeans manufacturing may not appear to be a major issue as far as sustainability goes, yet because of the role it has in strategies for the creation jeans, it will have a major effect.

Numerous denim organizations are endeavouring to find greener strategies and new procedures for delivering jeans as a piece of their business techniques to save the earth. They have understood the significance and the need to conduct a sustainable business.

Cotton prices are still high and there is rising interest from clients for jeans. In this precarious circumstance, makers need to hold fast to the idea of manageability. Sustainable jeans are not an idea that exists only in developed countries; it is also in developing nations such as India and China. Indian fashioners have come up with eco-accommodating denims for clients.

These jeans are fabricated from 100% natural cotton without using bleach. To achieve the coveted denim look, two procedures are used. One is the wet procedure, in which basic catalysts are employed to reduce water waste; the second is a dry technique. This strategy uses different methods such as scratching, whiskering, granulating and troubling. Rather than synthetic colours, vegetable colours are used to obtain a credible indigo shade of denim.

Some denim organizations have thought up new systems to fabricate 'green jeans'. A denim maker in Sweden has attempted to reuse 1600 tons of material into new articles of clothing and has saved 50 million litres of water while fabricating jeans. Another denim brand makes jeans completely from cotton and reuses water bottles. Plastic waste is a major worldwide issue. Around 66 billion water bottles are used each year just in the United States.

Every pair of these jeans contains eight to 10 used bottles and scrap cotton gathered from manufacturing plant floors which generally are tossed into landfills. This demonstrates the organization's dedication to saving the environment and figuring out how to maintain a gainful business. Although it looks straightforward, a considerable measure of innovative work has gone into assembling jeans from used water bottles and in making plastic popular.

New improvements such as waterless jeans and the development of the natural cotton industry have kept up the sustainability approach. One must make sustainable choices while purchasing denim, such as going for organic cotton, natural dyes and rivets and buttons made from natural or recycled materials. Avoid purchasing jeans with washes that require harsh chemicals, such as distressed, stonewashed and dyed jeans. Continuously purchase exemplary styles that will stay in fashion.

Denim organizations in have been offering some astounding green jeans alternatives to shoppers globally. Some make jeans that use 100% natural cotton, some employ characteristic indigo colours and others use bamboo catches for jeans. There are denim organizations that produce fashionable natural cotton jeans and give away some percentage of their profits to charity.

It is estimated that delivering a couple of pairs jeans uses more than 2500 gallons of water and nearly a pound of chemicals, and a huge amount of energy. If this quantity of jeans is increased exponentially, one can obtain a look at the gigantic problem of wastewater and unsafe gasses from the denim business for the earth.

Researchers have reported another procedure known as 'cutting-edge denim' that can significantly decrease these impacts. This green process can lessen 92% of water use and 30% of energy expenditure and create no wastewater while delivering a couple of pairs of jeans, compared with the usual strategies. In addition, the procedure lowers cotton waste by 87%, which is regularly burned and adds carbon dioxide and other greenhouse gasses into the atmosphere.

Moreover, the creation of conventional denim requires around 25 colouring barrels and scope of perilous chemicals. The Advanced Denim process needs just one barrel and advanced eco-safe, concentrated liquid sulphur dyes. The rest of the means of generation are evacuated. As per the researchers, if 25% of the denims delivered on the planet are coloured utilizing this innovation, water prerequisite of 1.7 million individuals can be met each year.

This innovation would also keep wastewater down to 8.3 million cubic metres and save 220 million kilowatt hours of power. Meanwhile, it would prevent a vast measure of carbon dioxide from being discharged into the air annually. Denim makers of the world are highly enthusiastic in embracing the Advanced Denim innovation. This sort of green process is a great example of saving water, energy and materials and reducing environmental contamination.

The denim industry will soon witness exceptional enhancements from its developments. Denim organizations need to quantify their environmental footprint to secure the environment as a response to the call for sustainable jeans, at the same time benefitting from the business.

## 1.11 Future trends

Worldwide interest in denim is moving toward refinement and inconspicuous design. Fabrics are becoming better with sublime shades and finishes and high-quality augmentations such as printing and hand painting. Developments are originating from mixes, specialized stretch denims, natural denims and green reused denims. The business sector is searching for something new all the time and mixes can bring originality. In this manner, fabrics are being created with numerous mixes such as cotton with thick, modular, cloth, bamboo fibre, soya fibre, hemp fibre, milk fibre and numerous other man-made and normal filaments. A considerable measure of polyester fibre has made advances into denim with different structures, prompting lower-valued items, whereas natural denim has gradually taken a back seat.

Sustainability is also adding to more current styles. A noteworthy issue identified with denim remains water use: 1200–1500 L of water is needed to deliver one pair of jeans. However, denim clothing washes are turning out to be generally greener with manageable washing techniques, waterless washing, and new fashion depending on laser prints, laser outlines, and so forth.

Indian fabric makers think that vivid denim has an essentially decent future. A more agreeable and milder feel for women's jeans and lighter weight with super stretchiness are among most recent trends. More closely woven and better yarns are also being used to create denim. Glossy silk weaves are famous as well. Denim fabrics with various hues and tints and covered denims are progressively being used by popular brands in India.

Because denim is an aggressive business sector item, cost and quality will always be important. One can gain the high ground in evaluating if the expense of assembly is low. One method for decreasing the expense of assembly is to lessen the cost of the crude material, minimizing the production damages and lessening the second's era.

Quality assurance in denim factories can accomplish these targets. Quality assurance may be characterized as 'the arranged and deliberate exercises actualized in a framework for satisfying the quality prerequisites of an item or administration'. The current chapter has briefly highlighted different inspection procedures utilized in Denim plant for capturing the non congruities in order to diminish the generation misfortunes and quality down degrees.

The spread of denim culture everywhere throughout the world has carried with it quickly changing styles and innovations in production. Regarding eco-accommodating solutions inside the denim business, there is no lack of inventive perspectives. Rather, the trouble lies in the ability to actualize a green approach into the clothing market. No dependable change happens overnight. Reasons change through the inventory network, from transient issues with evaluation to the absence of training concerning the strategies for production as well as the straightforward hesitance of organizations and purchasers to make a much-needed overhaul in the way they direct their business and lead their lives.

## 1.12 Conclusion

Denim is an adaptable fabric and has dominated the closets of individuals around the world. New patterns arrive each season, affirming that this famous fabric is imperishable.

Given the increase in the level of prosperity in the Indian population and the shift in preferences to the high-prices product, mostly directed toward the ready to wear segment, especially denim, the vast Indian consumer market offers a unique opportunity for domestic and international retailers and wholesalers of denim products. Because the cost of denim is falling, as a result of oversupply, it has turned into a bonanza for purchasers. The purchaser's business sector of denim is liable to remain stable for another 5–10 years. Subsequently, universal brands and retail chains are looking at India as a marvellous business sector. The coming years will see more denim items in the Indian markets, even with rivalry between retailers and suppliers.

Jeans made from denim are the most mainstream jeans in the field of apparel. However, the world of style has not kept to pants, and flexible finished are given to denim. No other fabric provides such a large number of choices of style and fashion with which to play. Many distinctive washes and clothing systems make several outlines out of the same fabric. Because denim is naturally a woven fabric, it sometimes characteristically fails to fulfill specific needs of the conscious and sophisticated customers who are more cognizant about comfort and flexibility. Advancement of weave denim delivered by existing sewing hardware has ended up being an extremely fruitful expansion in the denim world. Also, with both a woven and sewn structure, denim has truly progressed, in contrast to other fabric. It is a fabric that has had a magnificent past and it will probably have an incredible future as well.

## Further reading

Agarwal S. World denim market – a report on capacities, market size, forecasts etc. 2009.

Bellis M. Levi Strauss – the history of blue jeans. About.com.

Miller D, Woodward S. Global denim. Bloomsbury Publishing Plc; 2016.

Dartmann L, Hang K. The denim bible jeans encyclopedia. 2006.

Dascalu T, Acosta-Ortiz SE, Ortiz-Morates M, Compean I. Removal of the indigo colour by laser beam-denim interaction. Opt Laser Technol 2000;34:179–89.

Ghoranneviss M, Shahidi S, Moazzenchi B, Anvari A, Rashidi A, Hosseini H. Comparison between decolourization of denim fabrics with oxygen and argon glow discharge. Surf Coat Technol 2007;201:4926–30.

Hospes C. Article: dutch denim. JFK Magazine 2007. Nr 02 November/December.

Hung ON, Chan CK, Kan CW, Yuen CWM, Song LJ. Artificial neural network approach for predicting colour properties of laser-treated denim fabrics. Fiber Polym 2014;15(6):1330–6.

Kan CW. Washing techniques for denim jeans. Denim: manufacture, finishing and applications. 2015. p. 333.

Kan CW. Colour fading effect of indigo-dyed cotton denim fabric by $CO_2$ laser. Fiber Polym 2014;15(2):426–9.

Kan CW, Yuen CWM. Effect of atmospheric pressure plasma treatment on the desizing and subsequent colour fading process of cotton denim fabric. Color Technol 2012;128:356–63.

Ghoranneviss M, Moazzenchi B, Shahidi S, Anvari A, Rashidi A. Decolourization of denim fabrics with cold plasmas in the presence of magnetic fields. Plasma Process Polym 2006;3:316–21.

Kan CW, Yuen CWM, Cheng CW. A technical study of the effect of $CO_2$ laser surface engraving on some colour properties of denim fabric. Color Technol 2010;126:365–71.

Khalil E, Islam MM. Wrinkle finish on denim by resin treatment: a review. AASCIT Commun 2015;2(3):82–7.

Montazer M, Sadeghian V. Bio-washing of pigment dyed denim with laccases. Res J Text Appar 2014;18(3):35–41.

Vivekanadana MV, Raj S, Suffixeenivasan S, Nachane RP. Parameters affecting warm-cool feeling in cotton denim fabrics. Indian J Fibre Text Res 2011;36(2):117–21.

Mondal MIH, Khan MMR. Characterization and process optimization of indigo dyed cotton denim garments by enzymatic wash. Fash Text 2014;1(1):1–12.

Moore MA, Ankeny M. Performance of garment washed denim blue jeans. AATCC Rev 2005;5(6):23–7.

Ortiz-Morales M, Poterasu M, Acosta-Ortiz SE, Compan I, Hernandex-Alvarado MR. A comparison between characteristics of various laser-based denim fading processes. Opt Laser Technol 2003;39:15–24.

Ozguney AT, Ozcelik G, Ozkaya K. A study of specifying the effect of laser fading process on the colour and mechanical properties of the denim fabrics. Tekst ve Konfeksiyon 2009;2:133–8.

Roshan P, Sandeep NR, Jegadeesh T. Denim: the evergreen favorite. Textile Dyer & Printer; 2009.

Paul R. Denim and jeans: an overview. Denim: manufacture, finishing and applications. 2015. p. 1–5.

Radetic M, Jovancic P, Puac N, Petrovic ZL. Environmental impact of plasma application to textiles. J Phys Conf Ser 2007;71(1):012017.

Sarkar J, Khalil E. Effect of industrial bleach wash and softening on the physical, mechanical and color properties of denim garments. IOSR J Polym Text Eng 2014;1(3):46–9.

Sarkar J, Khalil E, Solaiman M. Effect of enzyme washing combined with pumice stone on the physical, mechanical and color properties of denim garments. Int J Res Advent Technol 2014;2(9):65–8.
Shalini N. Fabric and garment finishing: basic washes in denim fabric. India: Sanblue Enterprises Pvt. Ltd.; 2013. [Available].
Tarafder N. Denim fabrics – an overview. Man-Made Text India 2008;51(11):387–95.

## Websites

http://www.garmentsmerchandising.com/types-of-denim-fabric-used-in-garments-manufacturing/.
http://www.denverfabrics.com/pages/static/denim/denim.htm.
https://scilearn.sydney.edu.au/fychemistry/labmanual/e04.pdf.
http://textilelearner.blogspot.com/2014/08/overview-of-denim-fabric-types-.
http://www.tandfonline.com/doi/full/10.1080/10350331003722851.
http://www.fashionintime.org/history-jeans/.
http://textilelearner.blogspot.com/2016/03/latest-dry-washing-techniques-.
http://www.historyofjeans.com/jeans-history/history-of-denim/European Market information: The European market for denim jeans.
http://www.cbi.eu/marketinfo/cbi/docs/outerwear_the_eu_market_for_denim.
http://www.scribd.com/doc/23645858/Global-market-review-of-the-denim-andjeanswear-industries-%E2%80%93-forecasts-to-2012.
http://www.denimsandjeans.com/events/international-conference-on-denims/.
http://www.Thedailygreen.com.
http://www.Tmagazine.ey.com.
http://www.Urbantimes.co/magazine.
http://www.Sciencedaily.com.
http://www.Sustainablebrands.com.

# Environmental impacts of denim

**2**

*K. Amutha*
Bharathiar University, Coimbatore, India

*Denim is close to everybody, everywhere, anytime.*

***Denim Book by Clariant***

## 2.1 Introduction

Denim is a unique fabric which is woven with indigo-dyed cotton yarn as the warp and undyed cotton yarn as the weft in a warp-faced twill pattern. Hence, the face of the fabric is blue and the back is white. Denim can be called a fabric for all-time fashion. It is a very trendy fabric. It has undergone constant evolution to remain sustained in fast-changing fashion. Currently, denim is a fabric for an entire lifestyle rather than just a garment. The future of denim is also promising owing to new innovations. To create new fashion trends, denim continues to be revived through the fashion cycle. According to Clariant [1], adjectives that define denim are 'comfortable', 'risky', 'adventurous', 'relaxing', 'glamorous', 'attractive', 'aggressive', 'smart', 'casual', 'funny', 'dynamic', 'energetic', 'fashionable', 'practical', 'creative' and 'universal'. Denim jeans have created a blue revolution in the fashion world. It is a popular fashion because consumers have an everlasting love for denim. Advantages of denim are high durability, excellent strength, all-time fashion and appealing aesthetics. Disadvantages are higher stiffness, cloth damage caused by finishing treatments, and the inability to be ironed. When it is ironed the fabric becomes stiffer. In 1969 a writer for an American fabrics magazine declared, 'Denim is one of the world's oldest fabrics, yet it remains eternally young'. It is an amazing fact that even after 47 years the saying is pertinent.

In a news release on 18 June 2012, the American Chemical Society mentioned that the production of a pair of jeans consumes more than 2500 gallons of water (including cotton growing, dyeing and processing of denim), about a pound of chemicals (in the form of dyes, auxiliaries and finishing agents) and a vast amount of energy (including the irrigation of cotton and subsequent processes such as spinning, weaving, processing and sewing). These quantities need to be multiplied by 2 billion to arrive at the world's total jeans production for a year. Hence the question of sustainable production arises. This chapter discusses key sustainability issues of the denim industry, the environmental impacts of denim processing, human health hazards, sustainable production practices, trends followed around the world and some future trends.

**Sustainability in Denim. http://dx.doi.org/10.1016/B978-0-08-102043-2.00002-2**
Copyright © 2017 Elsevier Ltd. All rights reserved.

## 2.2 Denim industry

Denim is one of the major sectors of the textile industry. Denim is manufactured by more than 15 countries around the world; denim mills across the globe are listed in Table 2.1. Cotton is the key raw material for denim; hence the denim industry includes the cotton textile industry. The process begins with cotton farming, followed by ginning, spinning, warp dyeing, weaving, finishing and garment making.

Denim will probably never go out of fashion. Various garments and styles can be made with denim, including classic and vintage, chinos, slim fit, shorts, capris, skirts and dresses. Denim is also manufactured with special effects and finishes such as printed denim, water-repellent, recycled denim and so on.

Reports by the Securities and Exchange Board of India [3] estimated that the world's total denim manufacture is 7.7 billion metres. Asia is the largest contributor, with a production capacity of over 70%, and China is the leading producer, with a capacity of approximately 3 billion metres. Next to China, India is in second place, with a production capacity of over 1 billion metres. Other key denim manufacturing countries are Pakistan, Bangladesh, Turkey and the United States.

Wazir Advisors (an India-based business consultancy service), reported in 2012 that about 1.8 billion pairs of denim trousers, amounting to USD 51 billion, are sold annually worldwide. Europe and the United States are the biggest markets for denim fashion, even as newer economies such as China and India are witnessing a steady rise in demand. According to industry estimates, the per-capita denim consumption in the United States is eight pairs whereas it is one pair in China and 0.3 pairs in India.

According to *Business Standard* [4], in an article entitled 'Excess capacity hits denim industry', the overall denim industry capacity of India is estimated to be 1.2–1.3 billion metres a year with use of about 700–800 million metres. Also, 80–100 million metres of additional production is expected in a year.

During the 1970s and 1980s Texas was the denim capital of the world. The denim jeans market slowly extended from developed countries to developing countries. In the mid 1990s, there was a steep decline in the denim market, and Xintang began

**Table 2.1** **Denim mills worldwide**

| Region | Number of mills |
|---|---|
| Asia (China) | 297 |
| Asia (other countries) | 104 |
| North America | 9 |
| Europe | 41 |
| Latin America | 46 |
| Africa | 15 |
| Australia | 1 |
| Total | 513 |

Taken from DenimsandJeans. http://www.denimsandjeans.com/latest-denim-reports/denim-data-figures/world-denim-market-a-report-on-capacitiesmarket-size-forecasts-etc/919.

**Figure 2.1** Problems of denim industry.

producing denim jeans at the lowest cost. With free trade and increased production capacity, Xintang became the denim capital of the world, producing about 300 million pairs of jeans a year.

ISKO is the world's largest denim manufacturer under one roof. It is the only denim mill in the world that has the Nordic Swan Ecolabel certificate. A production capacity of 250 million metres of fabric per year is achieved with 1500 high-tech automated looms and the global distribution of employees.

Japanese denim is said to be world's best denim. Two important qualities that make it unique are that the fabric is woven on old looms that produce selvage fabric; and it uses only natural indigo dye derived from the *Indigofera tinctoria* plant. Hence it is called premium denim.

The denim industry is associated with certain problems, as shown in Fig. 2.1. Solid waste is an integral part of textile manufacturing, and denim waste is composed of preconsumer waste (clean waste) and postconsumer waste (secondhand trash). Metal accessories such as rivets and zippers used in jeans have to be recycled properly.

Overcapacity or overproduction leads to a huge inventory that cannot be sold in time. An example is India faced the problem of overcapacity in the first quarter of 2014. Except for a few leading manufacturers, others also faced huge losses in profit. When a new trend is introduced, stock that is already available becomes unsellable.

Finally, environmental pollution caused by the denim industry includes air, water and noise pollution, which varies according to the production stage.

## 2.3 Key sustainability issues of denim industry

With an increase in awareness among consumers, the denim industry is shifting towards sustainable production methods and eco-friendly jeans. Sustainability issues of the denim industry may be categorized into five different aspects, based on three pillars of sustainability: social, environment and economic, as shown in Fig. 2.2.

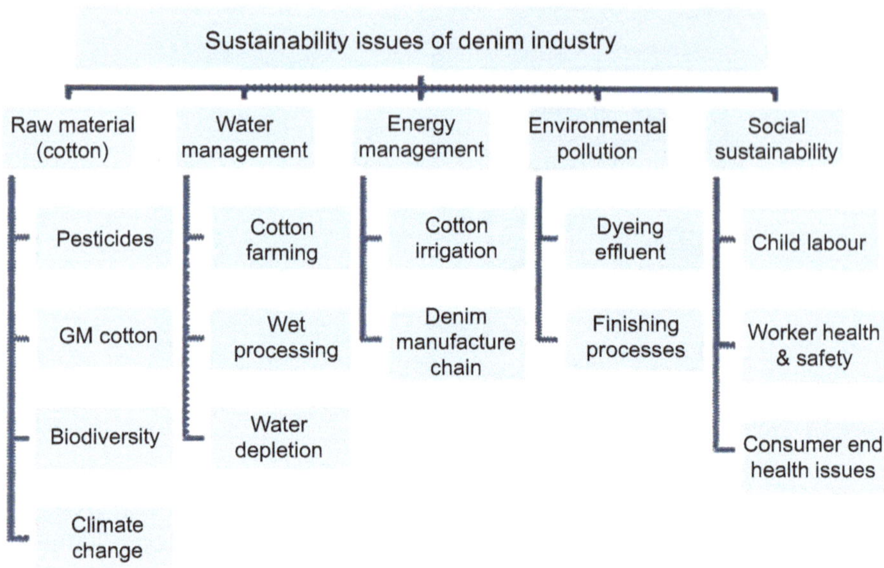


**Figure 2.2** Sustainability issues of denim industry. *GM*, genetically modified.

### 2.3.1 Raw material (cotton)

Cotton, a cellulosic natural fibre, is the key raw material for the denim industry. It belongs to the genus *Gossypium* and is the most widely cultivated crop in more than 90 countries. It is a soft, cool fibre with good strength and absorbency, which makes it suitable for a range of clothing and other home textile applications. The most important trait of cotton is its biodegradability. Common varieties of cotton are Upland, Pima, Sea Island and Egyptian. Egyptian cotton is an extra long–staple cotton used to produce fine fabrics. Cotton fibre could be described as shown in Fig. 2.3.

Problems associated with cotton cultivation are portrayed in Fig. 2.4. No definite benchmark has been reached for the sustainable production of cotton. Cotton production is affected by heat stress and the ability to withstand heat stress is becoming weaker.

Cotton growing requires the use of pesticides, which in turn impose hazards on human health (Fig. 2.5) and the environment. According to Allen Woodburn [5], cotton production uses approximately 25% of the world's total insecticides and more than 10% of pesticides. The insecticides and pesticides used are hazardous in nature and pose a threat to global freshwater resources.

The effects of using pesticides include:

*quantity*: the amount of pesticide
*the nature of the pesticide*: factors such as acute and chronic human toxicity, environmental toxicity. It is advisable not to use highly hazardous pesticides
*handling and use of pesticides*: factors such as having to use personal protective equipment while spraying; following label instructions for use, storage and disposal; application procedures; selection, use and cleaning of equipment used for pesticide application

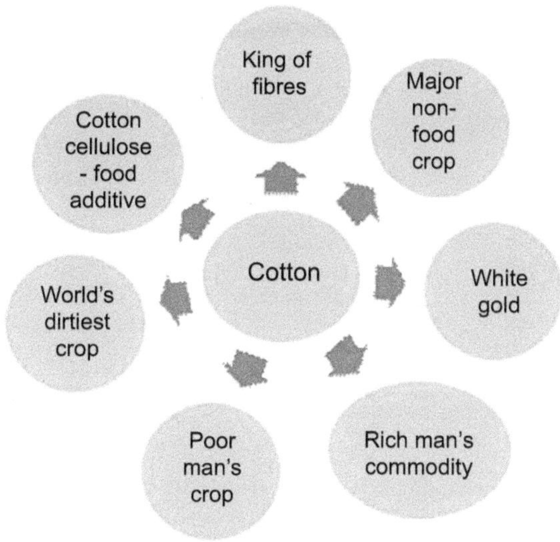


**Figure 2.3** Cotton fibre descriptions.

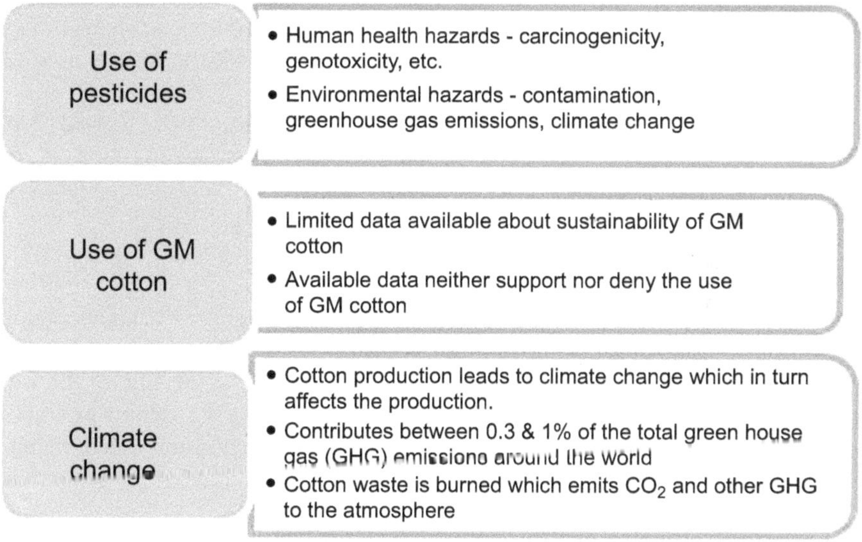


**Figure 2.4** Problems associated with cotton cultivation. *GM*, genetically modified.

The International Cotton Advisory Committee [6] pointed out the environmental impacts associated with the improper use of pesticides:

- contamination of drinking water, river systems, groundwater and aquifers
- poisoning of fish and other aquatic organisms and loss of biodiversity
- long-term persistence in soils affecting rotational crops and beneficial soil organisms and loss of ecosystem services

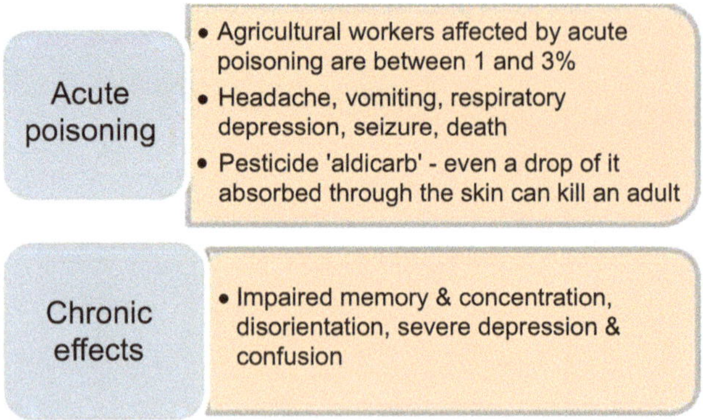


**Figure 2.5** Effects of pesticides on humans.

- poisoning of wildlife (including birds and bees) and loss of biodiversity
- poisoning or contamination of livestock: cotton seed fed to cattle; traces of pesticides detected in milk; entry into human food chain
- reduced populations of pollinating insects important for crop yield
- air pollution

In Egypt, a study by Farahat et al. [7] on 52 cotton labourers working in fields revealed that 88% had never used protective clothing; only 6% of workers used face masks to cover their nose and mouth.

According to a study by researchers at the Technical University of Lodz, Poland (2004), hazardous pesticides applied during cotton production can be detected in cotton clothing.

Traces of pesticide residues are found in blood samples of Indian cotton labourers.

### *2.3.2 Water management*

The virtual water content of a product (a commodity, good or service) is the volume of freshwater used to produce the product, measured at the place where the product was actually produced (production site definition). It refers to the sum of the water use in the various steps of the production chain. Water use for denim production includes cotton fibre production, wet processing and, at the consumer end, laundering, as shown in Table 2.2. In the pyramid of water-consuming textiles, denim holds the top position.

The water requirement for cotton farming varies with the cultivar, length of the growing season, temperature, hours of sunshine, amount and distribution of rainfall, irrigation method and characteristics of the soil. Cotton is identified as one among the nine of the thirstiest crops. It can survive drought and hot conditions.

A study on the relationship between cotton yield and the use of water was found to be linear. About 3% of global irrigation water is used for cotton cultivation. Water withdrawn for irrigation purposes is sustainable, provided an equal amount of water is replenished in time. Irrigation of cotton is associated with negative impacts on regional freshwater resources including eutrophication, salinization, pollution, wildlife contamination, a rise in water tables and habitat destruction.

Table 2.2 **Water use along the denim supply chain**

| Stage | Average water use (per kilogram of textiles) |
|---|---|
| Growing cotton fibre | 20,000 L (irrigated conventional crop) |
| Treating/dyeing/finishing fibres and fabrics (all fibres) | 100–150 L |
| Consumer laundering (all fibres) | 1650 L |

*Source:* GLASA_2015_StateofApparelSector_SpecialReport_Water_150624.pdf.

According to scientists and technicians, colour is an aesthetic pollutant. Warp yarn is indigo dyed for denim production. Conventional denim (indigo) dyeing employs up to 15 dyeing vats with potentially harmful chemicals. Use of synthetic sulphur dyes does not need water because they are water insoluble, but they are hazardous to the environment. Other processes such as sizing, bleaching and denim washing consume huge amounts of water.

1 pair of jeans = 2900 gallons/11, 000 litres of water

## 2.3.3 Energy management

Cotton farming consumes energy in the form of electricity or fossil fuels. It includes irrigation pumps, tractors, harvesting machines, etc. Ninety percent of energy that is employed in cotton production is used on the farm and only 10% is used in the ginning process. The subsequent processes of spinning, dyeing, weaving, finishing and sewing also consume a high amount of electrical energy.

## 2.3.4 Environmental pollution

Dyeing of warp yarn for denim manufacture employs either natural or synthetic indigo dyes and sulphur dyes. Environmental pollution caused by dyeing and finishing is discussed in Section 2.4.

Using core-spun cotton-elastane (spandex) for stretch denim poses problems such as disposal and recycling. Elastane is a synthetic rubber synthesized from polyurethane which is derived from petrochemicals. Hence it is nonbiodegradable and leads to pollution.

## 2.3.5 Social sustainability

Child labour is engaged in cotton fields for various work (Fig. 2.6) including pesticide applications in India and Uzbekistan. Children are often the first victims of pesticide poisoning, especially those who live in close proximity to cotton fields, sometimes owing to the reuse of empty pesticide containers.

**Figure 2.6** Sanliurfa, Turkey, 25 October, 2011: Unidentified seasonal child workers work in very difficult conditions. They cannot go to school to work in the cotton fields. Harran plain near Syria border of Turkey.
*Source:* http://campaign.worldvision.com.au/news-events/deaths-during-cotton-harvest-highlight-child-labour-dangers/.

**Figure 2.7** Amravati, Maharashtra, India, 19 August 2016: Unidentified farmers work in a soybean field. An Indian farming scene, Amravati, Maharashtra, India.
Used with permission from "Dirty White Gold" and Leah Borromeo.

Cotton farmers around the world:

- lack safety awareness
- lack the use of protective apparatus (Fig. 2.7)
- are illiterate
- use pesticides with poor labeling
- practice inadequate safeguards

Fig. 2.8 shows the engagement of child labour (along with their parents) in a jeans factory in Xintang Town; every day, each child cut threads for 200 pairs of jeans for 15 fen (about 2 cents). Fig. 2.9 shows a family-owned workshop in Gurao town. Children are often given part-time jobs there. The pay is based on the number of pieces; at most, a whole day may fetch about 20 to 30 Yuan (Chinese currency).

**Figure 2.8** Child labour along with family in jeans factory.
Image courtesy: Greenpeace.

**Figure 2.9** Child labour in jeans factory.
Image courtesy: Greenpeace.

The textile industry has been reported to pose the threat of various types of occupational diseases. **Byssinosis**, also called 'brown lung disease' (Fig. 2.10) or 'Monday fever', is an occupational lung disease caused by exposure to cotton dust in inadequately ventilated working environments. Byssinosis commonly occurs in workers who are employed in the yarn and fabric manufacturing industries. The risks of developing byssinosis is higher for smokers. The symptoms are tightening of the chest, coughing and wheezing. Symptoms worsen upon exposure to cotton. Ways to prevent byssinosis are avoiding exposure to cotton and providing appropriate ventilation in the workplace. Bronchodilators, steroids and nebulizers are used as treatment.

Sandblasting is a process by which the denim (jeans) surface is abraded using sand containing silica. This process imparts a 'worn-out' look to the denim.

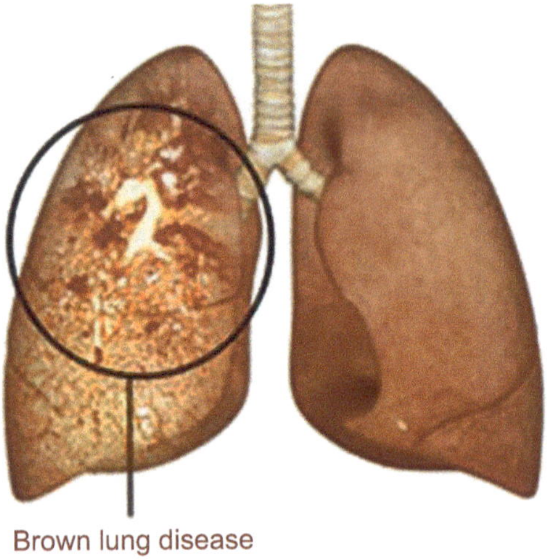


**Figure 2.10** Byssinosis.
*Source:* http://www.epainassist.com/chest-pain/lungs.

**Figure 2.11** Unsafe working conditions in a denim factory.
*Source:* Justin Jin/Panos.

Silicosis is a lung disease caused by the inhalation of dust that contains free silica in a crystalline form. Denim sandblasting was found to be a cause of silicosis, which is an incurable disease. The disease proliferates rapidly under extremely high-exposure conditions. The condition is permanent and progresses even after exposure is stopped. It is an occupational disease which kills thousands of people every year across the globe. Because silicosis is an occupational disease, it is preventable. Silicosis, caused by denim sandblasting (Fig. 2.11), was first diagnosed among workers involved in denim sandblasting in workshops in Turkey in 2004.

Reasons for silicosis are:

- lack of awareness about the dangers of silica
- inadequate protective measures
- poorly ventilated work areas
- longer working hours in a day (higher exposure)
- poor working conditions (smaller work area)

Alternative methods are:

- use of sandpaper
- chemical treatment using potassium permanganate
- laser technology

Sandblasting is preferred over these alternatives because it is:

- quicker
- more reliable
- cheaper

Potassium permanganate is largely used to lighten the colour of denim. The chemical is sprayed onto denim garments and then washed off, leaving the treated area with a lighter colour than the surrounding fabric. Workers spray the chemical onto the denim using a hose or sometimes a brush. After the chemical dries, it is neutralized with a second chemical and then washed off. The process exposes workers to harmful inhalation of the chemical's fumes. The odour is irritating and strong. The advised method usually involves spraying of denim garments in a sealed and ventilated booth, and if such a booth is unavailable, it is advisable to use respiratory masks and the potassium permanganate applied from a 'safe' distance.

On the consumer end, denim jeans also pose certain health threats, especially 'skinny' jeans or jeggings. Wearing skinny jeans and squatting for a prolonged time can damage muscle and nerve fibres in the legs, making it difficult to walk, a case study revealed that was published online in the *Journal of Neurology, Neurosurgery & Psychiatry*. A 35-year-old woman had severe weakness in both ankles. The doctors diagnosed the case and reported the reason to be the tight jeans she was wearing the previous day. She had been in a squatting position for a long time, which had made her calves swollen. The jeans had to be cut from her skin.

In the case of men, skinny jeans lead to urinary tract infections and infertility. TENA is a leading global brand of health-care products and services for individuals. According to a study carried out by TENA, one in ten men experienced unpleasant side effects as a result of wearing jeans that were too tight.

## 2.4 Environmental impacts of denim processing

Indigo dye is part of the numerous marketed organic colourants used to colour textiles, paper, leather and plastic and for specialized applications such as food, drugs, cosmetic and photochemical production. Textile effluents containing indigo dye (Fig. 2.12) and other dye types make water toxic and unfit for human and animal consumption, and cause an imbalance in different aquatic ecosystem food chains.

Indigo obtained from plant source (*Indigofera tinctoria*) have been used for more than a century. Synthetic indigo dye was introduced in 1897. During the 1980s indigo was combined with sulphur dye to acquire a new look and effects. Later, in the 1990s, to widen the denim colour range (apart from blue), sulphur dye was used separately to obtain colours such as black, brown, olive and burgundy.

The use of synthetic indigo and sulphur dyes pose serious effluent problem.

An archaeological study by Splitstoser et al. [8] found that indigoid dye (indigotin) was used to colour cotton fabric which was 6000 years old, in Peru. It was supposed to be derived from the *Indigofera* species native to South America. Indigo is one of the most valued and most globally widespread dyes of antiquity and of the current era (it is the blue of blue jeans).

Various researchers have identified the mutagenic effects of textile samples and wastewater from the textile industry. Their findings showed that dyes used for textile finishing are mainly responsible for the mutagenic effects observed. Indigo dye had a mutagenic effect on the bacterium *Salmonella typhimurium*.

A study by Mathias A Chia et al. [9] reported that indigo dye effluents exhibited acute toxic effects on the microalgae *Scenedesmus quadricauda* (Fig. 2.13). Therefore, there is risk of disrupting the primary aquatic food chain when untreated indigo dye effluents are discharged into aquatic systems.

A research by Rannug et al. [10] studied the mutagenic effect of pure cotton, jeans fabric extracts and synthetic indigo, indirubin and isatin on *S. typhimurium* strains TA98 and TA100. Mutagenic effects on TA98 ± S9 and TA100 ± S9 were seen on the extracts of both bleached and unbleached jeans. The mutagenic effects increased in the presence of S9. Moreover, synthetic indigo of a technical grade or 98% pure showed mutagenic effects, especially on TA98 + S9. The researchers reported that

**Figure 2.12** Denim dyeing effluent.
Image courtesy: Greenpeace.

**Figure 2.13** Denim dyeing effluent in water stream.
Image courtesy: Greenpeace.

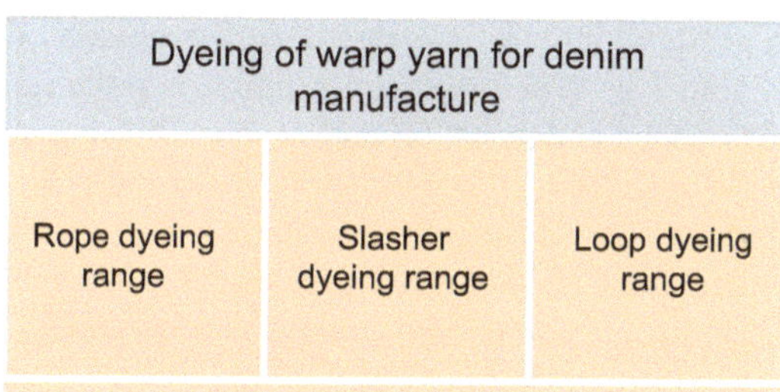


**Figure 2.14** Warp dyeing methods for denim manufacture.

indigo might cause a potential health risk either by revealing toxic effects of other compounds or by being a non-genotoxic carcinogen. If the effluent contains higher concentrations of indigo and indigoid dyes, it would be genotoxic.

Three different dyeing techniques are used to dye the warp beam for denim weaving (Fig. 2.14). More than 95% of warp-dyeing for denim employs rope and slasher dyeing ranges.

The environmental impacts of denim processing can be classified into the following categories:

- water pollution: dyeing and finishing effluent discharge in water bodies
- air pollution: cotton dust, abrasives and chemicals found in air
- solid waste (sludge)

Ecological and toxicological problems caused by the direct discharge of textile wastewater in natural water bodies have been one of the most important water pollution problems. The presence of colour (dye), dissolved solids (TDS), total suspended solids (TSS), toxic (heavy) metals and residual chlorine in the effluent results in a high chemical oxygen demand (COD). The presence of organic pollutants leads to a high biochemical oxygen demand (BOD).

Five heavy metals (cadmium, chromium, mercury, lead and copper) were found in 17 out of 21 water and sediment samples taken from throughout Xintang and Gurao in China. In one sample, cadmium exceeded China's national limits by 128 times.

Denim washing (Fig. 2.15) is an aesthetic finish that is imparted to fabric to improve the softness and comfort of the fabric. In addition, the fabric achieves a different look such as a faded or worn-out appearance.

Pumice stone is used to stonewash denim garments. The machine is loaded with water (1:5 material to liquor ratio), denim garments, pumice stone (half the volume of garments), bleaching powder (4%) and soda ash (8%) and processed for about 20–40 min. Then the liquor is drained and the garments are given a hot wash at 40–50°C for 5 minutes and dried. The stone gets abraded during the process and becomes powdered; part of it remains in the liquor and part of it sticks to the garment. A huge amount of water is required for repeated washing cycles to remove the deposited pumice from the denim. The effluent and pumice dust lead to environmental pollution.

Sandblasting is a mechanical finish which uses sand containing silica. The minute silica dust spreads in air and poses serious respiratory disease such as silicosis.

Microsanding is a finish which uses a series of rollers covered with abrasive paper or chemically coated abrasive. When the denim fabric is pulled over these rollers it results in a soft, raised surface and dulls the colour. The abrasive material pollutes the environment.

Chemical washing such as bleaching employs chemicals such as sodium hypochlorite or potassium permanganate. It is harmful to human health and corrodes the stainless-steel drum of the bleaching machine. The effluent contains chlorinated organic substances which cause severe pollution to the environment.

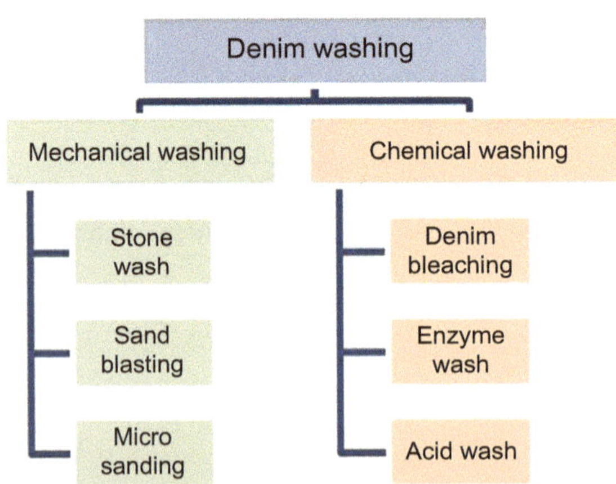


**Figure 2.15** Denim washing techniques.

Acid wash uses both pumice stone and chemicals (sodium hypochlorite or potassium permanganate) for washing. It does not require water but leads to pollution. Using a smaller amount of stones does not give the expected look (bleaching effect) and use of more stones damages the fabric and accessories and the machine itself. The acid damages the washing machine and the metal accessories used in the jeans. The effluent contains pumice dust and residual manganese, which are hazardous, and also cotton fibre dust as a result of abrasion caused by pumice stones. Acid-washed jeans have a tendency to yellow upon repeated use.

Denim dyeing effluent is heterogeneous in nature and demands a technologically sound system for the treatment and recovery of dyes/chemicals and reuse of water. Based on the nature and complexity of the dyes and chemicals present in the effluent, a suitable treatment process can be employed. The different processes adapted are precipitation, degradation (catalytic, bio and photo), adsorption, filtration (ultra, micro and nano) and ozonation, coagulation, electrolysis, etc.

Levi Strauss & Co. established Global Effluent Requirements for all of its supplier factories that finish or launder garments. It states the limits of pH, temperature, TSS, BOD, COD, colour, foam and heavy metals such as mercury, cadmium, lead, arsenic and copper.

DyStar and LoopTEC devised a closed-loop technology to reuse the wash water and recycle indigo on denim dyeing ranges, hence recovering and reusing 99% of the wash water and 99.9% indigo recovery from the wash water; eliminating 96% of the salt content use; and reusing recovered indigo liquor, which is highly concentrated.

Levi Strauss & Co., in the life-cycle assessment, studied the impacts of Levi's 501 jeans on climate change and calculated carbon dioxide emissions throughout the life of jeans by a cradle to grave approach. The results are shown in Fig. 2.16. The consumer phase emissions were found to be the highest, followed by the fabric production phase.

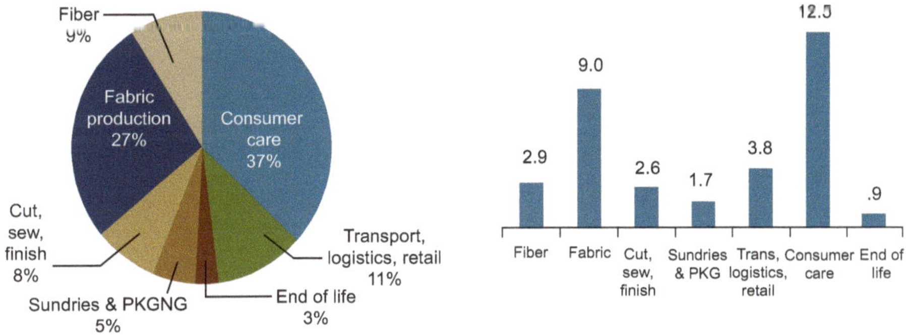


**Figure 2.16** Levi's 501 jeans: climate change impact. *PKGNG*, packaging.
*Source:* http://levistrauss.com/wp-content/uploads/2015/03/Full-LCA-Results-Deck-FINAL.pdf.

## 2.5 Sustainable denim production

Sustainable denim production can be achieved by reducing the use of water, energy and chemicals; this will significantly reduce the environmental impacts, as in the following cases.

*Advanced Denim*: A process developed by Clariant, Switzerland uses 92% less water and 32% less energy than conventional denim manufacturing methods. Also, cotton waste generation is reduced up to 87%; in other companies it is burnt, which leads to greenhouse gas emissions.

Schimper et al. [11] found an eco-friendly alternative to sandblasting. It was found to be cheaper and more efficient than sandblasting. The process is called 'Surface activation'. The concentration of indigo dye is greater on the surface than in the core of the yarn, and this property of indigo-dyed yarn takes advantage of surface activation to achieve a worn or faded look. Advantages are that the time taken for the wash process is shorter, the fabric strength does not decrease, and it consumes fewer chemicals that are costly.

In 2014, Levi's made 100,000 pairs of women's jeans with 100% recycled water, saving about 12 million litres of water (Fig. 2.17).

As of September 2010, Levi Strauss & Co. implemented a global ban on sandblasting in all of its manufacturing units.

Enzymes can be used in place of corrosive chemicals such as caustic soda, potassium permanganate and hypochlorite. Enzymes are obtained from renewable resources; they are biodegradable, they are manufactured with the highest efficiency, they do not damage fabric, unlike other chemicals, and they are flexible. Hence they are sustainable.

Denim products named *Waste Less* by Levi's use recycled plastic bottles and food trays for denim manufacture, using less water. The plastics are first sorted based

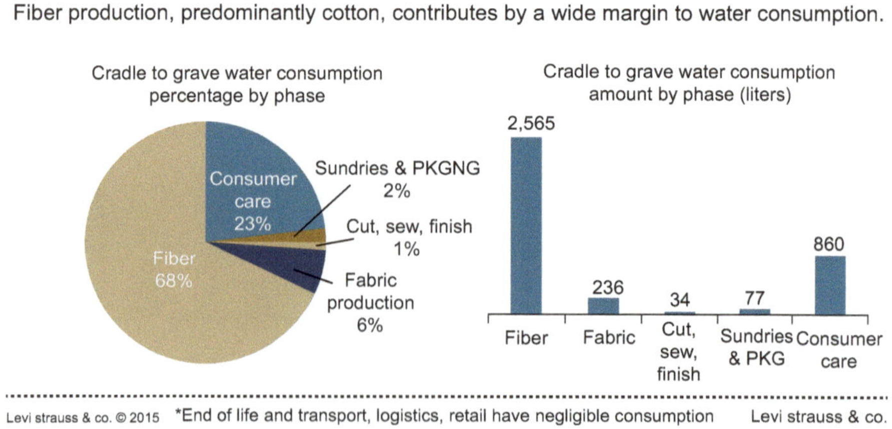


**Figure 2.17** Water consumption of Levi's 501 jeans. *PKG*, packaging.
*Source:* http://levistrauss.com/wp-content/uploads/2015/03/Full-LCA-Results-Deck-FINAL.pdf.

on colour, crushed into flakes and made into polyester fibre. This polyester fibre is blended with cotton and spun into yarn, which in turn is woven into denim. By 2020, Levi's aims to reduce its greenhouse gas emissions by 25%.

In 2010, Levi's initiated *Waterless* jeans with a 96% reduction in water use in the finishing process. In 2012, it saved 360 million litres of water through Waterless jeans.

Jack & Jones, a leading denim brand, introduced Low Impact Denim in 2011. It is an initiative to save water and energy. In 2013, 55 million litres of water and 5 million kilowatts of energy had been saved by means of Low Impact Denim production. It uses laser and ozone to achieve different visual effects on the finished garments.

MUD Jeans is an attempt to manufacture jeans in a closed loop (circular economy). It is made with 30% recycled content and 70% organic cotton. The consumer can either purchase or lease the jeans. The lease span is 1 year, after which the consumer can swap the old jeans for new ones, continue to lease them for another year or extend the lease for another 4 months. The offer includes free repairs. The company won awards such as the 'Sustainability Leadership Award' and the 'Peta Vegan Award'.

A case study by *Nobody Denim*, supported by Sustainability Victoria, carried out a resource assessment. Based on the recommendations, the process and equipment were upgraded, as a result of which energy and the cost of materials could be saved significantly. It was found that the purchase price of waste fabric is 20 times the cost of disposal; the actual cost (includes energy, labour and other costs involved in production) of the finished fabric that is wasted is 425 times the cost of disposal.

The New Denim Project by Iris Textiles focuses on a Zero Waste Industrial System by maximizing recycling; waste minimization; proper reuse and repair and redesigning. Zero Waste makes upcycling a powerful tool to manage consumption, waste, environmental destruction and to picking up corporate responsibility.

Preconsumer denim waste is collected, ground into fibre, spun into yarn and woven/knitted into new sustainable fabrics. This process is chemical-free and uses minimal water and energy, saving up to 20,000 litres of water per kilogram of upcycled material. Even the waste generated by this manufacturing process is donated to farmers for use as compost waste, which turns to be an organic fertilizer (Fig. 2.18).

Textile artist Yvonne Dalton of Australia upcycled used denim garments into floor rugs that are warm and machine washable.

Jeanius is upcycled denim developed by students at a recycling event at the University of Salford. Major attractions include a waterproof coat made from shredded denim and denim-made sunglasses.

Conventional chemical bleaching (to achieve a faded effect) of denim may be replaced by one-bath enzymatic fading, which saves water and energy.

Denim finishing processes such as bleaching and sandblasting could be replaced by the application of lasers and ozone.

Pumice stones used for denim washing may be replaced by synthetic stones. Even small quantities of enzyme can replace the effect produced with several kilograms of pumice stones.

Applegate Insulation is a world leader and private manufacturer of cellulose insulation materials. They introduced cotton batt insulation made of cotton and polyester mill scraps (especially denim) through pollution-free manufacturing.

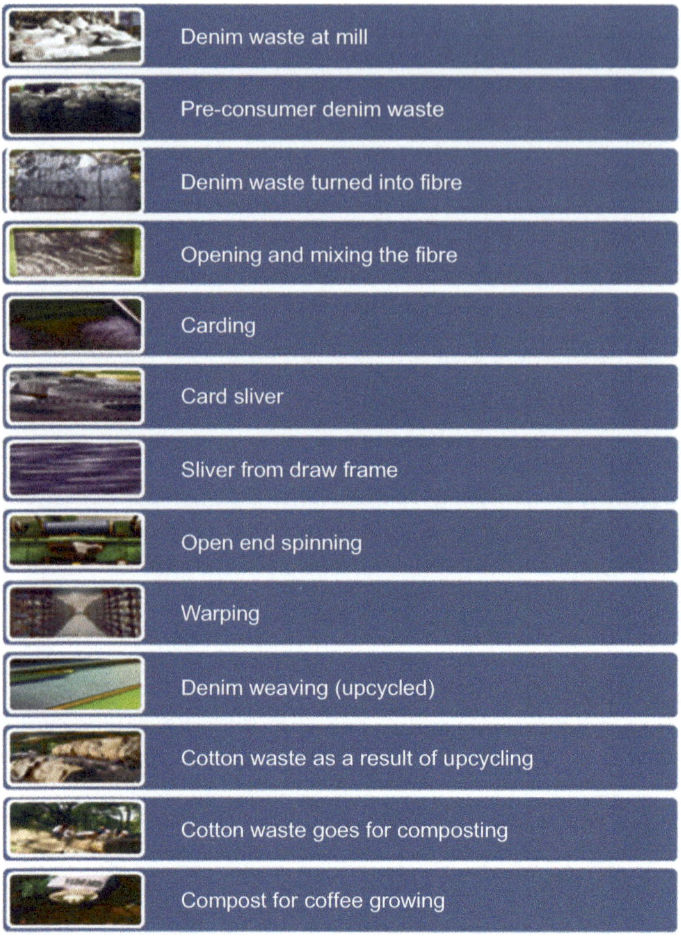


**Figure 2.18** The process sequence of the new denim project.
*Source:* http://www.thenewdenimproject.com/process.

A study by Mendoza et al. [12] evaluated the application of denim fibre scraps for the synthesis of an effective and low-cost absorbent for water treatment (defluoridation) and showed that denim-based absorbents are promising for water treatment and purification. They also help in solid waste management in the denim industry.

Mexico has one of the world's largest privately owned wastewater treatment plants with state-of-the-art technology that recycles about 75% of its water and is used for other production processes, rest rooms and landscape irrigation.

The net emissions of greenhouse gases by cotton fields are negative because cotton plants store more carbon than is released by the inputs used during production or growth. Hence, the carbon footprint of cotton farming is better than neutral. The Australian cotton industry recorded minor greenhouse gas emissions, which is less than one-third of 1% of total emissions by the country's agriculture, in the range of 0.16%–0.29%.

Invista, one of the world's largest integrated producers of polymers and fibres, and owner of the Lycra brand, introduced the first commercial offering of a bioderived spandex, the stretchy fibre commonly found in tights, jeggings, swimwear, sportswear and dance wear.

Research by Annu Kumari et al. [13] used regenerated cellulosic fibres such as viscose, modal, Tencel and bamboo as weft yarn and indigo-dyed cotton as warp yarn for denim manufacture and studied the comfort properties using the Kawabata Evaluation System.

## 2.6 Future trends

Led by innovations in manufacturing, new shopping patterns and an increase in the use of sustainable production methods, the global premium denim market is predicted to grow 8.09% by 2020, according to a new report by analyst team Technavio. It is estimated that by 2020 the global denim industry will be worth USD 64.1 billion. The future of denim looks promising, with new trends and innovations.

The future trends of denim are briefly described below:

According to Technavio, the demand for stretch denim will be more compared with nonstretch denim. Although the cost of stretch denim is 10%–15% higher, it is preferred by consumers owing to its comfort and style.

The use of organic cotton is preferred for making organic denim; also, potato starch, natural indigo, etc. are preferred over tapioca starch and synthetic indigo. Indian designers have developed eco-jeans using 100% organic cotton, skipping the bleaching process.

CORDURA denim is made from a blend of cotton and T420 nylon 6.6 fibre. This particular blend possess the look and feel of cotton denim but improved functional properties such as abrasion resistance and toughness. Jeans made with CORDURA Denim fabric are exceptionally durable.

Denim fabric is also manufactured using polyamide, lycra, polypropylene or polyester and a special bonding with 100% nylon net for a more lively look. This fabric is actually a mix of nylon, cotton and elastane blend yarns.

A granular softening technology is used to manufacture denim fabric to increase its productivity. The specialty of this technique is that it potentially reduces the use of amount of water, energy and labour required for manufacturing.

Storm denim is a treatment by Cotton Incorporated which offers a superrepellent finish to denim. It protects the wearer from moderate rain, snow and wet conditions without compromising the natural comfort and breathability of cotton.

A patina effect gives a look of corroded metal to denim. A bright orange pigment is added to an adhesive and applied manually over the fabric surface with the aid of a sponge. Then a turquoise foil is applied to the surface, which finally gives a metallic patina effect.

Black foil is applied to the denim surface in an irregular pattern to impart a leather-like look.

Bio-indigo, a natural dye extracted from the plant *I. tinctoria*, can be used to dye denim. It is the only powder form of indigo certified by Global Organic Textile Standard International Working Group Version III by the Control Union of the Netherlands.

Hemp is a more sustainable fibre than cotton, and hence hemp can be blended with cotton for denim manufacture. Hemp jeans hits denim fashion trends with numerous advantages.

In 2013, Levi's introduced wool denim (a blend of Australian Merino wool and fine-grade cotton) which is more comfortable than 100% cotton denim. Wool denim is appropriate for winter.

Printed denim is trendy and suitable for kids and women's wear.

Lenzing introduced the 4S Denim Capsule Collection ('4S' stands for 'Softness, Stretch, Science and Sustainability').

## 2.7 Conclusion

*Success is the sum of small efforts, repeated day in and day out.*

***Robert Collier***

Sustainability cannot be achieved at once. It needs careful research and understanding. Then it has to be implemented at every stage of the supply chain. It needs the participation and cooperation of all stakeholders along the entire supply chain. The future of denim seems to be sustainable. The need for sustainable production has to be communicated to manufacturers; similarly, sustainable use and disposal need to be communicated to the consumer. Recycling, upcycling, use of substances that does not harm the environment, and water and energy savings will be the key elements of sustainable denim production. Some leading manufacturers and brands of jeans have already begun research into and implementation of sustainable production.

Sustainable practices have to be:

- grown along with the fibre,
- spun into the yarn,
- woven along with fabric and
- imparted while dyeing and finishing.

Above all, consumer awareness regarding the sustainable use and disposal of denim is important.

## Sources of further information and advice

1. 'World Denim Production and Consumption Report: 2012' at www.denimsandjeans.com
2. 'US Denim Jeans Market Report: 2014' by Koncept Analytics
3. 'Indian Denim Market: Trends and Opportunities (2012–17)' by Daedal Research
4. 'Global Premium Denim Jeans Market 2015–19' by Technavio
5. www.levistrauss.com – production – sustainability
6. The life cycle of a jean: Understanding the environmental impact of a pair of Levi's 501 jeans
7. '2016 Market Research Report on Global Denim Fabric' at www.deepresearchreports.com

## References

[1] Denim Book. Barcelona: Clariant; 2012.
[2] http://www.suryalakshmi.com/Uploads/PDF/whitepaper001.pdf.
[3] http://www.sebi.gov.in/cms/sebi_data/attachdocs/1430568646805.pdf.
[4] http://www.business-standard.com/article/markets/excess-capacity-hits-denim-industry-114082100969_1.html.
[5] Allen Woodburn Associates Ltd./Managing Resources Ltd.. Cotton: the crop and its agrochemicals market. 1995.
[6] https://www.icac.org/getattachment/Home-International-Cotton-Advisory-Committee-ICAC/measuring-sustainability-cotton-farming-full-english.pdf.
[7] Farahat TM, Shaheen HM, Sanad ZF, Frag NA. Knowledge, attitudes, and practices of organophosphorus pesticide exposure among women affiliated to the Manshat Sultan Family Health Center (rural area) in Menoufia governorate: an intervention study. Menoufia Med J 2016;29:115–20.
[8] Splitstoser JC, Dillehay TD, Wouters J, Claro A. Early pre-hispanic use of indigo blue in Peru. Sci Adv 2016;2:e1501623.
[9] Chia MA, Musa RI. Effect of indigo dye effluent on the growth, biomass production and phenotypic plasticity of *Scenedesmus quadricauda* (Chlorococcales). Ann Braz Acad Sci 2014;86(1):419–28.
[10] Rannug U, Bramstedt H, Nilsson U. The presence of genotoxic and bioactive components in indigo dyed fabrics – a possible health risk? Mutat Res 1992;282:219–25.
[11] Schimper C, Ibanescu C, Bechtold T, Doppler C. Surface activation of dyed fabric for cellulase treatment. Biotechnol J April 2011. http://dx.doi.org/10.1002/biot.201100002. Wiley-Blackwell http://doi.wiley.com/10.1002/biot.201100002.
[12] Mendoza-Castillo DI, Reynel-Ávila HE, Bonilla-Petriciolet A, Silvestre-Albero J. Synthesis of denim waste-based adsorbents and their application in water defluoridation. J Mol Liq 2016;221:469–78.
[13] Kumari A, Khurana K. Regenerated cellulose-based denim fabric for tropical regions:an analytical study on making denim comfortable. J Text 2016;2016:4614168.

## Further reading

[1] Palabiyik SS, et al. Immunomodulation and oxidative stress caused by silicosis. Arh Hig Rada Toksikol 2013;64:431–7.
[2] Sadeghian Maryan A, Montazer M, Rashidi AS. Nanoclay as new tool for discolouration of dyed denim garment with indigo. Prog Color Color Coat 2013;6:25–36.
[3] http://www.levistrauss.com/wp-content/uploads/2014/01/A-Short-History-of-Denim2.pdf.
[4] https://en.wikipedia.org/wiki/Jeans.
[5] www.iswa.org/index.php?eID=tx_iswaknowledgebase_download&documentUid.
[6] http://www.historyofjeans.com.
[7] http://www.denimassociation.com/knowledge-centre.aspx?mpgid=12&pgid=15.
[8] http://waterfootprint.org/media/downloads/Report18.pdf.
[9] http://cottonaustralia.com.au/uploads/factsheets/INNOVATION__TECH_FACT_SHEET_TEMPLATE_OL_Climate_challenges_and_cotton.pdf.

[10] http://www.cottoncampaign.org/uploads/3/9/4/7/39474145/2007_ejf_deadlychemicalsin-cotton.pdf.
[11] http://www.engineerlive.com/content/new-concept-sustainable-denim-production.
[12] http://levistrauss.com/unzipped-blog/2014/02/recycling-water-to-make-your-jeans-info-graphic/.
[13] http://medicalxpress.com/news/2015-06-squatting-skinny-jeans-nerve-muscle.html.
[14] https://www.buildinggreen.com/blog/new-cotton-insulation-enters-market.
[15] http://www.fashionintime.org/history-jeans/.
[16] http://wwd.com/markets-news/denim/india-denim-market-growth-10388888/.

# Environmental impacts of denim washing

# 3

*A.K.R. Choudhury*
Retired Professor, Government College of Engineering and Textile Technology, Serampore, India

## 3.1 Introduction

A number of technological factors have contributed to making denim the fashion icon that it is today, including vast improvements in spinning, weaving and finishing. One of the most important parts of creating beautiful denim jeans is washing. Washing has such an important part in the denim chain because of the umpteen effects that consumers look for in their jeans. Denim is a firm 2/1 or 3/1 cotton twill-weave fabric (work-wear twill) with a dyed blue warp and raw white weft. Coarse yarns are used in warp and weft. An attempt was made to produce a cotton denim style using 100% polyester threads in the 1980 period. However, denim is the most preferred dress material in cotton and the ensuing period may witness the role of blends in denim. The warp is dyed in a solid colour, usually with indigo. Because it is warp faced, denim fabric is blue on the surface and almost white on the back. The warp is sometimes dyed during sizing.

As early as 1873, after reaching North America, Bavarian immigrant Levi Strauss realized that an urgent need could be satisfied with this material. At the time of the California Gold Rush, tough, hard-wearing trousers were needed by hard-working gold prospectors (miners). Strauss designed a pair of trousers from denim material provided with ample pockets. He achieved rapid success and was soon able to build a factory in San Francisco for their production. These trousers were given the name 'jeans'. The story goes that 'jean' is derived from the word 'Genoa'. It refers to the material that sailors from Genoa used in their pants. This was a coarse cotton wool and/or linen blend. It originally came from Italy and is evidence of the custom of naming a material for its place of origin. By the late 16th century, jean was already being produced in Lancashire, England. The composition eventually evolved to 100% cotton by the 18th century. Today, 'jeans' usually refer to a garment that has five pockets (two in the front, two in the back and a small change pocket on the front right pocket); this style can be made using any kind of fabric, be it corduroy, twill, or bull denim.

The origin of the term 'denim' can be traced to late 16th-century France, where a fabric known as *serge de Nimes* (twill from Nimes) was popular. Some doubt remains as to whether the contraction *denim* actually came from this French fabric or twill called *nim*, also used in France at the time. *Serge de Nimes* was a blend of silk and wool, which has led some historians to doubt whether this was truly the origin of modern-day denim. Either way, the history of jeans goes this far back in history. Both

Sustainability in Denim. http://dx.doi.org/10.1016/B978-0-08-102043-2.00003-4
Copyright © 2017 Elsevier Ltd. All rights reserved.

fabrics (jeans and denim) grew in popularity, with denim being the stronger and more expensive of the two. The major difference between them was that denim was woven with one coloured thread (the warp) and the other white (the weft), whereas jean was woven with two coloured threads.

Jean and denim remained two different fabrics and were used for different types of clothing. Denim was used mainly for workers' clothes and jean for lighter clothes that did not have such a high need for durability.

A big problem with miners' clothes was the pockets, which easily tore from the jeans. Jacob Davis had the idea of using metal rivets (fasteners) to hold the pockets and the jeans together so that they would not tear. Davis wanted to patent his idea but he did not have enough money, so in 1872 he wrote to Levi Strauss and offered Strauss a deal if Strauss would pay for the patent. Strauss accepted.

In 1886 Levi sewed a leather label onto the jeans. The label was printed with a picture of a pair of jeans being pulled between two horses. In 1963 Levi Strauss first marketed prewashed/preshrunk jeans which did not shrink further after purchase, allowing the consumer to buy his or her correct size.

Fig. 3.1 shows a flowchart of denim fabric production. Ring-spun or open-end spun yarn is sent for warping, and then warp yarns are dyed by the rope dyeing, sheet dyeing or sheet dyeing method (discussed in Section 3.2). The dyed warp yarns are sized and woven into fabric, which is subsequently finished, inspected for faults, folded and dispatched for garment making. Denim garments have a lot of demand in the market of regular garments as well as in the fashion market. People of all ages, especially youth, have great interest in denim. Different value-adding processes such as industrial washing not only make denim look beautiful but impart some functional properties to the garments. Ready-made garments are made from dyed or pigment printed fabrics and then the garments are washed by different garment washing techniques. Thus the color, look and comfort of the garments are modified and a new appearance is produced in the garments which is not possible using any other method. Moreover, starch present in the garments is removed by washing. Washed garments can be worn after purchase directly from the store or shop. Some garments shrink after washing; but, washed garments can be purchased as per the required size as further shrinkage is nil or negligible.

The versatility, creativity, and uniqueness of denim are the virtues of denim washing. Denim washing has totally changed the market potential and demands of denim fashion around the world. From stonewashing to the concepts of emerging enzyme washing and biopolishing, the effects and shades achieved are a great breakthrough in the denim clothing industry.

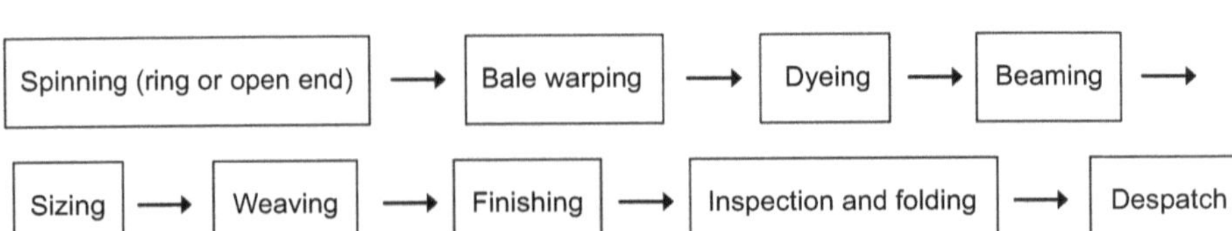


**Figure 3.1** Flowsheet for denim production.

## 3.2 Denim dyeing

Jeans made from blue denim have been popular over a longer period than any other item of apparel. They are inexpensive, durable and versatile. The blue-dyed warps wash down to an attractive blue without staining the white weft.

Indigo dye was originally extracted from the *Indigofera tinctoria* plant, but today's indigo dyes are man-made or synthetic. The dye was synthesized 14 years after its chemical structure was identified by Adolf Bayer in 1897. Indigo is a vat dye:

- It is water-insoluble and made soluble by reduction with caustic soda and sodium hydrosulphite and oxidized after application to textile yarn or fabric.
- It has poor affinity for cellulose fibre.
- Its colour fastness to light and washing are moderate.
- SDC (The Society of Dyers and Colourists, UK) colour index number of the dye is C.I. Vat Blue 1 and chemical formula is $C_{16}H_{10}O_2$.

Insoluble indigo dye is first made water soluble by converting into leuco compound by reducing it with sodium hydrosulphite in the presence of caustic soda at around 60–65°C (Eq. 3.1). Leuco salts are sparingly soluble in water. In the presence of excess alkali, leuco indigo is converted into sodium salt of leuco indigo (Eq. 3.2). After impregnation in this solution, the dyed yarn is run through air for a short duration when the soluble leuco salt is oxidized back into the original insoluble indigo.

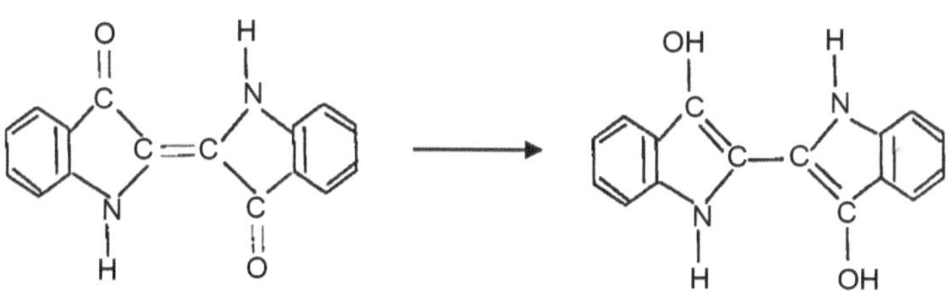


(3.1)

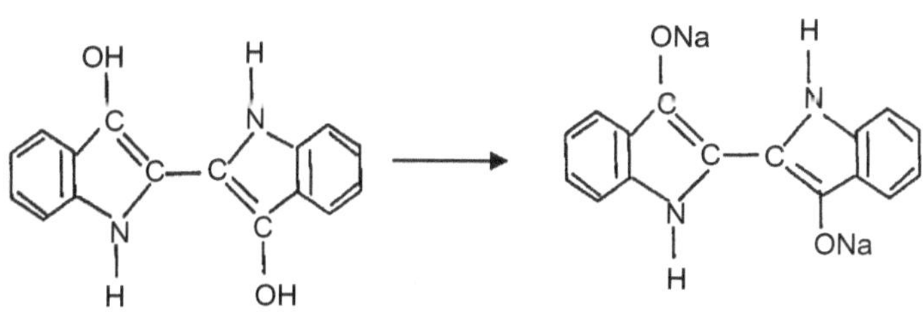


(3.2)

Leuco indigo has poor substantivity for cellulose. Hence, only pale depths are obtainable by exhaust dyeing procedures. Indigo is applied by a series of dips, with intermediate squeezing and atmospheric oxidation. By repeating the process, dye is applied to the substrate layer by layer to give deep dyeing with relatively low rubbing fastness. Denim clothes (dyed warp and natural white weft) are dyed in the form of warp sheets or ball warps, usually continuously. The batch-wise dyeing of indigo is of little importance. The batch-wise dyeing of hanks of yarn may be carried out in becks with repeated short dips, followed by squeezing and atmospheric oxidation. Wound packages cannot be dyed in full depth with indigo.

In the continuous indigo dyeing of warp yarn, the yarn is in the form of a full-width warp beam or in ball form, containing 300–400 individual threads. It passes through several (five to six) vats, each followed by passage through air for oxidation. The processing speed is 20–30 m/min with an immersion time of 20–30 s in each vat. Upon leaving the vat, the material is squeezed to a liquor pickup of approximately 100%, after which the dyeing requires around 2 min for oxidation before it passes into the next vat. Rinsing at ambient temperature in two or three rinsing baths after oxidation is normally adequate. The dip tanks are coupled to keep the dye bath composition exactly the same. A good volume of flow with minimum turbulence is essential to ensure uniformity.

The stock vat is prepared by mixing indigo dye with 0.7–0.8 times caustic soda (solid) and 0.8–0.9 times sodium hydrosulphite. The vatting is done at 60–65°C for 15–30 min.

A typical recipe for dye-liquor is as follows [29]:

1. caustic soda (27% by weight), 5 mL/L
2. sodium hydrosulphite, 1.5 g/L
3. stock vat, 62.5 mL/L

The stock vat is composed of:

1. indigo pure (BASF, C.I. Vat Blue 1), 80 g/L
2. dispersing agent (Setamol WS), 4 g/L
3. wetting agent, 1 g/L
4. caustic soda (27% by weight), 130 mL/L
5. sodium hydrosulphite, 60 g/L

Two grams per litre sequestering agent (sodium hexametaphosphate) and antifoaming agent may be added.

The content of caustic soda and hydrosulphite must be checked at regular intervals to determine what quantities are necessary. The amount of hydrosulphite may be monitored by controlling the potential difference (milli Volt) across the solution. The pH is kept near 11.

Careful buffering of the dye bath pH can be an effective means to conserve indigo and achieve more reproducible dyeing. Depending on the dye bath's pH, reduced indigo can exist in three forms: as a nonionic enolic acid leuco compound, a monoenolate anion or a bis-enolate anion. A monoenolate moiety is much more soluble than the acid leuco form and more substantive than doubly charged bis-enolate. Its amount reaches a maximum at pH 10.5–11.5 and the colour yield correlates closely with its fractional amount.

Warp preparation, dyeing and sizing procedures are of great significance in the manufacture of denim. Dyeing with indigo requires a combination of practical experience and precise knowledge of the physical and chemical backgrounds. The various dyeing and sizing methods, recipes and concentrations, and the reaction and oxidation times not only influence the associated weaving process but also determine to a substantial extent the appearance, characteristics and quality of the denim.

There is a fundamental distinction between the classic indigo rope dyeing method and the slasher, sheet dyeing or double-sheet dyeing methods, in which the dyeing and sizing processes are carried out in a single cycle or separately. Another variation is the loop dyeing process.

### 3.2.1 Rope dyeing method

First, 350–400 warp threads are combined by the ball warper to form a rope that is 10,000–15,000 m long. Twelve to 36 ropes at a time are conducted through the continuous dyeing plant, dried on cylinders after dyeing and deposited in cans. The ropes are converted into warps on the long chain beamer. These are introduced to the sizing machine, sized and assembled to form the weaver's warp with the required number of ends. The sequence for rope dyeing is:

ball warping→rope dyeing→rope opening (rebeaming)→sizing→weaving

Advantages of rope dyeing are:

- better squeezing takes place, resulting uniform and better oxidation, because it is squeezing in rope form
- better fastness
- no stripe effect
- less side–centre–side variation

A disadvantage is:

- the process sequence is high, leading to more space and a higher worker requirement

### 3.2.2 Slasher or sheet dyeing method

Instead of ropes, warps are fed into the continuous dyeing/sizing machine. The warps are dyed, oxidized, dried and sized, and after drying again, assembled to form the warp. The sequence for rope dyeing is:

beam warping→dyeing and sizing combined→weaving

### 3.2.3 Double-sheet dyeing

A further development of the sheet dyeing method, the dyeing and sizing processes are carried out separately here. Double numbers of warps are fed into the dyeing machine and then dyed, oxidized and dried. The sequence for double-sheet dyeing is:

beam warping→dyeing→sizing combined→weaving

After weaving, the denim fabric is finished. The sequence of finishing is:

grey fabric→opening→skew imparting→drying→rubber belt unit (sanforizing)→Palmer unit→finished fabric

## 3.3 Selection of denim fabric

The right selection of fabric can minimize the cost of treatment and solve environment-related issues. There are several variations in dyed denim fabric, such as (http://www.facebook.com/topic.php?uid=118135278228528&topic=179):

- varying colour depth
- quick-wash denim
- tinted denim
- soft-feel denim.

### 3.3.1 *Fabrics with varying colour depth*

To cut the processing time and effluent load and achieve minimum damage to the fabric and the minimum use of chemicals, there are different depths of indigo that can be used with denim. For example, in the case of an ice wash, in which more than half the dye is removed during washing, one can use a lighter shade of fabric, which will cut the process time, chemical consumption and effluent load. It will help the garment processor work more economically and with minimum error.

### 3.3.2 *Flat-look denim*

Different chemicals and processes are used to obtain a flatter look to denim garments. To meet this requirement special denim fabric has been developed which offers a flat look after washing.

### 3.3.3 *Quick-wash denim*

Quick-wash denim fabric is dyed with a modified technique of dyeing, so that during the wash cycle, indigo dye can be removed quickly, giving a washed look at a shorter washing cycle. This results in more economical washes, i.e., low water consumption, less use of chemicals, less time and retained fabric strength.

### 3.3.4 *Tinted denim*

With an increase in demand for a tinted/overdyed look to garments, garment processors are using an additional process of tinting/overdyeing, which is time-consuming. It consumes large quantity of water and chemicals. It is also associated with the risk of patches and unevenness in garments. Denim fabric is also available in a tinted form, which saves processors time and risk.

### 3.3.5 *Grainy look denim*

Different chemicals and processes are used to obtain a grain look in the denim garments. To meet this requirement, special denim fabric has been developed which gives a grainy look after the denim garment is processed.

### 3.3.6 Soft-feel denim

To meet such a need, different varieties of denim fabric are available which have inherent softness. These fabrics require a minimum application of softener at the garment stage.

## 3.4 Denim washing

Denim washing is the aesthetic finish given to the denim garment to enhance its appeal and provide strength. The emphasis is on comfort and softness. Fashion trends favour the broken-in look and worn/faded seams that can only be achieved through garment processing. Much of the appeal of dry denim lies in the fact that with time the fabric will fade in a manner similar to that which artificially distressed denim attempts to replicate.

The indigo dye remains mostly at the surface; hence the dyeing process may be called *ring dyeing*. Such ring-dyed materials are subjected to treatment with stones or enzymes, which removes dye randomly from portions of the fabric, exposing the white surface. This popular style is used in *faded jeans*. Microscopy reveals that for indigo dyeing, the cross-section of the resulting dyed cotton yarn depends on the pH of the bath. When the pH of the dye bath is decreased from 13 to 11, the denim yarn progressively becomes more ring-dyed. Associated with increased ring dyeing, more colour yield is obtained, making the wash-down process easier. The highest colour yield was observed within the pH range of 10.8–11.2 [28].

Denim comes in all forms, looks and washes to match every type of dress. In denim, garment washing is done to produce effects such as colour fading with or without patchiness, crinkles, seam puckering, hairiness, depiling, softened hand feel and stabilized dimensions.

Every small step in denim washing makes a big difference because indigo has poor wet and dry rubbing fastness. All parameters (e.g., pH, material:liquor ratio, rpm of the machine) are critical to maintain repetitive results.

### 3.4.1 Garment washing/dyeing machines

A garment dyeing and washing factory needs the following machines:

- garment dyeing/washing machine for samples and bulk
- hydroextractor
- dryer

Basically there are two types of machines used to dye garments:

- pedal
- rotary

The pedal-type garment dyeing machine is the first and oldest such machine still used in European countries. Pedal-type garment dyeing machines are either of the overhead or side pedal type.

Some important steps in the processing of denim garments are:

1. pretreatment (desizing, rinsing, scouring, etc.)
2. enzyme or stonewash
3. cleaning up to adjust the desire effect
4. bleaching
5. tinting/dyeing
6. softening

## 3.5 Desizing

Desizing is the process of removing size applied during denim fabric making in warp yarn. The size is applied to all woven fabrics to strengthen the yarn for weaving. There are many types of sizes available on the market; they can be divided into two major groups:

- water-soluble carboxymethyl cellulose- or poly(vinyl alcohol)-based sizes
- water-insoluble starch-based sizes. Starch-based sizes are most commonly used because of their low prices and ready availability.

Methods for removing sizes from denim jeans are:

- washing with high alkaline agents (i.e., soda ash)
- washing with highly acidic agents (i.e., acetic acid)
- washing with oxidative chemicals (i.e., hydrogen peroxide)
- desizing with $\alpha$-enzymes; this is eco-friendly and convenient.

## 3.6 Garment washing

There are several objects of garment washing [20]:

1. to develop softness in garments: Size materials applied during manufacturing that are present in the fabric are removed, which enhances the soft hand feel. Additional softness may be attained by using softeners.
2. to introduce a fading effect: Dyes or pigments present in the fabric, used during colouration, are washed out locally or partially, resulting in fading or worn-out effects in the garments.
3. to create new fashion: The washing process of garments creates a different outlook (faded, colour tinted, etc.), thus creating new fashions for new generations especially for teenagers.
4. to satisfy the consumer: Because contraction or extraction (shrinkage) does not occur during washing, the wearer can use the garments satisfactorily after purchasing them.

There are advantages to washing garments [20]:

1. Removal of starch or size materials gives the fabric a soft hand feel.
2. Softness can be increased by adding a softener at the very last stage of washing.
3. If dirt, spots, impurities, gum, etc., have accumulated in the garments during manufacturing, they can be removed.

4. During washing shrinkage may take place; therefore no shrinkage occur during washing of garments.
5. Washed garments can be worn directly after purchase.
6. Faded or worn effects can be introduced to the garments, creating a new fashion.
7. A similar look can be obtained by different washing techniques.
8. Comparatively lower capital is required to set up a washing plant.
9. A smaller amount of property and fewer paid workers are required to run a washing plant.

Limitations of garments washing are that:

1. garment sizes change: Size change takes place owing to the shrinkage properties of fabric. The amount of shrinkage of the fabric determines the size change of the garments during washing.
2. size material is partly removed: Unfixed dyes may remain on the surface of the garments and it is necessary to remove them completely after washing.

## 3.7 Types of garment washing

Several types of garment washing for denim fabric are carried out to create varied effects in denim garments to meet the needs of today's denim fashion trends (http://www.burton.co.uk/blog/denim-washes-and-finishes/?geoip=home). However each washing technique has its own advantages and limitations. Denim is either rubbed or worn with stones and other abrasive materials, called mechanical washing, or treated with bleach and other kinds of colour-altering substances, known as chemical washing.

Mechanical washes include:

1. rinse wash
2. water jet fading
3. stonewash
4. whiskering
5. microsanding including sandblasting
6. mechanical abrasion
7. laser marking

Chemical washing includes:

8. acid wash, ice or snow wash
9. hydrogen peroxide wash
10. enzyme fading
11. ozone fading
12. spray techniques
13. overdyeing and tinting

The effects desired for denim goods change every year owing to fashion. Various washing effects have been popularly obtained by laser, sandblasting and enzymatic washing processes. In addition to these processes, hypochlorite bleaching (acid wash) is often preferred, especially for summer denims [30].

### 3.7.1 Rinse wash or normal wash

This is the simplest type of industrial garment wash. The main aim is to remove starch, dust and dirt from garments. Sometimes reasons for a normal wash include softening and giving the garment a used look. For colour garments, removal of unfixed dyes can be achieved to improve colour fastness. In special cases, intentional shrinkage in garments can be achieved. Precautions need to be taken against unwanted shrinkage issue and back staining.

It is most basic wash for denim garments. Typically named a dark wash, the only purpose of this wash is to make the garment wearable. During the construction of denim fabric, starch is applied to strengthen the warp. This makes the fabric pretty stiff and harsh to skin. In a rinse wash the garment is desized and some softening agent is applied to improve the hand feel of the garment. In some cases, potassium permanganate (PP) is applied to the blasted area with a spray gun or paintbrush. This is commonly used where the fading effect is desired to have high intensity.

A sample recipe for a normal wash is:

If the lot size is 100 kg cotton shirts,
Material: liquor ratio (M: L ratio)
Water = 500 L
Detergent = 1% (based on the weight of the garments)
Temperature = 40–60°C
Time = 15–30 min
Back-staining chemical is used if required

### 3.7.2 Water jet fading

Water jet or hydrojet treatment was developed to create patterns and/or to enhance the surface finish, texture and durability, and other characteristics of the denim garment. Hydrojet treatment generally involves exposing one or both surfaces of the garment to hydrojet nozzles. The degree of colour washout, clarity of patterns, and softness of the resulting fabric are related to the type of dye in the fabric and the amount and manner of fluid impact energy applied to the fabric. Particularly good results are obtained with blue indigo-dyed denim. Because this process does not involve chemicals, it is pollution free. Using the water recycling system, the technique can be employed as economically and environmentally friendly denim processing. Colour washout of dye in the striped areas produces a faded effect without blurring, loss of fabric strength or durability, or excessive warp shrinkage [9].

### 3.7.3 Stonewashing

A number of people and organizations have claimed to have invented stonewashing:

- According to Levi Strauss & Co., Donald Freeland, an employee of the Great Western Garment Company (later acquired by Levi's), invented 'stone-washing' denim in the 1950s.
- Inventor Claude Blankiet has also been credited with having invented the technique in the 1970s.
- The jeans company Edwin claims to have invented the technique in the 1980s.
- In 1982, Lee introduced 'stone washed' jeans. This was followed by 'acid-washed' denim, which produced an even more faded look.

Stonewashed denim gained popularity in the 1960s and remains popular to this day. This type of denim is characterized by a lightly distressed, vintage look. The name comes from the fact that the original processing method involved rubbing pumice stones over the denim to wear down the fabric. Pumice stones are the silica-rich, high-temperature melted product of explosive eruptions during the ascent and expulsion of the earth. These are lightweight, highly porous, rocky substances which will float on water. Major pumice stone–supplying countries are the United States, Turkey, Italy, Iceland, New Zealand, Japan, Indonesia and the Philippines. It is an important component of stonewashed apparels and is used to abrade the surface of the fabric to obtain a surface pattern effect with colour contrast and a soft hand.

Because the degree of abrasion may vary in different parts of garments, such as the trouser leg, button slays and seaming parts, a number of neutral patterns can be formed. The degree of the wash effect depends on the stone size, stone ratio, liquor ratio, duration of treatment, amount of garments loaded, garment's grams per square metre, etc. Stone sizes vary from 1 to 7 cm in diameter. Other washes such as sand, micro and microsand wash are refer to the use of very small size pumice stones.

*Selection of stone*: Stones should be selected for their proper hardness, shape and size for the particular end product. Large, hard stones last longer and may be suited only for heavyweight fabrics. Smaller, softer stones can be used for lightweight fabrics and more delicate items (stone weight/fabric weight=0.5 to 3/1). It depends on the degree of abrasion needed to achieve the desired result. Stones can be reused until they completely disintegrate or wash down the drain.

However that method proves problematic owing to environmental concerns regarding the use of pumice and the fact that the stone often weakens the fabric too much, causing wear and tear. In the process of stonewashing, freshly dyed jeans are loaded into large washing machines and tumbled with pumice stone or volcanic rock to achieve a soft hand and desirable look. Variations in composition, hardness, size, shape and porosity make these stone multifunctional. The process is expensive and requires a high capital investment. Pumice stone give the additional effect of a faded or worn look as it abrades the surface of the jeans like sandpaper, removing some dye particles from the surface of the yarn.

The procedure is as follows:

1. Load stones into the machine.
2. Load garments into the machine (the ratio is usually 0.5–3.0 part weight stones to 1 part weight garments).
3. Desize with α-amylase enzyme and detergent. The liquor ratio is approximately 5–8:1.
4. Rinse.
5. Refill and tumble with stones for 30–90 min, depending on the desired effect. The Liquor ratio is 5–8:1 at 50–70°C. Scouring additives can also be used.
6. Drain. Separate garments from stones (garments can be transferred to another machine).
7. Rinse.
8. Apply a softener (garments can be transferred to another machine for softening).
9. Extract and unload.
10. Destone and tumble dry.
11. Press, if required.

Softeners and/or lubricants can be added during Steps 3 and 5 to reduce the potential for creasing.

Stonewashing the denim with pumice stones has some disadvantages:

- Stones could cause wear and tear of the fabric.
- The quality of the abrasion process is difficult to control and the outcome of a load of jeans is never uniform. A slight percentage always gets ruined by too much abrasion.
- The process is nonselective.
- Metal buttons and rivets on the jeans in the washing machines get abraded.
- It creates the problem of the environmental disposition of waste for the grit produced by the stones.
- High labour costs become necessary because the pumice stones and the dust particles they produce need to be physically removed from the pockets of the garments and machines by labourers.
- Denim needs to be washed several times to get rid of the stones completely. The process of stonewashing also harms big, expensive laundry machines.

#### 3.7.3.1 Substitution of pumice stone

To overcome the shortcomings of pumice stones, synthetic stones were developed. These are made of abrasive materials such as silicate, plastic, rubber and Portland cement.

Major problems associated with the use of volcanic-grade pumice stone can be overcome by using these products. Advantages of using these synthetic stones are:

1. The durability of such a product is much higher and can be used repeatedly from 50 to 300 cycles, depending on the type of synthetic stone.
2. Reproducibility of washing is manageable.
3. Because there is much less stone discharge in the process, the process is economical and ecological.
4. There is less damage to the machine and garments.

Perlite is a form of naturally occurring silicon rock. It has the distinctive property of expanding to 4 to 20 times its initial volume when heated at a particular temperature. This happens because the raw Perlite rock consists of 26% water. When it is heated above 870°C, crude perlite rock becomes swollen and tiny sealed glass bubbles form. Its original black or grey colour changes to greyish white or else white. This heated form of perlite is used for stonewash purposes.

It has the same function of stonewashing as stones. Perlite treatment reduces the rate of harm caused to large washing machines by pumice stones and gives denim better suppleness and a softer finish. Many jeans manufacturing companies use Perlite in the place of enzymatic treatment, which reduces the rate at which jeans wear out when used. It gives a uniform worn and old look throughout to the denim, not just the upper part of the garment. There are many grades of Perlite, differing in size. Each is used to giving the right stonewash finish to denim, from the largest to the finest grades; some are very tiny, just like ground earth [14].

### 3.7.4 Simultaneous stonewashing and tinting

Tinting of denim garments is usually done after the stonewash process. In this procedure, the garment has been lightly coloured to give the final denim appearance a slight shift. This is not true overdyeing but merely gives the impression of a change in the overall colour of the fabric. This process consumes a large quantity of water and chemicals. To make this process economical and ecologically friendly, some novel colour-based enzymes have been introduced to the market. Using this technique, tinting and stonewashing can be achieved in a single bath, resulting in:

- less process time to achieve a tinted look
- no extra chemical required, which makes the process more economical
- less water consumption
- less energy consumption
- less chance of patchiness or unevenness

Main disadvantages of this procedure are:

- the difficulty of removing residual pumice from processed clothing items
- damage to the equipment by an overload of tumbling stones
- pumice stones and particulate material can also clog machine drains and the drains and sewer lines at the machine site [1].

Despite these disadvantages, pumice stone is still used on its own in some factories and for certain types of research [7].

### 3.7.5 Whiskering

Whiskering is the technique of artificially aging the look of denim by washing areas prone to bunching and fading over time. This is primarily done by hand with rods and other tools to simulate years of wear within minutes. The name comes from the signature look of whiskering: several rows of thin white lines sprawled across the thigh section.

Whiskers are one of the most important designs of a used look garment. The idea of whiskers is taken from the worn lines and impression patterns generated by natural wear on the hips and front thigh area. On old jeans, a number of patterns can be finding consequential to the fabric, the body shape of the user or the sitting posture. It is also known as cat's whiskers [15].

Rubber balloons are available with different pattern designs. Garments are mounted on balloons and filled with air to obtain impressions of whiskers. The garment is carefully scrubbed over the pattern with sandpaper on engraved pattern lines. This operation requires highly skilled labour that can handle uniformity and sequence in whisker lines. Fabric may be damaged during the rubbing process if care is not taken. This method is famous for high quality and cost-effectiveness.

Sometimes separate patterns are used and fixed on the balloons to make moustaches. These patterns are made of thin rubber sheets and electric wires or rubber cords and pasted on the balloons for pattern lines. Thin line patterns are easy to work with in that way, but it is not as successful for broad lines.

Engraved patterns on thick boards such as hard rubber sheets are widely used to make whisker impressions. This idea is simple for drawing lines on rubber sheets and engraving them with blades. The garment is placed on the sheet and scraped in a specific area to draw the impression. For low-cost patterns, it is most frequently used in small industries, especially where production is not consistent in style.

### 3.7.6 Microsanding

A fabric finishing process in which fabrics are sanded (using actual sandpaper) to make the surface soft without hairiness can be performed before or after dyeing. In this fabric treatment process, a series of cylindrical rolls in a horizontal arrangement, either wrapped with an abrasive paper or chemically coated with an abrasive, is used to create a soft suede hand. The denim is pulled over the face of the sand rollers, creating a raised surface finishing. Some colour reduction is experienced.

There are three parts to this technique:

1. sandblasting
2. machine sanding
3. hand sanding or hand brushing

Microsanding is used in various ways: on flat surfaces (tables and ironing boards) or inflatable dummies (standing, flat or seated). Various templates can be used to create a three-dimensional (3D) effect.

#### 3.7.6.1 Sandblasting

Sandblasting involves blasting an abrasive material in a granular, powdered or other form through a nozzle at very high speed and pressure onto specific areas of the garment's surface to give a desired distressed, abraded and used look (Fig. 3.2).

**Figure 3.2** Sandblasting process.

As the name implies, sandblasting creates light wash denim patterns by blasting areas of the fabric with sand or other abrasive materials at high velocity. The result is a subtle fade across the front and back of the pants leg.

Sandblasting is a mechanical process that creates localized abrasion or colour changes on the denim garment. The process involves blasting an abrasive material in granular, powdered form at a very high speed and pressure through a nozzle onto certain areas of the garment such as knees and elbows. The garment treated surface results in a distressed, abraded and used look. Common blasting materials used are sand and metal granules.

Before sandblasting the garment is first subjected to a stonewash to the desired degree of washing. It is then sandblasted.

A solution of sodium hypochlorite or PP is often sprayed in the desired area of the garment to obtain the same look. The garment is then neutralized, rinsed, softened and dried.

Sandblasting is a water-free process; therefore no drying is required.

Advantages of sandblasting process are that:

- It is a purely mechanical process that uses no chemicals.
- It is a water-free process; therefore no drying is required.
- A variety of distressed or abraded looks is possible.
- Any number of designs can be created by special techniques.

The procedure of sandblasting is as follows [15]:

- The individual garment to be processed is put on a table.
- An air compressor is started and pressure is maintained at 3–4 kg/cm$^2$.
- The air compressor then stores air in the machine's cylinder.
- Aluminium sand or silicon sand [$Al_2(SiO_4)_3$] is then put in the chamber of the blasting machine.
- Sand and compressed air are then sprayed over the garments at specific points at a 10–200 degrees angle by a hand spray pipe.
- Sand passes into the machine by a regulating system.
- The machine is composed of a blasting regulator and a hand regulator to control the amount of sand and compressed air, respectively.
- The angle of sand blowing is important. The higher the blowing angle, the higher the fading effect is and higher the risk is of fabric damage.
- Fading is done on premarked areas.
- Sandblasting is a hazardous technique; therefore the area should be isolated from other work areas.
- Handling the sandblasting gun is risky; therefore operators should use safety equipment such as hand gloves, masks and gowns. Production/hours/guns vary from 25 to 45 garments, depending on the degree of fading and the extent of fading to be attained.
- The faded garments are cleaned by shaking them manually. After sandblasting, garments are washed. The technique used may be normal washing or bleach or bio-stone washing.

### *3.7.7 Mechanical abrasion*

To give a worn effect or abraded or used look, some mechanical processes were developed. These are based on mechanical abrasion, which removes the indigo. Some of these processes are sueding, raising, immersing, peaching and brushing.

Advantages of these processes are:
Abrasion can be controlled.

- A different look to the garment can be achieved.
- All are dry processes.
- The processes are economically, ecologically and environmentally friendly.

### 3.7.8 Laser marking

A laser is a device that emits light through a process of optical amplification based on the stimulated emission of electromagnetic radiation. The term 'laser' originated as an acronym for 'light amplification by stimulated emission of radiation'.

Treatment with a $CO_2$ laser has been used in different areas of the textile industry for several years because it allows the brief surface design of patterns with good precision, desirable effects, and various sizes and intensity without causing much damage to the bulk properties of the textile materials [8]. Popular uses of lasers are laser marking (only the surface of the fabric is processed and faded), laser engraving (controlled cutting to a depth), laser welding (a molten material joins or welds two or more layers of fabric together) and laser cutting (cutting through a fabric) [27]. Among these lasers, fading (marking) is a popular dry process for denim. It has been used extensively to replace some conventional dry processes such as sandblasting, hand sanding, destroying, and grinding, which are potentially harmful and disadvantageous.

Laser marking is a computer-controlled process for denim fading. This technique enables patterns to be created, such as lines and/or dots, images, text or even pictures. In one version of this concept, a mask is used to give a desired shape to be applied to the fabric. The laser projects through a lens system, which expands the beam. This beam is passed through the shaped mask, which is composed of an aperture of the desired shape, and then is deflected by a mirror to strike the textile substrate. The duration of exposure determines the final effect on the fabric.

Novelties of this system are that:

- It results in the water-free fading of denim.
- It is an eco-friendly and economical process.
- It can create local abrasion and fabric breaks, a used look and a moustache with excellent reproducibility and high productivity.
- Because it is an automatic system, the chances of human error are slim.
- The design is electronically translated onto the fabric, which avoids the need for photographic method of making screen for screen printing or serigraphy. The machine is simple and compact, it requires low maintenance and cleaning, and is extremely safe and reliable.

The laser is a source of energy which can be directed onto desired objects; its power and intensity can easily be controlled. Use of the laser enables great variety of materials to be cut from metal to fabric. It is then possible to transfer certain designs onto the surface of textile material by changing the dye molecules in the fabric and creating alterations in its colour quality values by directing the laser to the material at a reduced intensity. In designed systems, computer-controlled laser beams are used to change the colour of the dye material on the textile surface by directing the laser beams at a desired wavelength and intensity onto various textile surfaces selected for application. For this purpose, a laser beam source that can reach the initial level of power and can direct this beam at two axes, a galvanometer composed of an optical aperture, and a computer program that can transfer images obtained in standard formats for the galvanometer control card have been used [17].

Ortiz-Morales et al. [18] conducted a comparative study of laser fading using different types of lasers (Nd:YAG, CTH:YAG and $CO_2$) and varying parameters. The study showed that $CO_2$ laser-based processes were the most suitable ones because of their higher efficiency compared with other lasers. Pezelj et al. [23] studied the danger of fabric or fiber damage in laser treatment. Sample strength was tested and compared, as well as the average degree of cellulose polymerization as a measure of possible macromolecule degradation. Morphological changes in fibres also contributed to worn effects on the fabric. The macromolecular degradation of the fibres was moderate (maximum of 14%) and the reduction in strength was acceptable (maximum of 10%). Öndogan et al. [17] investigated patterning methods of denim goods by laser and handwork, and Özgüney [19] presented a comparative study of laser patterning and pigment printing methods.

Sarkar et al. [27] compared laser fading with manual dry processes and observed that the:

1. effect quality is comparatively finer in the manual system than the laser
2. hand feel is comparatively finer in the manual system than the laser
3. hairiness is comparatively less in the laser system
4. production cost is higher in the laser
5. work space required is less in the laser system
6. design consistency is higher in the laser system
7. tearing strength is higher in laser-faded denim (although it may differ in the case of stretch denim)
8. rejection rate is 0% in the laser system whereas it around 5% in the manual system
9. staff ratio required for laser to manual = 1:3 (approximately)
10. power requirement is higher in the laser system (approximately 7.5 kW)
11. laser beam is harmful to skin and eyes

### 3.7.9 Acid, ice and snow wash

This finish gives indigo jeans sharp contrasts. The process was created in Italy and was patented in 1986. Interestingly, acid wash denim does not use actual acid in the wash process. Instead the process uses pumice stones soaked in bleach to alter the colour in a nonuniform manner. The result is often splotchy yellowish patterns. This finish provides jeans with sharp contrasts in tonal colour. This is also known as ice or snow washing. This is achieved by dry tumbling with pumice stones soaked with bleaching agents to produce a 'snow' pattern effect on denim. Stones are used to deposit a chemical on garments to strip off the colour. This chemical deposit removes colour only from the outer surface of the garments and produces a frosted appearance. Indigo and selected sulphur dyes are the most popular candidates for this procedure.

Acid wash on denim jeans has become popular owing to its significant contrasts and attractive appearance in colour. Acid wash can be carried out on indigo- and sulphur-based fabric garments. Acid wash is a chemical wash process on denim which strips the top layer of colour and makes a white surface while the colour remain seen the lower layers of the material, giving it a faded look.

Acid was first launched in the 1980s as an innovative finish on denim garments. This wash is carried out by soaking stones in bleach, followed by neutralization.

Acid wash of denim garments normally carried out by tumbling the garments with pumice stones or towels presoaked in a solution containing sodium hypochlorite (5%–10%) or PP (3%–6%) for about 20 min to an hour. The longer the pumice or towel is soaked, the more prominent the effects will be. When the soaked stone or towel is in contact with denim, localized bleaching occurs, producing nonuniform a sharp blue-white contrast.

Acid-washed denim fabric sometimes become yellowish after washing. This is the result of incomplete neutralization, washing or rinsing, so that manganese is not removed from the garment. However the manganese can be removed by washing additional ethylenediamine-tetra-acetic acid as a chelating agent.

Acid washing proceeds thus:

- Desize with α-amylase enzyme and detergent. The liquor ratio is approximately 5–8:1. After 10–15 min, the material is washed, squeezed and taken out of the machine.
- The stones are saturated with chlorine/PP, excess water is removed from stones or towel and porous stones are dried in the open air for 60–90 min.
- The stones followed by the jeans are put in a dry washing machine and treated for about 20 min. Make sure machine does not have single drop of water in it. Open the drain over the complete process.
- At this time, the jeans are checked to ensure that they look acceptable. Too long a time in the first machine can totally bleach or ruin the jeans; too little time may not produce the desired results or can create a patchy appearance. Check on the jeans once every 10 min or so during the second wash neutralization process to ensure that the desired look is achieved.
- Move them to another washing machine for a regular washing process.
- Antichlor or neutralize with sodium bisulphate.
- Cool by adding water.
- Dry tumble.
- Use optical brightening.
- Use softener.

The process of acid washing jeans involves soaking porous pumice stones and chlorine or PP bleach. Sodium bisulfate is used for neutralization. It is recommended to use two separate washing machines for acid washing, neutralization, etc.

Porous pumice stones are soaked in a chlorine or PP bleach solution for about 20–60 min. After complete the stones are completely soaked in the bleaching agent, excess water is removed from them. Then the stones and denim garments are put into a washing machine for about 20 min. In this process there should not be a single drop of water. The drain should be open during the complete process.

When the required look is achieved, the fabric is transferred into another washing machine for regular washing for neutralization. After neutralization, the garments are dried.

#### 3.7.9.1 *Precautions*

During acid washing, the workers should wear masks and aprons. The operator should use protective rubber gloves and safety glasses.

### 3.7.10 *Hydrogen peroxide wash*

Hydrogen peroxide is rarely used as a bleaching agent only when very little loss of colour is required or if the fabric is coated with sulphur, because it takes longer to achieve the desired effect. Hydrogen peroxide has a prime role in the bleach wash technique. In an alkaline medium, hydrogen peroxide breaks up and gives off some perhydroxyl ions, which destroy the colouring matter and result in a fading effect. Hydrogen peroxide is used in a scouring, bleaching bath for white textile material or "ready for dyeing" form of garments made from grey fabrics.

### 3.7.11 *Enzyme wash*

There has been increasing interest in the use of environmentally friendly, nontoxic, fully biodegradable enzymes in modern textile technology finishing processes. Enzymatic treatment can replace a number of mechanical and chemical operations which are applied to improve the comfort and quality of fabrics.

Because of the problems with pumice, alternative methods for stonewashing of denim fabrics were developed. Cellulases have been used in denim washing for many years and it is estimated that about 80% of denim washing is done in this way [7]. The most important cellulase-producing organisms are fungi of the genera *Trichoderma*, *Penicillium* and *Fusarium*. The cellulases used are chemically complex and consist of at least three enzyme systems working together synergistically. Endo-β-(1,4)-gluconase or endocellulase hydrolyse cellulose chain molecules randomly, preferentially attacking the internal regions of the polymer. They generate more chain ends which are further hydrolysed by exo-β-(1,4)-gluconases (also termed cellobiohydrolases) to produce cellobiose. These exo-cellulases may assist in the disintegration of crystalline regions, making the region more susceptible to hydrolysis by endocellulases. The cellobiose and other small oligomers are then hydrolysed by the third enzyme, β-(1,4)-glucosidase, into glucose. All of these enzyme components act synergistically during the degradation of cellulose [25].

During enzymatic treatment, the indigo that is removed can redeposit on the white yarn of denim fabric. This process is called back staining and it can destroy the look of a denim garment. An ideal bio-stoning enzyme possesses highly abrasive activity as well as low back staining. Although there have been some studies on back staining which focused on the mechanism of indigo redeposition on denim garments and the reduction of the effect of back staining, it continues to be a problem for manufacturers.

It is claimed that a significant reduction in back staining can be achieved at a much lower temperature than a conventional process by using a small amount of protease either at the end of the cellulase washing step or during the rinsing step. By adding the protease at the end of the cellulase wash step, one rinse step is eliminated, offering savings in time and energy. This process at a lower temperature also claims to achieve a significant reduction in residual cellulose [26].

The finishing and washing effects delivered by cellulases are always applied in processes in which a strong mechanical action on the fabric is provided [1].

During enzymatic treatment, the indigo that is removed can redeposit on the white yarn of denim fabric. This process is called back staining and it can destroy the look of a denim garment. An ideal bio-stoning enzyme possesses highly abrasive activity as well as low back staining. Although there have been some studies on back staining which focused on the mechanism of indigo redeposition on denim garments and the reduction of the effect of back staining, it continues to be a problem for manufacturers [21,22].

Cellulase are available in three categories:

- neutral
- acidic
- hybrid enzymes

Acid cellulases exhibit the greatest activity, generally in the pH range 4.5–5.5 at 45–55°C, whereas neutral cellulases require a pH of 5.5–8.0 at 50–60°C [25]. Neutral enzymes give a better salt and pepper effect with much less back staining; they are generally available in powder form and the fabric retains better strength than with acidic cellulase. Acidic cellulase gives faster results but with too much back staining; it cuts down the indigo colour and affects the strength of the fabrics.

Laundry workers need faster results in less time and less expensively; hence chemical suppliers combine neutral and acid cellulase in such way that it works faster and with better results than acid cellulase. It is a cost-effectiveness solution known as a hybrid enzyme.

Enzymes, which are proteins present in all living cells, speed up chemical processes that would run slowly if at all. They are nontoxic and are readily broken down. Enzymes are used in textile processing, mainly in the finishing of fabrics and garments. Enzymes such as yeast are used to eat away cellulose physically in cotton. The colour in denim fabric is actually on the outer surface of the yarn owing to poor exhaustion. When the denim is washed in a cellulose enzyme bath, the indigo is removed along with the fibre. When the desired colour is achieved, enzymes are stopped from reacting, either by changing the alkalinity of the bath or by heating the bath. A rinsing and softening cycle follows. This process is more environmentally friendly than stonewashing because strip-mined pumice stones are not used.

Laccases (benzenediol:oxygen oxidoreductase, EC 1.10.3.2) have a broad substrate specificity with respect to the electron donor. They catalyze the removal of a hydrogen atom from the hydroxyl group of *ortho-* and *para-*substituted mono- and polyphenolic substrates and from aromatic amines by one electron abstraction to form free radicals capable of undergoing further depolymerization, repolymerization, demethylation or quinone formation [30].

The group of enzymes called laccases, or phenol oxidases, has the ability to catalyze the oxidation of a wide range of phenolic substances, including indigo. Laccases are extracellular, multicopper enzymes that use molecular oxygen to oxidize phenols, and various aromatic and nonaromatic compounds by a radical-catalysed reaction mechanism.

Application of laccases in the textile industry is growing quickly. The first commercial use of laccases in the textile industry was in the denim washing process. Laccases in combination with redox mediators were used to reduce back staining, enhance abrasion levels and bleach indigo. In 2005, Pazarlıoglu et al. [21,22] showed that a phenol-induced laccase from *Trametes versicolour* was an effective agent for creating stonewashing effects in denim fabric without using a mediator.

The enzymatic treatment uses a chemoenzymatic approach to remove indigo dye selectively from denim fabric. Laccase is applied together with a suitable mediator to create a bleached look to jeans. A better contrast between white and blue yarns can be achieved. In this process, the indigo chromophore is transformed into Isatin and back staining is reduced or avoided [12]. Laccases have been reported to bleach indigo fabrics and laccase-based systems are able to degrade indigo both in solution and on denim, leading to various bleaching effects on the fabric. There have been many studies regarding indigo degradation with laccase from *Polyporus* sp., *Sclerotium rolfsii, Trametes hirsute* and *T. versicolour* [21,22]. However because of the high cost and short shelf life, laundries do not prefer it.

Most denim garments are abraded using enzymes or a combination of enzymes and pumice stones. A wash processing concept from Novozymes, called Novozymes Denimax ore, makes it possible to integrate the abrasion process and the preceding desizing process, with abrasion results equal to or better than conventional processing. Nielsen [16] assessed the environmental impacts of moving from traditional enzymatic denim washing processing to the combined denim wash process. A savings of 50% of the water used for desizing and abrasion in the traditional process was reported. Heat use was reduced by 50% and electricity by 12%–15%. The environmental benefits of shifting to the combined denim wash process can be expected to be even larger for textile laundries that shift from a pumice stone treatment.

### 3.7.12 Ozone fading

Using this technique, the garment can be bleached. Bleaching of denim garment is done in a washing machine with ozone dissolved in water. Denim garments can also be bleached or faded by using ozone gas in a closed chamber. The advantages associated with this process are that:

- Colour removal is possible without losing strength.
- This method is simple and environmentally friendly because, after laundering, ozonized water can easily be deozonized by UV radiation.

Ozone typically acts as a mild bleaching agent as well as a sterilizing agent. In this technique of denim washing, the garment is bleached with ozone dissolved in water in a washing machine. However, this technique can also be carried out in a closed chamber using ozone gas. Advantages of this method are:

- a minimum loss of strength
- simplicity
- water- and chemical-free; environmentally friendly
- low energy costs for processing
- short treatment time

After laundering the ozonized water can easily be deozonized by UV radiation. Ozone fading can also be achieved by plasma equipment. Under plasma treatment, high-energy electrons are formed. Some of the high-energy electrons react with moisture in air and a mixture of radicals is generated. During the generation of ozone plasma, a combination of charged particles, free radicals and UV light is generated. The UV light, which is the by-product of the plasma treatment process, also contributes to production of the •OH radical. The hydroxyl radical •OH is the most oxidative radical among radicals generated in the plasma process and is the main radical responsible for degradation of indigo dye in textile materials. The •OH can oxidize indigo dye molecules (RH), producing organic radicals, R•, which are highly reactive and can be further oxidized. As a result, the colour fading effect of the indigo-dyed textile is achieved [11].

### 3.7.13 Spray technique

This technique is based on spraying chemicals or pigments to get different effects on the garments, as in case of laser marking. This can be done by using a robotic spraying gun or by a manual spray, followed by curing the garment.

- It a water-free process and therefore has no effluent discharge.
- It is less time-consuming.
- Different designs are possible.
- It is easy to use.
- There are no adverse effects on fabric strength.
- A single bath stonewashing and tinting technique is used.

### 3.7.14 Overdyeing and tinting

Tinting of denim garments is usually done after the stonewash process. In this procedure, the garment is lightly coloured to give the final denim appearance a slight shift. This is not true overdyeing but merely gives the impression of a change in the overall colour of the fabric. This process consumes a large quantity of water and chemicals. To make this process economical and ecologically friendly, some novel colour-based enzymes were introduced to the market. Using this technique, tinting and stonewashing effects can be achieved in a single bath. Advantages are:

- Less process time to achieve a tinted look
- No extra chemicals are required, which makes the process more economical
- less water consumption
- less energy consumption
- less chance of patchiness or unevenness

Tinting is a process in which a much lesser amount of tint is involved and mainly direct dye is used. This is being done to change hue, cast or tone of indigo. Tinting is used to give garments a used, vintage and muddy look.

Overdyeing or tinting of denim is an additional dyeing treatment normally carried out on jeans after they are sewn. This adds another colour to the jeans. Normally denim garments are overdyed with yellowish dye to appear dirty.

Denim garment tinting is normally carried out after the stonewash process. During tinting, a small amount of tint or colour is added to the garment to change the hue, cast or tone of indigo shade. However when the quantity of tint colour is increased, it covers up the indigo. Various types of dyes are used for overdyeing and tinting:

- direct dyes
- reactive dyes
- pigment colours
- sulphur dyes

These processes takes 5–25 min for better results, followed by dye fixing and cleanup of superficial dye.

Tinting of denim garments is normally carried out after the stonewash process. However this process consumes more water and chemicals. In an alternative method, enzymes are used. This process allows tinting and stonewashing to be carried out in a single bath. Advantages of the technique are that:

1. It consumes less water and energy and takes less process time to achieve a tinted look.
2. No extra chemical is used in this process.
3. The chances of patchiness or unevenness in the garment are almost zero.

In a study by [30], the laser and sandblasting procedures were examined because these procedures are popular and preferred in fusion trends. After the laser and sandblasting procedures were applied, various washing and bleaching processes were used as the chemical methods.

As a result of fading tests involving sandblasting, laser and washing, the strength and weight values of fabrics decreased depending on the increase in pressure in sandblasting, the burning intensity of the laser, and the increase duration and number of washings.

Significant colour loss was observed in fabric colour in line with the increase in these parameters. Back-staining values obtained from washings increased in desizing and other washing processes. However they decreased and were not even observed in bleaching and laccase enzyme washings and the colour of pocketing moved toward a yellow shade.

Significant losses were observed in the strength and weight of the fabric as a result of physical fading methods involving laser and sandblasting. To decrease these losses in production conditions, physical wear must be applied to the lowest degree. The required colour loss effect must be obtained later by regional bleaching to be directed to the area that is to be worn out physically.

On the other hand, to prevent strength and weight losses after hypochlorite washing, the pH of bleaching liquor must be around 11–12, at which hypochlorite least damages the cotton, or the required colour must be obtained by alternative bleaching processes (eco-bleach, laccase, organic peroxide, etc.). Certainly, the selection of these processes is not easy because of the required colour shade.

Back staining, which is one of the most important problems in denim washing, can be removed with the aid of laccase washing applied after normal washings for a short time and at a low concentration, because it fairly lessens after laccase washing, as seen from the results of these trials [30].

### 3.7.15 Plasma

Plasma has been used in the treatment of textiles over the past 2 decades, but a generally available and implemented technology has yet to be developed. Nevertheless, several industrial devices are available as well as numerous examples of custom-designed plasma processing stages for specific industrial production lines. Two directions for development are atmospheric pressure devices and low-pressure plasmas. There are numerous solutions for atmospheric pressure (or pressures of the order of several 100 Torr) devices, which include glow regime of dielectric barrier discharges and corona discharges. Low-pressure devices, in particular radio-frequency (RF)-powered plasma sources, allow easier control of properties and provide greater stability and uniformity at the cost of more complex handling of the fabric. Corona and RF low-pressure plasma treatments brought about an increase in denim lightness, indicating a removal of indigo dye from the fabric surface. The increased yellowness of the treated denim was more pronounced in case of corona and low-pressure plasma in air in comparison with low-pressure RF plasma in argon. This can be associated with the production of chemically active molecules and radicals in gas mixtures containing oxygen, which consequently leads to an oxidation of dyes.

Low-pressure plasma and corona treatments can be a viable alternative to conventional bio-stoning to obtaining the 'worn' look of indigo-dyed denim fabric [3]. In addition to satisfactory colour change effects, the main advantages of these treatments are the lack of water consumption and shorter process duration. However, it requires further research to avoid an increase in yellowness and the appearance of a harsh fabric handle.

Corona and RF low-pressure plasma treatments brought about an increase in denim lightness, indicating a removal of indigo dye from the fabric's surface. The increased yellowness of the treated denim was more pronounced in the case of corona and low-pressure plasma in air compared with low-pressure RF plasma in argon. The formation of pits and striations on the fibre's surface was more prominent after low-pressure RF plasma than after corona treatment [24].

## 3.8 Denim finishing

Denim finishing includes chemical shrink resistance, sanforizing, liquid ammonia treatment and the permanent press process.

The finishing and sanforizing of denim usually takes the form of a continuous (in-line) method, but depending on the plant and machinery it may be carried out in several stages.

The finishing process is made up of the following stages [6]:

- brushing (on one or both sides)
- singeing (on one or both sides)
- desizing (where necessary)
- padding (using one to two padding mangles)
- stretching to the desired cloth width (where necessary) using a stretch meter (compensator),
- weft skew correction (antitwister) at 6%–16%, excluding 3/1 and 2/1 twill, to prevent twisting of fabric
- drying (12–24 cylinders) to a residual moisture content of 15%–20%

- stentering (using a clip or clamp stenter)
- sanforizing (moistening, shrinking or calendering) with a shrinkage of 11%–15% to a residual shrinkage value of 2%–2.5%

### 3.8.1 Wrinkle finish

Resin is applied on the finished garment, in which wrinkles have to be created manually or semiautomatically. Low-formaldehyde or formaldehyde-free catalyst integrated cross-linking *N*,*N*-dimethylol-4,5-dihydroxy-ethylene urea–type resins along with support auxiliaries are used on a large scale in the denim industry. Support auxiliaries are used for various purposes such as to improve resin penetration into thick fabrics and seams, create creases at low temperatures, improve the handle when used as a softener in last bath, impart an extrashiny look, etc. Different types of resin application systems may be employed:

- In the garment-dip method, garments are impregnated with a resin solution, extracted to about 65% wet pickup and then tumble-dried to 8%–10% moisture content.
- In the spray method, a technology of wrinkle finishing, the resin is applied by spraying it onto the garment during tumbling in an enclosed rotational device, or in a spray booth where garments are hung onto inflated balloons and sprayed with a measured quantity of resin solution.
- Spray cabinets with a conveyor allow chemicals to be applied to robots dressed up with jeans while they are moving on the conveyor. It is an easy system that makes increase the workflow and daily capacity.

There is an industrial wrinkle-making technique:

- After resin application, tying is done with thread at the specific area where a crinkle effect is desired. The garment is then sent to an oven to cure.
- In this process small areas of the garment are bound using clips on for the desired effect and then the garment is put in the oven to cure.
- This machine consists of three operation heads and three stations. Variable pressure can produce unlimited crease designs which can easily be altered by changing the wire configuration. The operation time is around 30 s.
- It is a scrunch/crushing machine that uses steam to achieve different levels of wrinkles and/or pressing. This machine is used for scrunch applications and makes wrinkle all over the garment. The effects can be adjusted depending on the levels of pressure, steam and chemicals.
- This machine is used to give some crinkle effects to denim jeans during the washing process. The flexible hanger consists of one station and one flexible hanger. There is no energy consumption in this system. It works with just air.
- The crinkle press consists of two application stands and one head that presses the trousers to be given an effect on these stands.
- The system works with air and electricity. Air is used to creating force to the air pistons whereas electricity is used to feed resistances to the heating group. The system slides and the product can be prepared on one stand while the pressing procedure is carried out on the other.
- Curing is the process of placing the fabric into high temperature to allow the chemical to carry out the reaction process. For high-quality 3D crunching/wrinkles, the two most important criteria are temperature and a controlled cycle. Oven curing is needed to reach the right temperature (145–160°C) and time (7–22 min) as per resin technological bulletin advice. Some typical industrial ovens, e.g., automatic type ovens with a conveyor, box-type curing ovens with trolleys, semiprofessional curing ovens with conveyors, are being widely used.

Wrinkle effects on denims create a unique and vintage look that makes the garments more fashionable, thus adding value to denim garments and improving its salability in the market. It prevents intermolecular slippage in the fibre core and it may not harmful if formaldehyde-free resin is used.

It decreases the tensile and tear strength and abrasion resistance. It has an unpleasant odor and an unwanted harsh and stiff feel. It may turn the fabric yellow after chlorine bleaching. Sometimes it is difficult to remove unfixed resin [10].

## 3.9 Impact on environment

To study the impact of denim production on the environment, in 2007 and again in 2015, Levi Strauss & Co. conducted a life-cycle analysis of a pair of Levi's 501 jeans [33]. A pair of 501 jeans uses 3781 L of water in its full life cycle, from growing cotton through manufacturing, consumer care at home and end-of-life disposal.

A life-cycle assessment (LCA) is a systems-based, quantitative method for evaluating the environmental impact of a product. This tool is used to assess the stages and impact of a product's entire life from raw material extraction (cradle) to waste treatment (grave). LCAs focus on environmental impacts and typically do not consider social or economic impacts. The key findings are [33]:

Water consumption
Nearly 3800L of water is used to make a pair of jeans. Fiber production, predominantly cotton (68%), consumes the most water, followed by consumer care (23%).
Climate change
Of the 33.4kg of carbon dioxide produced during the life cycle of a pair of jeans, consumer care (37%) and fabric production (27%) generate the most significant impact on climate change and energy use.
Expanded scope
The amount of water used to grow cotton varies significantly across the world. Also, by including new consumer markets it was found that washing and drying habits vary by region.
Impact
By wearing jeans 10 times before washing, American consumers can reduce their impact on water and climate change by 77%, United Kingdom and French consumers by 75% and Chinese consumers by 61%.

Current average washing frequency was determined as follows:

- In the United States, consumers typically wear their jeans twice before washing them.
- In the United Kingdom and France, consumers typically wear their jeans 2.5 times before washing them.
- In China, consumers typically wear their jeans four times before washing them.

Consumer jeans-washing habits vary by region:

- Americans use more water and energy to wash and dry their jeans than do consumers in the other markets.
- Consumers in the United Kingdom and France mostly air-dry their jeans and use more hot water than do others.
- Consumers in China mostly wash their jeans in cold water and air-dry them.

Around 1.5 lb of cotton is needed to produce one pair of jeans. A total of 1500 gallons of water is needed to grow 1.5 lb of cotton required for one pair of jeans. So, 1500 gallons of water is going into one pair of jeans. Wow! Keep in mind that on a yearly basis over 450 million pairs of jeans are sold in the United States. That is just the United States, not including other jeans-clad countries such as Canada and those in Europe.

For a given weight of cotton harvested, a farmer uses one-third of that weight in chemical fertilizers. Cotton plants are highly susceptible to pests, especially in humid areas [5]. Clay [2] reported that whereas cotton production is restricted to 2.4% of the cultivatable land globally, an estimated 25% of insecticides and 11% of global pesticide production is consumed in cotton cultivation.

The makers of some 'green' jeans sidestep pesticides and oil-guzzling machinery by opting for hand-picked organic cotton. Because pesticides are not applied to organic cotton, yields can be as much as 50% lower than those of conventionally produced cotton, which means that more land may be needed to make organic jeans. Some manufacturers also use bamboo. It is a fast-growing, water-efficient crop that can be cultivated without pesticides and fertilizers, but turning it into a comfortable pair of jeans requires some nasty chemical treatments.

Cotton yarn is typically 'sized' with starch to increase its strength for weaving, and in some cases it is 'mercerized' in caustic soda, which gives it a worn look. Starch biodegrades, but when it is dumped in waterways the microbes that eat it also consume oxygen. Aquatic life depends on that oxygen, and starch is just one of many chemical treatments, including dyes, that deplete it. Caustic soda can kill aquatic life and burn workers. Bamboo eco-jeans are made by 'cooking' shoots in caustic soda before they are turned into fibre.

To obtain the right shade of blue, cotton yarns may be dipped a dozen or more times in enormous vats of synthetic indigo, which is often made from coal or oil. Some factories have machines that measure the concentration of dye in solution precisely, enabling a manufacturer to recycle spent liquid by adding just the right amount of fresh dye. However in developing countries, where water and dyes are cheap and environmental regulations are not strict, factories that lack modern equipment often dump the unexhausted dye into nearby waterways. Water samples taken from downstream of textile plants in Tehuacan, Mexico, a major denim-producing region, have been shown to contain lead, mercury, cadmium, and selenium. Local farmers report chemically burned seedlings and sterile soil [4].

Eco-jeans may be dyed with pigment from the *I. tinctoria* plant, but these natural pigments are found in less than 1% of all indigo dyes. Natural or not, indigo pigments do not dissolve in water, so creating a liquid dye requires chemicals. The upshot is that most eco-jeans are dyed conventionally.

### 3.9.1 Danger from stonewashing

'Stone washing' is a euphemism for a number of processes that give denim a worn look. Each pair of jeans is washed and rinsed multiple times, consuming more water and energy at every turn. Denim may be physically sanded or blasted with silica, or dye may be stripped using chemicals such as PP, which is highly toxic and contains heavy metals. Various enzymes may also be used; when improperly dumped, they deplete oxygen from waterways. Scrubbing jeans releases denim and silica dust, which can inflame workers' lungs and cause silicosis. In a study of denim sandblasters in Turkey, more than half of the workers showed signs of the condition; the average worker was 23 years old and had been on the job just 36 months [4].

Because municipal discharge criteria are restrictive and the cost of water is high, producers will need to reuse wastewater. Many techniques are readily available for a processor to treat wastewater. The basic limitation of a garment processor is the high installation cost associated with such techniques. Technocrats need to realize that they have to develop simple and cost-effective garment processing technologies that aim to conserve energy, water and time, and are environmentally friendly.

Jeans are weathered using volcanic pumice stones. These pumice stones have to be mined and shipped to denim companies all over the world, creating a huge carbon footprint. As that pumice is used to 'weather' denim, it breaks down. Most companies simply put this pumice dust into their local watershed. Large quantities of pumice dust pollute local streams and rivers. Distressing denim also puts small particles of denim and silica into the air, which is then breathed in by factory workers.

The entire textile industry is extremely destructive to the planet. Manufacturers are known for dumping wastewater and harmful chemicals into nearby waterways. The denim industry is the worst, because the chemicals used to create denim are the most destructive. Denim manufacturers dump chemicals such as cadmium, mercury and lead into water sources; these same chemicals are used to make the popular jeans.

One can only imagine the destruction these chemicals have on the environment as well as the quality of water. Nearly 1 billion people do not have direct access to clean water. We are polluting water sources further to create fashion. Water pollution is at an all-time high. If this pollution continues, Kumi Naidoo predicts a grim future. Naidoo, executive director of Greenpeace, says, 'The wars of the future will not be fought over oil. The wars of the future will be fought over water'.

Keep in mind that countries that produce the highest quantity of textiles, including denim, are China and India. These countries are battling the serious consequence of such successful textile industries: dangerously polluted water sources. In areas where clean drinking water is already compromised, locals may find even greater difficulty in obtaining the essential life source.

A popular trend in the past several years poses different threats to the environment. These articles of clothing are made through colouring with toxic dyes, acid baths and sandblasting and then are chemically bathed, all which produces the popular distressed look to make a pair of jeans look old. This distressed look requires multiple treatments, which further causes a negative impact on the environment because repeated washing of jeans results in the repeated introduction of dangerous chemicals in the area's water supply.

A denim disaster can be seen in Tehucan, Mexico, also known as the 'heartland of Mexico's denim industry'. Tehucan is known for housing dozens of industrial laundries which are responsible for completing the finishing touches on denim jeans. Laundering of these jeans includes using bleach, dye and detergents, all of which are dumped as wastewater into the area's rivers. This wastewater is then used to irrigate corn fields. Unfortunately, these laundries are unregulated and Mexico does not manage the contamination of the water source. This results in Mexican rivers literally being dyed blue. This has led to some serious questions regarding the impact the textile industry is having on developing nations [32].

### 3.9.2 Danger from sandblasting

A survey was conducted by the Alternative Movement for Resources and Freedom Society in association with the Clean Clothes Campaign about sandblasting units in Bangladesh [31]. Sandblasting removes dark indigo pigmentation from a garment that is usually made of denim, giving it a popular preworn look. The process involves smoothing, shaping and cleaning a hard surface by forcing abrasive particles across that surface at high speeds using special types of sands. These are sprayed onto selected parts of the garments at high pressures through air compressors to remove colour from the areas to create the desired design. Sandblasting may be of two types: manual and mechanical. In manual sandblasting, compressors are used to blow out sand under pressure through a gun to strike the denim fabric. This process is done in the absence of sealed blasting cabinets and ventilation, exposing operators directly to silica particles (tiny particles of blasted sand) that are released from the guns.

If inhaled, this silica dust can cause severe respiratory problems in workers. In cases of intense or long-term exposure, it may even lead to the contraction of fatal diseases such as silicosis and lung cancer. Although the most common form of sandblasting is manual blasting, sandblasting can also be performed mechanically in blasting cabinets where the process is supposed to be more controlled. However the report showed that mechanical sandblasting as done in Bangladesh actually continues to expose workers to silica dust. The research showed that mechanical sandblasting is largely carried out in unsealed environments with little protection for workers, who are provided with inadequate safety equipment. As a result, the use of this technique continues to expose workers to a potentially fatal risk.

After the imposition of strict regulations on sandblasting in many European countries, the clothing industry has largely outsourced production to as yet unregulated regions such as Turkey, Bangladesh and China.

Conservative estimates suggest that Bangladesh has over 2000 full-time sandblasters who produce garments for export, and almost half of the 200 million pairs of jeans exported from Bangladesh each year are sandblasted [31].

It was in Turkey that the negative health effects of this process in the garment industry were recognized; Turkish doctors were the first to sound the alarm over silicosis among garment sandblasters. In 2005 the first major study to link sandblasting jeans with silicosis was published. Since Turkey implemented a ban on sandblasting in 2009, pressure on brands to stop using manual sandblasting has increased. In autumn 2010, the Killer Jeans campaign was launched, public calling for abolishing the practice in the industry, and many brands have announced a voluntary ban on sandblasting. Yet few if any brands have provided clear information regarding how these bans are being implemented and no brand has yet agreed to take responsibility for identifying and treating affected workers in their supply chain [31].

Some countries, such as the Netherlands and Denmark, have classified silica as a carcinogen. In 1987, the International Agency for Research on Cancer (IARC), an agency of the World Health Organization, concluded that crystalline silica (but not noncrystalline, or amorphous, silica) is a 2A substance (a probable carcinogen for humans). In October 1996, an IARC panel concluded that crystalline silica inhaled in the form of quartz or cristobalite from occupational sources should be classified as carcinogenic to humans (Group 1).

Sandblasting itself is not prohibited in most countries, and restrictions are instead placed on the type of sand used. Regarding the practice of sandblasting itself, the US Department of Labor, Occupational Safety and Health Administration states that 'the most severe worker exposures to crystalline silica results from sandblasting'.

The use of crystalline silica was banned for most blast-cleaning operations in Great Britain in 1950 (Factories Act of 1949) and in other European countries in 1966. In 1974, the US National Institute for Occupational Safety and Health recommended that silica sand be prohibited for use as an abrasive blasting material and that 'less hazardous materials be substituted for silica during abrasive blasting'. Under European Union directives and national legislation, sandblasting is allowed provided that the abrasive materials contain less than 1% silica; in the United States, the figure is less than 0.5% silica. Silica sand used in denim sandblasting can often contain 90%–95% crystalline silica.

The report recommended that the brands should immediately stop using sandblasting throughout their supply chains. To that end, we demand that companies continue to publicly support a complete ban on sandblasting in their supply chain and ensure that this ban be enforced by using adequate monitoring processes.

In the processing of jeans there are a number of opportunities to reduce the environmental impact. Starting from dyeing, less dye could be used, especially in cases in which most of the dye is bleached or washed away. Alternative processes such as inkjet printing can save enormous amounts of dyes, water and energy, but this technology is not ready for the mass production of jeans. The advantage of using inkjet printing is that 'worn' designs can be printed directly on the jeans, greatly reducing the finishing treatments. Dyeing the warp yarns in a lighter shade has the possibility of reducing dye consumption by 50%.

The finishing process is the most damaging part of the whole production cycle. Finishing processes may reduce the strength of the fabric by 50% or more, resulting in a much lower technical lifetime. To apply the finishing effects by a printing technology (conventional or inkjet) would not reduce the strength of the fabric and hence would make it possible for jeans to withstand more than 100 washing cycles. Of course, from a fashion point of view, this could change the appearance of the jeans. Such changes in processing have to be accompanied by a marketing strategy, with increased sustainability as the central theme.

### 3.9.3 Fewer washings during use

Special attention should be given to find ways to reduce the frequency of wash and care cycles. Because the use phase has the most environmental impact, opportunities have to be worked out to reduce wash and care. Jeans are washed for two reasons: they are soiled and it is hygienic. Soiling can be reduced through antisoiling finishes; hygienic properties can be improved using a nano-silver finish. However, antisoiling finishes are mostly based on fluoro compounds, which are heavily criticized by Greenpeace. Alternative antisoiling compounds are based on silicon chemistry, which have little environmental impact.

### 3.9.4 Recycling

A huge reduction in environmental impact can be achieved by using recycled fibres originating from discarded jeans and other cotton waste sources. Although fibres of discarded jeans might have reduced strength, they can be used to spin a good yarn. About 50% of recycled fibres can be added.

The use of such yarns in jeans (50% of the yarns in the warp direction and 100% in the weft direction) will result in a fabric with nearly 40% recycled content. In processing, these fibres do not have to be bleached and can be used directly in the dying process. Because these yarns already have a certain blueish colour, the amount of dye needed will be somewhat reduced. It is believed that the environmental impact of a pair of jeans can be reduced by about 30%.

Environmental performance can be improved by the selection of materials, improved processing using (inkjet) printing technologies, a reduction in mechanical finishing, application of an antisoiling and antimicrobial finish to reduce the frequency of wash and care cycles and the use of a design for recycling principles in the selection, processing and production of a pair of jeans Luiken A [13].

One of the leading innovation companies, Jeanologia (http://www.jeanologia.com/), launched a water-free bleaching process machine which works on ozone, called G2 technology, that produces various shades of denim using air technology as opposed to water and chemicals. Jeanologia, the owner of the G2, estimates that if every company were to adopt their machine, the amount of water saved could supply the entire country of Spain with clean drinking water for 8 months. This machine saves both water and energy and has no impact on the environment.

There are many up-and-coming companies producing friendly denim using organic cotton and more eco-friendly ways to distress the fabric, such as using ozone to fade the denim. It is the responsibility of the consumer to search for these enlightened businesses by checking for labels such as Fair Trade and Global Organic Standard certification, and the recycling logo.

## 3.10 Future trends

Denim sales in the United States have fallen a whopping 6%. That is pretty significant! Executives in the denim industry are surprised about the rapid decline in denim sales. Unfortunately the downward trend has not been so significant in Europe, which has only seen a 3% decrease. VF Corp., which represents the brands Lee and Wrangler, stated, 'In the Americas, the jeans wear business was down to a low-single digit rate due to continued challenges in the US mid-tier channel and the ongoing unfavourable women's denim trend'. Thus if denim sales are on the downward trend, jeans are going out of style.

The focus of the current fashion trend is all about ease of movement and comfort. Denim is having a hard time adjusting to meet the needs of this new trend. As a traditionally tight, inflexible fabric, denim has taken second place to the more flexible, comfortable fabrics seen in yoga pants and athletic pants. This new look and style is called 'athleisure' because of the underlying athletic design of most of these products. These trends are worn by people of all ages and sizes, from leggings-clad teens to yoga pants–sporting moms and grandmas.

Although denim and jeans have a prominent role in the fashion trade, it is time to think about jeans' impact on the environment. Pumice stones have to be mined and shipped to denim companies all over the world. That is a huge carbon footprint. Their disposal is also a problem: pumice dust is sometimes disposed of into the local watershed.

A wash-processing concept from Novozymes, called Novozymes Denimax Core, makes it possible to integrate the abrasion process and the preceding desizing process, with abrasion results equal to or better than conventional processing. There is another option: The eco-minded can invest in a pair of jeans woven from organic cotton, dyed with natural indigo and faded with nontoxic ozone.

Once consumers are aware of the detrimental effects denim production has on the environment, they will be motivated to identify more ethically produced clothing brands. There are also many eco-friendly companies that make eco-friendly jeans from recycled denim.

## References

[1] Cavaco-Paulo A. Mechanism of cellulase action in textile processes. Carbohydr Polym 1998;37:273–7.

[2] Clay J. Cotton, in world agriculture and the environment. Washington, DC (USA): World Wildlife Fund, Island Press; 2004. p. 283.

[3] Ghoranneviss M, Shahidi S, Moazzenchi B, Anvari A, Rashidi A, Hosseini H. Comparison between decolorization of denim fabrics with oxygen and argon glow discharge. Surf Coat Technol 2007;201:4926–30.

[4] Glausiusz J. How green are your jeans?. 2008. http://archive.onearth.org/article/how-green-are-your-jeans.

[5] Grose L. Sustainable wool production and processing. In: Blackburn RS, editor. Sustainable textiles: life cycle and environmental impact. Cambridge (UK): Woodhead; 2009. p. 33.

[6] Rouette H-K. Encyclopedia of textile finishing. Aachen (Germany): Springer; 2000.

[7] Heikinheimo L, Buchert J, Oinonen AM, Suominen P. Treating denim fabrics with *Trichoderma reesei* cellulases. Text Res J 2000;70(11):969–73.

[8] Hung ON, Song LJ, Chan CK, Kan CW, Yuen CWM. Using artificial neural network to predict colour properties of laser-treated 100% cotton fabric. Fiber Polym 2011;12:1069–76.

[9] Kan CW. Washing techniques for denim jeans. Denim: manufacture. Finish Appl 2015:333.

[10] Khalil E, Islam M. Wrinkle finish on denim by resin treatment: a review. AASCIT Commun May 10, 2015. ISSN: 2375-3803;2(3). Online.
[11] Khalil E. Sustainable and ecological finishing technology for denim jeans. AASCIT Commun July 10, 2015. ISSN: 2375-3803;2(5). Online.
[12] Körlü AE, Bahtiyari MI, Perincek S, Duran K. Usage of laccase in denim washing. Tekstil ve Konfeksiyon 2013;23(4):364–8.
[13] Luiken A. Blue jeans environmental aspects and opportunities to reduce the environmental impact. 2016. www.iswa.org/.
[14] Mahfuz M. All about garments washing Part4: acid & wet wash. 2013. http://textilesociety.blogspot.in.
[15] Rahman Md M. Different types of dry process for denim wash. 2013. http://textilelearner.blogspot.in.
[16] Nielsen AM. Combined denim washing process. Int Dyer October 2012:16–8.
[17] Öndogan Z, Pamuk O, Öndogan EN, Özgüney A. Improving the appearance of all textile products from clothing to home textile using laser technology. Opt Laser Technol 2005;37:631–7.
[18] Ortiz-Morales M, Poterasu M, Acosta-Ortiz SE, Compean I, Hernandez-Alvarado MR. A comparison between characteristics of various laser-based denim fading processes. Opt Lasers Eng 2003;39:15–24.
[19] Özgüney AT. The comparison of laser surface designing and pigment printing methods for the product quality. Opt Laser Technol 2007;39(5):1054–8.
[20] Pareek V, Eklahare S. Denim washing: its process cycle. Indian Text J May 2014.
[21] Pazarlioglu N, Sariisik M, Telefoncu A. Treating denim fabrics with immobilized commercial cellulases. Process Biochem 2005;40:767–71.
[22] Pazarlioglu N, Sariisik M, Telefoncu A. Laccase: production by *Trametes versicolor* and application to denim washing. Process Biochem 2005;40:1673–8.
[23] Pezelj E, Cunko R, Andrassy M. Modification of denim surface using laser. In: World textile conference, 4th AUTEX conference, Roubaix, France. 2004.
[24] Radetić R, Jovančić P, Puač N, Petrović ZL, Šaponjić Z. Plasma-induced decolorization of indigo-dyed denim fabrics related to mechanical properties and fiber surface morphology. Text Res J 2009;79(6):558–65. http://dx.doi.org/10.1177/0040517508095612.
[25] Roy Choudhury AK. Textile preparation and dyeing. 2nd ed. USA: Science Publishers; 2006. ISBN: 1-57808-402-4. by The Society of Dyers and Colourists Education Charity, India. www.sdc.org.in. in 2010.
[26] Sangita S, Siva Kumar P, Chandran MR. Types of stone wash & their effects on the denim fabric. Indian Text J February 2010.
[27] Sarkar J, Rashaduzzaman M. Laser fading technology: facts and opportunities. Bangladesh Text Today 2014;7(8).
[28] Schmitt B. Colourage. ;XlV. October 1998. p. 20–4.
[29] Shore J. Cellulosics dyeing. Bradford: SDC; 1995.
[30] Tarhan M, Sariisik M. A comparison among performance characteristics of various denim fading processes. Text Res J 2009;79(4):301–9. http://dx.doi.org/10.1177/0040517508090889.
[31] Deadly denim, sandblasting in the Bangladesh garment industry. March 2012. www.cleanclothes.com.
[32] Environmental aspects of blue jeans. 2016. www.creeklife.com/blog.
[33] Sustainability: planet. March 2015. www.levistrauss.com.

# Life-cycle assessment of denim

4

*A.P. Periyasamy, J. Wiener, J. Militky*
Technical University of Liberec, Liberec, Czech Republic

## 4.1 Introduction

Sustainable fashion started before the Industrial Revolution, when people conserved resources and the cost and labour it took to produce the basic things necessary for life [1]. The raw material for textile fabrics came from nature (fibers and colourants) and it required a long time to transform these into garments. At that time, the requirements and cost of labour were huge; hence it was affordable only for the wealthy. By the end of the 18th century, industrial machines were introduced to spin yarns and weave fabrics, which accelerated the production process of textiles. In later years, synthetic dyes, synthetic fibres and other new products of the Industrial Revolution allowed the textile industry to grow quickly. Around the mid 19th century, textile products became affordable for all classes of people, and apparel was more abundant and easily accessible, which led to the development of consumerism. Therefore, the purchasing capacity of consumers increased the rate of production of textiles and garments. With the growth in the textile and apparel industries came many factors, including consumer awareness, government intervention by legislation, and the diversification of textile production. Since 1980, owing to the development of information technology, consumers have been able to access data showing the impacts of this industry. Thus, the attention of consumers has forced manufacturers and retailers to take action regarding the environmental, economic and social impacts of various processes of the textile value chain [2,3].

During the Industrial Revolution, textile manufacturers were neither aware nor concerned about the environmental effects of the production of textiles, and they disposed of chemicals into nearby rivers and watersheds. Subsequently, the emergence of environmentalism and social consciousness took place. Consumers began to consider new ways of disposing, recycling and reusing their clothes. They became concerned about how their clothing was produced and the impact this process had on the environment, and thus developed a conscientious lifestyle. As a response to conscientious consumerism, eco-labeling arose as a way to inform consumers that companies cared and were on board with a sustainable viewpoint. Over the past decade, the global society has begun to focus more on environmental issues (i.e., global warming) and its impact on all fields. Therefore, manufacturers have had to develop sustainable products rather than disposable ones. This chapter focuses on the life-cycle assessment (LCA) of denim jeans, because we are aware that denim is a significant part of the fashion industry as a result of its popularity. According to the report by Cotton, Inc., US consumers typically own seven pairs of denim at a time [4].

**Sustainability in Denim. http://dx.doi.org/10.1016/B978-0-08-102043-2.00004-6**
Copyright © 2017 Elsevier Ltd. All rights reserved.

The LCA is a detailed analysis of a product; it describes the entire life span of products, from fibres to the retailing of denim. Therefore, it encompasses the production of raw materials, the manufacturing process of denim garments, logistics and the garment's use, and ends with the denim's reuse, recycling and disposal. The product's use phase and end-of-life disposal phase depend on consumer behaviour; nevertheless they are critical stages in a product's life cycle. During the manufacturing phase of denim, at every stage there is an impact on the environment, such as energy and water use, $CO_2$ emission and waste [5,6].

### 4.1.1 Stages of a life-cycle assessment

Generally the LCA can be divided into four stages. The first stage includes defining the goals of the study and describing the product (denim) to be assessed. The second stage is inventory; data from all processes of the product's life cycle are collected and processed, from the acquisition of raw materials to the development of the final product. During this stage, energy, raw material requirements, environmental emissions and discharges relating to the product can be computed and presented. These data can be used to calculate the discharge from all of the processes of the product's life cycle.

The third stage is called the 'impact assessment stage', in which the inventory data are translated into their effects on human health, ecological health and resource depletion. The last stage of this process may be called 'recommendations' based on an analysis of the impact assessment results. These are the basic stages of LCA assessment, but they may vary for different raw materials, modified production processes and the choice of one product over another [6–9]. An LCA is an important procedure which helps us assess the environmental impact of any product. Hence, we can easily identify the most polluting stages of the process and use them as a guide to develop enhanced methods and processes. An LCA is a burgeoning topic and is studied by many researchers [10–12].

## 4.2 Product life cycle

The main purpose of this assessment is to reduce resource use and emissions into the environment as well as improve performance throughout its life cycle. Fig. 4.1 is an illustration of all phases in a product's life cycle [13].

### 4.2.1 Raw material procurement

Denim has always been made from cotton [14]. Sometimes synthetic materials are used to produce denim to provide additional properties such as comfortable fit and durability. We know that cotton is one of the most popular natural fibres used in the world. Three percent of the world's cultivated land is used for cotton production. Based on previous studies on the environmental, natural fibres are good (environmentally friendly) and synthetic fibres are bad (harmful). However, each fibre is different, with its own challenges to sustainability in the production process that cannot be

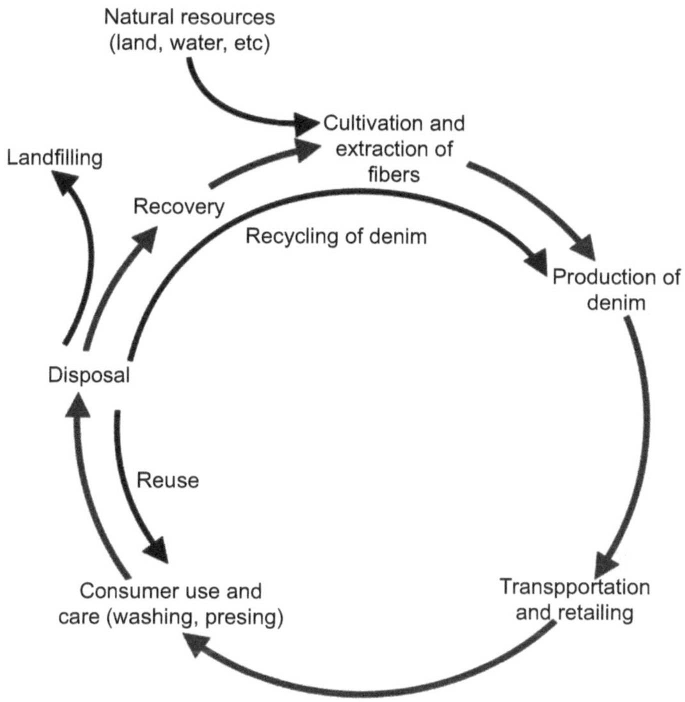


**Figure 4.1** Phases of a product's life cycle (for denim).

ignored [13]. Generally, cotton is a natural fibre which is grown and not manufactured. Although cotton is an environmentally friendly fibre, it consumes huge quantities of water and requires pesticides and fertilizers to grow well for the expected quality and quantity [15]. Based on a data sheet published by the US Department of Agriculture and the Organic Consumers Foundation, cotton is the most toxic crop in the world. Cotton uses more than 25% of all insecticides in the world and 12% of all pesticides [16,17]. The majority of these insecticides are considered the most toxic chemicals in the world. The health risks of pesticide exposure include birth defects, reproductive disorders and weaker immune systems [18–22]. Also, cotton requires 7–29 tons of water to produce 1 kg of raw cotton [23–25]. As a result of these environmental and health issues, most denim brands are committed to organic cotton materials [26,27]. As to synthetic materials, polyester and polyurethane are the most used in denim manufacturing. Although these fibres require less water and land to manufacture, they have effects on the environment. Generally, synthetic polymers are obtained from nonrenewable resources (e.g., fossil fuels) and the production of these fibres requires an enormous amount of energy, causing higher greenhouse gas (GHG) emissions. Environmental problems in waste management such as nonbiodegradability pose various health and toxicological threats that also have negative impacts created during the production of synthetic fibres [7].

### 4.2.2 Manufacturing

Manufacturing is the process in which fibres are converted into yarn and then fabric, and finally are converted into garments (three-dimensional structures). Fig. 4.2 shows

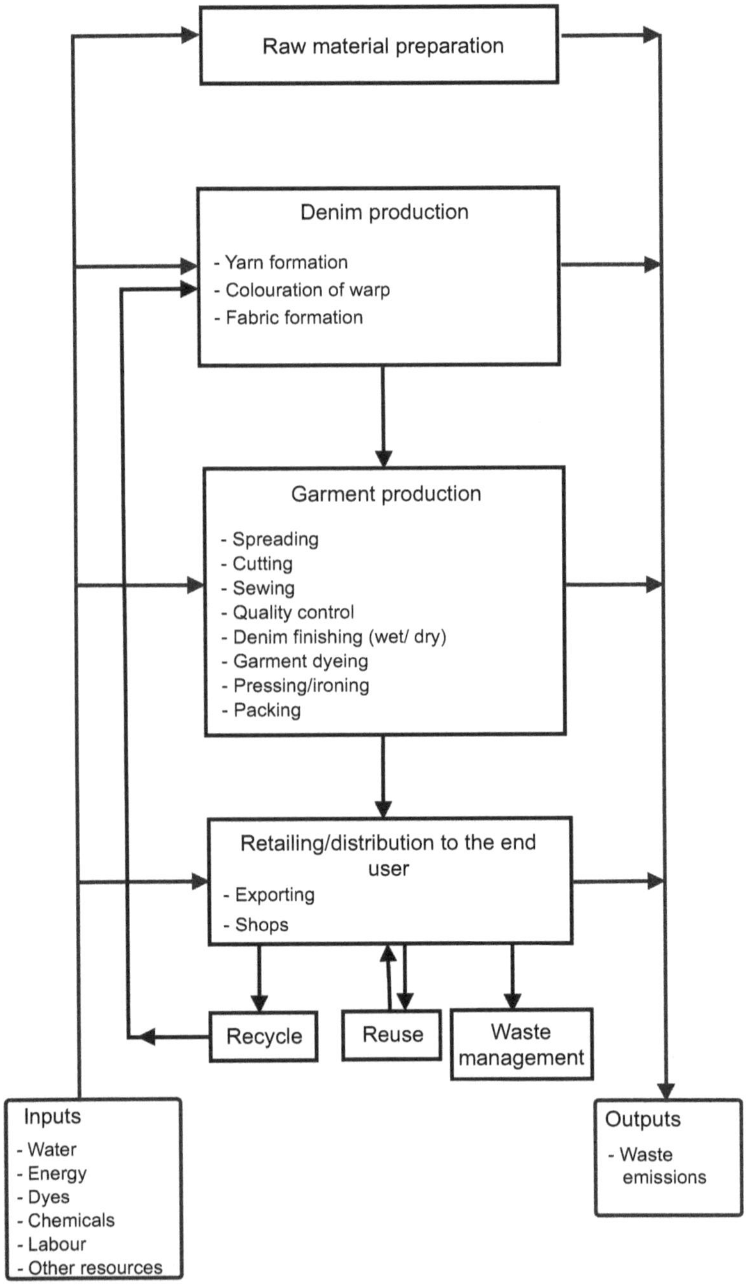


**Figure 4.2** Production life cycle of denim.

a map of the each process related to denim manufacturing with respect to the inputs and outputs of the production chain for the product life cycles.

### 4.2.3 Spinning

Yarn formation for denim fabrics is carried out by two methods: ring spinning or open-end spinning. Over the past 3 decades, open-end yarns have dominated in the field of denim fabric production. Generally, cotton and other fibres are produced or manufactured in different places and/or countries, and the first phase of yarn formation is to transport fibre from the place of harvest to the spinning factory. Several processes are carried out (i.e., opening, carding, combing, drawing, roving, spinning and winding) to form yarn, which require a high level of energy use. Dust, fibre and yarn waste production are main issues related to environmental protection during spinning [7,28]. Based on the condition of the fibre, there is a possibility for the use of lubricants, chemicals, water, packaging materials and other disposables during the various spinning processes, including *blowing, carding, drawing, combing, roving, ring spinning* and *open-end spinning*. The highest environmental and health risk in this process is the inhalation of the dust, which may cause a dangerous disease called byssinosis (commonly called 'brown lung') [7,29].

### 4.2.4 Fabric formation

Denim fabric is obtained by weaving twill structures. After the warping process (beaming), it is necessary for the warp sheets to undergo a sizing process in which a sizing box contains natural or synthetic sizing agents such as polyvinyl alcohol or carboxymethyl cellulose. The fabric manufacturing process requires high levels of energy consumption and produces waste from various chemicals, water and packaging materials. The disposal of this waste, as well as excessive noise, can cause issues [30–33].

### 4.2.5 Colouration of denim

Colouration is a process that is accompanied by major environmental impacts owing to the production of dyes, the use of water and chemicals and the fixation of the dyes and the drying of denim as well as nondenim fabrics. These textile materials need pretreatment before colouration. During pretreatment, they are cleaned with sodium hydroxide and detergents to remove natural hydrophobic substances. Washing of these process lines may cause water pollution [34]. For denim fabrics, only the warp yarns are dyed. Generally, indigo and sulphur dyes are used to colour warp sheets of denim. After dyeing, the fabric is rinsed with water, excess dye is removed and dried (water pollution results from coloured wastewater) and then sized. In the dyeing bath, the indigo dye has to be chemically reduced using sodium hydrosulphite as a reduction agent. The reduced indigo (with a yellow colour) will adhere to the fibre. The reduced dye will oxidize back to the characteristic blue colour when exposed to air [35–37].

### 4.2.6 Garmenting

The final stage of the denim supply chain is 'garment manufacturing', also known as the cut–make–trim (CMT) stage. During this process, various operations such as spreading, cutting, sewing, heat pressing, ironing and packaging are involved. Unlike yarn and fabric manufacturing, the CMT process involves mostly manual human labour to complete the garment. Important considerations here in the environmental impact assessment are the high level of energy use; fabric waste during the cutting process (often already recycled); accessory waste; rejected garments; lubricant, water and chemical use; use of packaging and hanging materials; and the production of high-carbon emissions during transport operations [7,13].

As mentioned earlier, the production stages from raw materials to finished textile goods are generally located in different places and countries, which make logistical activities an important component along the denim supply chain. Transportation of goods is not related only to manufacturing; once manufacturing is completed, the denim is usually sent to wholesalers, retailers, distribution centers and shops using different kinds of transportation (i.e., road, maritime or air), all of which have different amounts of energy consumption values and $CO_2$ emission levels. $CO_2$ emissions from freight transport are estimated at around 8%. In reference to the LCA, transportation to consumers is a huge factor which consumes large quantities of oil or gas, and the price of oil and gas has been steadily increasing, thus increasing transport costs. It is important to reduce the amount and distance of the transport of goods, popularly known as 'green logistics', to ensure the management of a green supply chain [38–41].

### 4.2.7 Consumer use and disposal

Well-meaning people who truly want to protect the environment can be passionate about what should be done with what they consider to be a generic material, called waste. The LCA process is now geared toward the consumer and is no longer solely within the manufacturer's control (e.g., consumers can pass valued garments on to family, friends or charity or send to for reselling or recycling) and varies among and within regions and cultures. This phase encourages consumers to buy clothes only when they are needed and to take care of them to prolong their life. However, most denim clothing is washed in water every 2 weeks and then disposed of into a landfill after 1 year. Most companies put much effort into the care of denim; for example Levi Strauss & Co.'s 'Care Tag for the Planet' reminds consumers to wash in cold water and wash less. However, you cannot make an egg-omelet without having eggshells to dispose of. The main motto of the disposal stage, ensuring recycling and donating to reduce the amount of waste going into landfills, helps to affect the environment positively. The most commonly known waste management strategies are the 3 R's: *reduce*, *reuse* and *recycle*. The meaning of reuse is simply using the product for the same purpose or for something completely different. Reusing includes donating unwanted items to charity groups or even to a family member or friend. The main benefits of reducing and reusing are to prevent pollution caused by developing new materials (i.e., producing or cultivating new fibres such as cotton or polyester). Also, they save energy and money, reduce GHG emission and global climatic change, and most important, help sustain the environment for future generations. The meaning of recycle is to collect and process materials that are considered trash and turn them into new products. Recycling saves resources and uses less energy than producing new items. Levi Strauss & Co. started the '8 Bottles. 1 Jean' concept, recycling eight polyethylene terephthalate bottles (typically beverage bottles) to produce one pair of jeans [42]. A few companies extracted the fibre from old denim (considered trash) for technical applications such as insulation for homes and civic buildings.

## 4.3 Life-cycle assessment

An LCA is an analytical tool for the systematic and quantitative evaluation of the environmental impacts of a product or service system throughout all stages of its life. It is an undeniable fact that denim products require the use of raw materials, energy, equipment and labour. Generally, these key points determine design and product development. Thus an LCA is necessary for an analysis of details about product design by assessing the environmental and economic impacts during the product's life cycle, with an aim of ensuring sustainable development. An LCA studies environmental aspects and potential impacts throughout a product's life (i.e., cradle to grave) from raw material acquisition through production, use and disposal. LCAs are critical for understanding the global progress of reducing negative impacts on the environment. Currently, an LCA is the most comprehensive approach to assessing the environmental impacts of textile products and is graded by the International Organization for Standardization (ISO) 14040-14043 industry standard [43–46]. An LCA does not include social or economic impacts of sustainability. The LCA framework has described as having four phases, shown in Fig. 4.3. This assessment is based on the definition of its goal and scope, inventory analysis, impact assessment and interpretation of results [13,47,48].

### *4.3.1 Goal and scope*

The goal and scope are the first phase of an LCA, which sets the context of the study. In them, the following items should be considered and clearly described:

- the functions of the product system or, in the case of comparative studies, product systems
- the functional unit
- the product system to be studied
- the system boundaries
- allocation procedures
- types of impact and the methodology of impact assessment, and subsequent interpretation to be used
- data requirements
- assumptions
- limitations
- initial data quality requirements
- type of critical review, if any
- type and format of the report required for the study

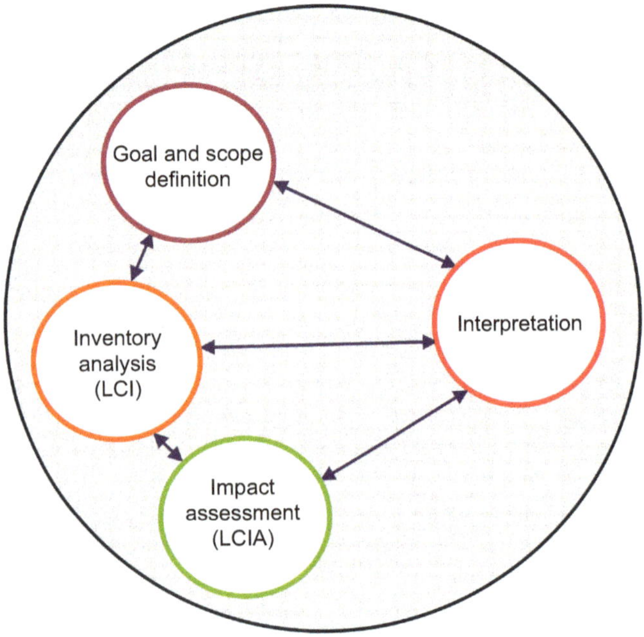


**Figure 4.3** The life-cycle assessment framework [47,49].
Adapted from sources – Ali S, Life cycle assessment, submitt. as coursework phys. 240, Stanford Univ. Fall 2010; 2010. http://large.stanford.edu/courses/2010/ph240/ali2/; Rebitzer G, Ekvall T, Frischknecht R, Hunkeler D, Norris G, Rydberg T, Schmidt W-P, Suh S, Weidema BP, Pennington DW, Life cycle assessment: part 1: framework, goal and scope definition, inventory analysis, and applications, Environ Int 2004; 30:701–20. doi:10.1016/j.envint.2003.11.005.

The LCA is an iterative technique; therefore, the scope of the study may require modification according to the product design and other parameters of the study being conducted as additional information is collected.

### 4.3.2 Life-cycle inventory analysis

A life-cycle inventory (LCI) analysis is the second phase of an LCA; it is the data collection portion. It is straightforward accounting and involves the 'system' of interest, consisting of the detailed tracking of all process flows for the entire system (i.e., from raw materials to final products, energy use, water, carbon, and other substance emissions) based on ISO 14040:2006. Therefore, it is a complex analysis which may involve dozens of individual unit processes in the supply chain (e.g., extraction of cotton or other fibres, yarn formation, colouration, fabric formation, garmenting, finishing, transportation, retailing and distribution) as well as hundreds of tracked substances.

### 4.3.3 Life-cycle impact assessment

The impact assessment is done after the inventory analysis. In general, this phase can focus on the results of the inputs and outputs and strategically categorize them, helping to analyse the environmental impact. Various methods are available globally to categorize and characterize the life-cycle impact of the flows to and from the environment, but it can complicate the comparability of different LCA studies. Apart from this, there are other variables in LCI assessment (LCIA) including the system boundary, the functional unit and specific LCIA methods such as allocation. Mandatory elements for the LCIA are:

- the selection of impact categories, category indicators, and characterization models
- the classification stage, in which inventory parameters are sorted and assigned
- impact measurement, indicating where the inventory in categories is characterized, using many possible LCA methodologies

### 4.3.4 Life-cycle interpretation

This is the last phase of LCA, in which we 'identify, quantify, check and evaluate' information from the results of the previous phases such as LCI and or the LCIA [50]. This study aids in ensuring the validity of the entire study. ISO 14040:2006 requires the interpretation phase to:

- identify significant issues based on the results of the LCI and LCIA phases of an LCA
- evaluate the study, considering completeness, sensitivity and consistency checks
- have conclusions, limitations and recommendations [13,50].

## 4.4 Life-cycle assessment of denim

In 2004, Levi Strauss & Co. began a green campaign involving sustainable fashion. They initiated many case studies with respect to their product. In 2007, they conducted an LCA analysis to assess the environmental impact of a pair of Levi's jeans, which are made from cotton. The analysis was made from cradle to grave (seed sowing to landfill). The study opened up a new era for awareness regarding the environmental issues caused by one pair of jeans. The environmental impact was assessed in the following categories:

1. cotton production
2. fabric production
3. garment manufacturing
4. transportation and distribution
5. consumer use
6. recycling, which then goes back to Step 2, or
7. waste stream in a landfill

From this work, an idea was born to create sustainable fashion in their supply chain (design, production, merchandise and marketing), which ensured the reduction of environmental impacts. Therefore, the primary objective of the LCA approach was to provide designers and developers with the information they needed to produce more sustainable products. A secondary objective was to provide a scientific method to support claims of the environmental improvement of products. Although not an initial objective of this effort, it was later realized that the methodology also provided a rigorous means to communicate environmental performance to suppliers (Levi Strauss & Co.) [51]. The LCA study on denim conducted by Levi Strauss & Co. was based on the ISO 14040 standards, in which all significant processes from 'cradle to gate' was analysed based on one pair of Levi's 501 jeans. The phases are:

- fibre cultivation (sowing to harvesting)
- yarn formation and fabric formation (ginning, spinning, preparation for weaving and weaving)
- colouration (colouration of warp)
- cut, sew and finish (garmenting and garment washing)
- transport and retail (transportation at each stage and distribution to retail stores)

Table 4.1 shows the impact categories that were investigated to carry out an LCA.

Commercially huge number of software is available to measure LCA, but Levi's used ReCiPe 2008 for this study [53]. This assessment helped Levi's and other brands in the apparel sector to use more effective and holistic approaches to address the environmental impact caused by this industry.

Several data were included in cycle mentioned earlier. For example, in the spinning phase, the type of fibre, origin of the fibre, and fibre loss during processing were taken into account. In the case of colouration and weaving, the type of dye (indigo, sulphur, pigment, etc.), weave structure, and sizing efficiency were considered. During the cutting and sewing phase, the cutting efficiency (i.e., marker efficiency), materials, sundry material/weight, and packaging material/weight were included. The types of chemicals used during the finishing phase were considered; finally, the transport mode and distance were considered and collected from all phases of the entire supply chain.

**Table 4.1** Impact categories and descriptions [52]

| Environmental impact categories | | |
|---|---|---|
| **Category** | **Description** | **Units** |
| Climate change | Global warming potential of greenhouse gases released into environment | $CO_2$ emissions (kg) |
| Water consumption | Net freshwater taken from environment minus water returned to the same watershed at the same quality or better | Litres |
| Eutrophication potential | Oxygen depletion as a result of nitrogen and phosphorous deposit into freshwater or marine environments | $PO_4$ emission (g) |
| Land occupation | Total land occupied to support product system assessed | $m^2$/year |
| Abiotic depletion | Measure of depletion of nonrenewable sources including fossil energy, metals and minerals | Sb emissions (mg) |

Levi Strauss & CO. The life cycle of a jean; 2015. p. 16–50. http://levistrauss.com/wp-content/uploads/2015/03/Full-LCA-Results-Deck-FINAL.pdf.

### 4.4.1 Functional unit

The functional units are the number of items to be analysed for this study. Therefore the functional unit for this study is the production of one pair of Levi's 501 medium stonewash jeans (women's) with a weight of 340g. Details of the functional unit are listed in Table 4.2.

### 4.4.2 System boundaries

The boundaries of denim jeans systems from the cradle-to-gate phase of the life cycle are: raw materials extraction, fabric formation, colouration, garment manufacturing, transportation, retailing and distribution.

### 4.4.3 Material phase

The material phase covers how much cotton cultivation is required to produce one pair of denims. We know that cotton cultivation requires a huge quantity of fertilizers, water and pesticides to prevent insects and worms, and herbicides to prevent weeds. This study phase includes procedures used to extract cotton (harvesting, transportation and ginning) to produce one pair of denim jeans.

### 4.4.4 Production phase

The production phase consists of more process lines which convert the raw materials (fibre) into a finished product (denim jeans). The number of processes is directly proportional to the higher amount of environmental impact. In this phase, different processes are involved in making final products, i.e., yarn formation, fabric formation, colouration, garment manufacturing and the finishing process.

Table 4.2 **Functional unit of life-cycle assessment (LCA)**

| Functional unit | Pair of Levi's 501 women's jeans (one pair) |
|---|---|
| Material | 100% cotton |
| Raw material | Cotton made in United States, Brazil, India, China, Pakistan and Australia |
| Dyeing | Indigo blue |
| Finishing | Medium stonewash |
| Weight of jeans | 340 g |
| Year of production | 2012 |
| LCA conducted | 2013 |
| LCA published | 2015 |
| Consumer phase | United States, United Kingdom, France and China |
| Study conducted by | Industrial Ecology Consultants and Levi Strauss & Co. |
| International Organization for Standardization (ISO) | ISO 14040 |
| Number of studies conducted | Not disclosed |

### 4.4.5 Environmental impact categories

The environmental impact potential is expressed at the midpoint and end-point levels. A description of environmental impact categories is shown in Table 4.1. The following categories were adapted from the LCA of a pair of Levi's 501 denim jeans conducted by Levi Strauss & Co., based on the ReCiPe 2008 methodology. The essential environmental impact categories are: ozone depletion, terrestrial acidification, human toxicity, photochemical oxidant formation, particulate matter formation, terrestrial eco-toxicity, freshwater eco-toxicity, marine eco-toxicity, ionizing radiation and natural land transformation [52].

### 4.4.6 Climate change

Climate change is one of the major concerns of the 21st century. Because it is one of the major polluting industries of the world, the textile industry has drawn intense attention to its affects on climate change, which results in global warming, a rise in sea levels and extreme weather. Increasing consumption of textiles for clothing is causing the biggest impact on the environment. In the case of denim, from fibre production to finished products, meaning throughout the supply chain, the textile industry emits a huge amount of carbon, initiating a change in the climate. These changes usually affect the ecosystem as well as human health, and they increase every year. This study focuses on the global warming potential of radioactive forcing, and measuring the effect of emissions of $CO_2$ and other GHGs into the environment. The final result can be expressed in terms of how it affects human health and the ecosystem. Some facts about cotton cultivation are that:

- The production of 1 kg of cotton (equivalent to a pair of jeans) can take more than 20,000 L of water.
- As much as 1.5% of the world's annual energy is consumed by industrial fertilizer production. Because it is a key consumer of fertilizer, cotton has a huge role in global climate change.

### 4.4.7 Water consumption

The World Bank estimates that almost 20% of all freshwater pollution is made by textile production, in particular wet processing. We are aware that the global textile (including denim) industry releases pollutants continuously, causing unimaginable harm to the environment. Generally, it pollutes the land and makes it useless. A survey showed that cotton crops consume the highest amount of harmful pesticides and fertilizers, which pollute and spoil the water and land. Thus water consumption is an important factor affecting human health and the ecosystem. The life cycle for LCA is expressed at the midpoint level simply because no model can provide the end-point level. As per the ReCiPe 2008 LCA tool, water consumption can be calculated from raw material procurement to the end product. Therefore, this study focused mostly on water that is lost and damaged rather than recycled. There are some facts which highly correlate cotton cultivation and water consumption:

- The Aral Sea, which is the world's fourth largest lake, has been reduced to 15% of its original size as a result of irrigation for the cotton industry.
- A total of 73% of the global cotton harvest comes from irrigated land.

### 4.4.8 Eutrophication

Eutrophication is an environmental problem that affects water bodies. Eutrophication occurs as the result of a process in which high nutrient levels stimulate the growth of aquatic plants, especially algae, on the surface of still or slow-moving water bodies. The growth of algae on the water surface reduces the penetration of sunlight, reducing oxygen levels. Also, as the result of photosynthesis, algae reduce oxygen levels, so both processes affect the water bodies. The resulting level of deoxygenation is increased by microbial action upon dead algae on the floor of a lake or estuary. Ultimately these limitations in the available levels of sunlight and oxygen reduce the diversity of plant and fish life that the affected waters can sustain. In LCA, eutrophication is expressed at the end-point level for damage to the ecosystem. Generally phosphorus is an essential nutrient for plants; however elevated amounts of phosphorus cause plants to soak up more nitrogen before all phosphorus is depleted, causing excessive algae growth in bodies of freshwater [13].

### 4.4.9 Land occupation

This study assessed how much damage to the ecosystem is caused by land occupation. In general, it concentrates on the area of land used for raw materials, cultivation and production, and the transformation of a certain area of land. Denim production requires larger areas of land than do other apparel products, owing to cotton cultivation. Therefore, in this study results are expressed at the midpoint and end-point levels, because both systems are affected [13].

### 4.4.10 Abiotic depletion

Abiotic depletion is the loss of nonliving natural resources such as naturally occurring minerals, fossil fuel energy, and metals. These resources, such as fossil fuels and minerals, are essential to our everyday lives and most are being extracted at an unsustainable rate, which seems to be leading to a shortage of resources for future generations. The results for denim regarding the overall costs of resources lost during cultivation and production are calculated, expressed at the midpoint and end-point levels.

### 4.4.11 Inventory analysis

LCA data for denim jeans was collected from Levi Strauss & Co. The results express the environmental impacts of a pair of Levi's 501 jeans explained in Section 4.4. These

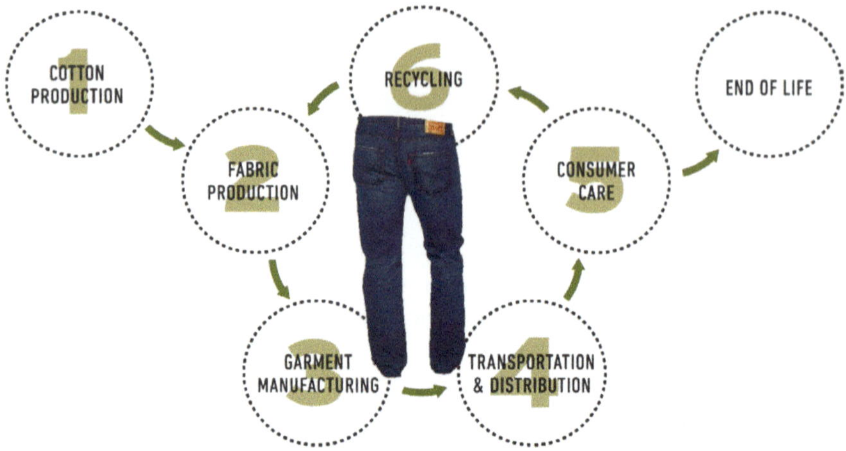


**Figure 4.4** Life cycle of a pair Levi's 501 jeans [36].
Adapted from The life cycle of a jean, by Levi Strauss & Co., 2015; Copyright 2015 by Levi Strauss & Co.

analyses were made with the help of the ReCiPe 2008 tool. Based on this tool, the input and output inventory was calculated with respect to environmental impacts. Fig. 4.4 illustrates the life-cycle phase of denim from cradle to grave [13,52].

## 4.5 Life-cycle assessment results

### 4.5.1 Climate change

Generally, climate change results from the emission of GHG and is followed by impacts on global warming. The emission of GHGs in textile production is substantial, resulting in an increase in the temperature of the system and damage to human health, animals and entire ecosystems. To produce one pair of denim jeans, consumer care was considered to have the highest impact on global warming; it measured 37% on the life cycle. Fabric production had the second largest impact on global warming (27%). The third biggest impact on global warming was transport and retail; its impact was 11% on the life cycle. Climate changes (expressed in kilograms and percentage) to the environment with respect to the different phases of denim production are shown in Fig. 4.5. Consumer care contributes the largest impact on climate change. $CO_2$ emissions are increased by processing factors such as washing with warm water, using a dryer, washing weekly and using a conventional washing machine. Results are shown in Figs 4.6 and 4.7.

#### 4.5.1.1 Water consumption

Water is important because it is needed for life to exist; it is an abundant resource on earth. Water is used for many purposes include agricultural, industrial, household, recreational and environmental activities. Most important, water is essential for sustaining the ecosystem. Climate change will have significant impacts on water resources

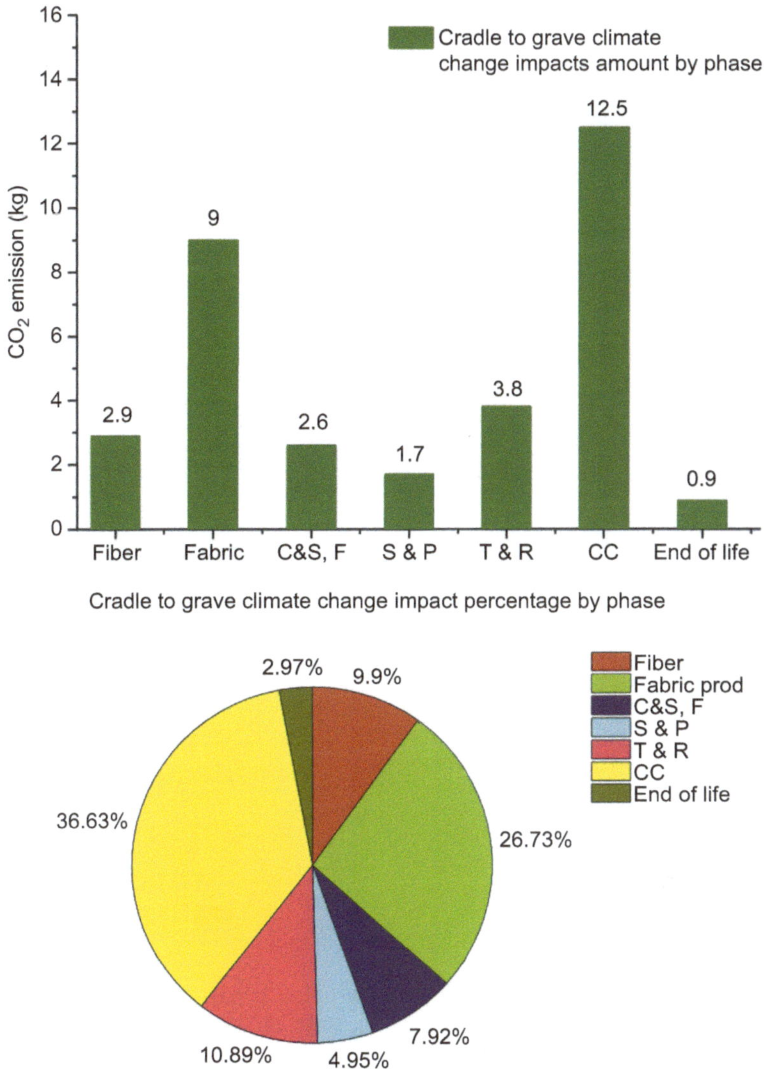


**Figure 4.5** Climate changes to the environment with respect to different phases of denim production. *C & S, F*, cut and sew, finishing; *CC*, consumer care; *Fabric Prod*, yarn formation, fabric formation, colouration; *Fiber*, fibre cultivation to harvesting; *S & P*, sundries and packing; *T & R*, transport logistics and retail [52].
Adapted from The life cycle of a jean, by Levi Strauss & Co., 2015; Copyright 2015 by Levi Strauss & Co.

around the world because of the close connections between the climate and the hydrologic cycle. Because of the expanding human population, competition for water is growing so that many of the world's major aquifers are becoming depleted. Thus water resource management has become and will remain important. Levi Strauss & Co. conducted water consumption studies in major cotton producing countries such as

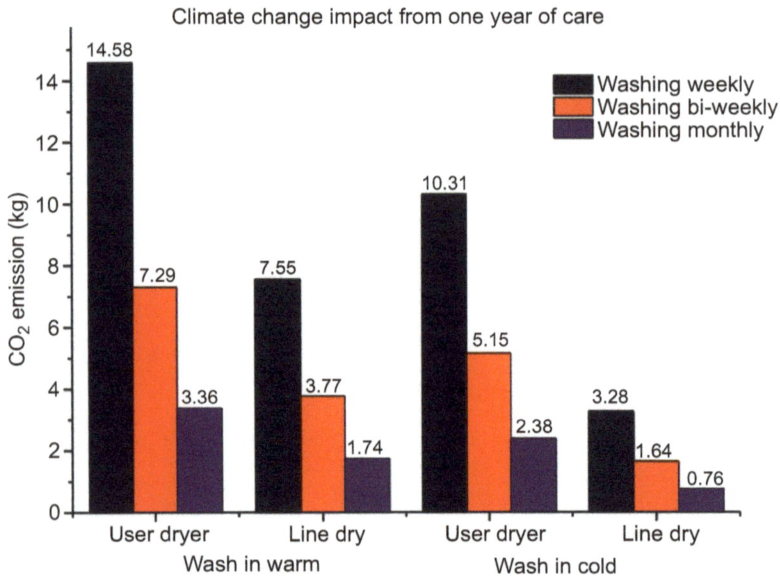


**Figure 4.6** Climate change impact of 1 year of care with a conventional washing machine [52]. Adapted from The life cycle of a jean, by Levi Strauss & Co., 2015; Copyright 2015 by Levi Strauss & Co.

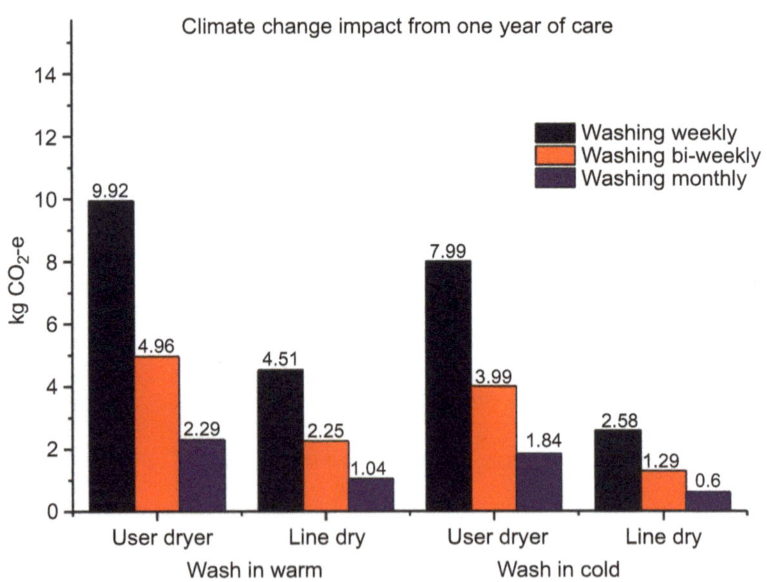


**Figure 4.7** Climate change impact of 1 year of care with an efficient washing machine. [52]. Adapted from The life cycle of a jean, by Levi Strauss & Co., 2015; Copyright 2015 by Levi Strauss & Co.

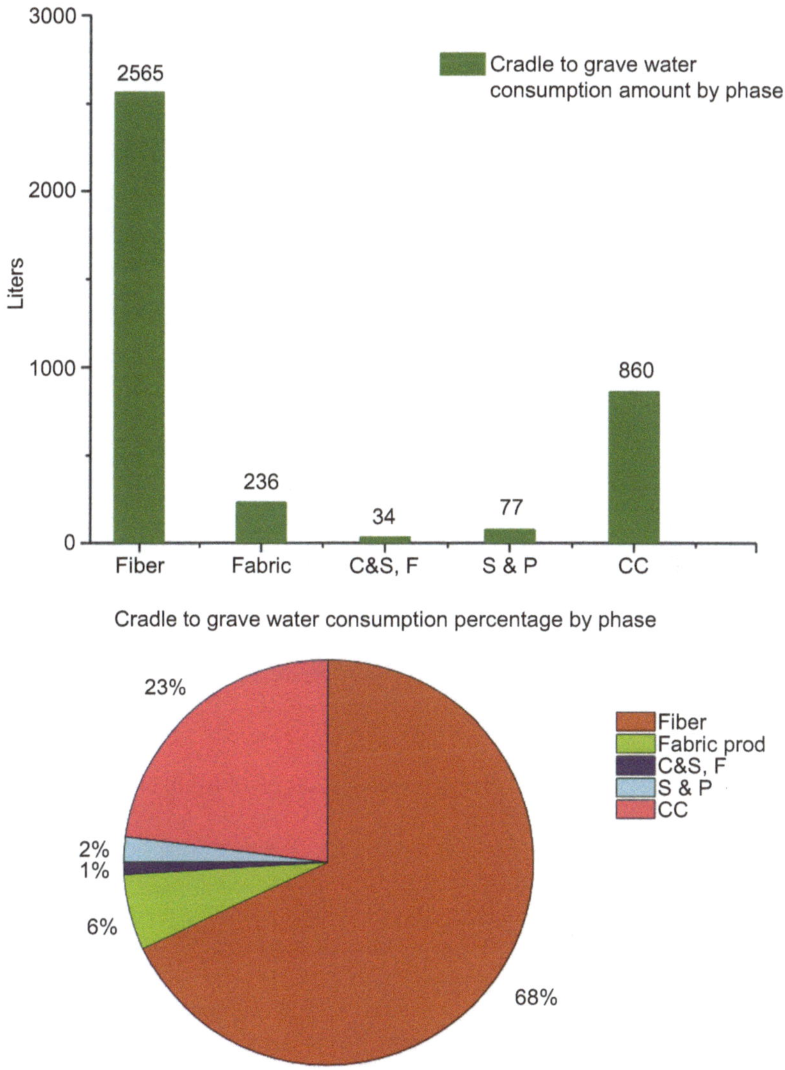


**Figure 4.8** Water consumption for different phases of denim production. *C & S, F*, cut and sew, finishing; *CC*, consumer care; *Fabric Prod*, yarn formation, fabric formation, colouration; *Fiber*, fibre cultivation to harvesting; *S & P*, sundries and packing; *T & R*, transport logistics and retail [52].
Adapted from The life cycle of a jean, by Levi Strauss & Co., 2015; Copyright 2015 by Levi Strauss & Co.

the United States, India, China, Brazil, Australia and Pakistan. Cotton crops consume a large quantity of water from sowing to harvesting. The study proved that fibre production consumes 68% of the entire LCA. The second largest consumption of water is during consumer care activities such as cleaning and washing, at 23% (Fig. 4.8).

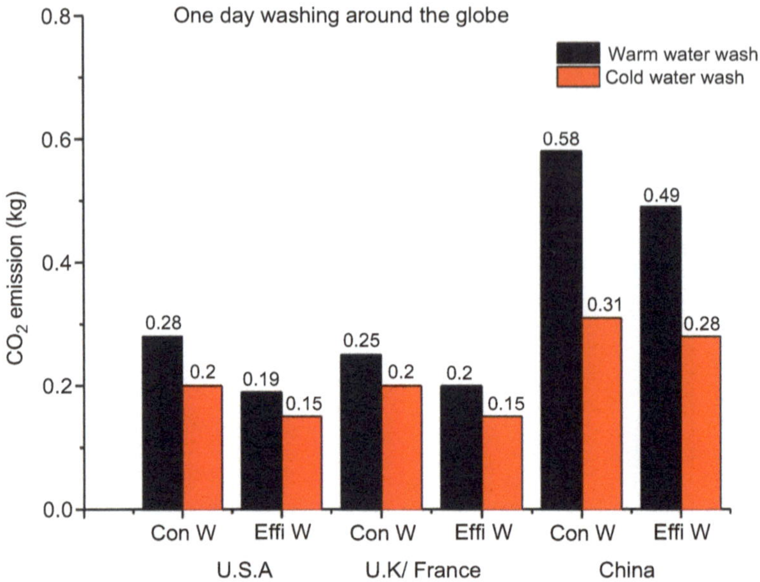


**Figure 4.9** Impact of emission resulting from 1 day of washing, in different countries. *U.K.*, United Kingdom; *U.S.A.*, United States [52].
Adapted from The life cycle of a jean, by Levi Strauss & Co., 2015; Copyright 2015 by Levi Strauss & Co.

Fabric production consumes 6% during the processes involved in converting fibre into yarn and yarn into fabric.

Concerning consumer care, Levi Strauss & Co. conducted various studies. From their studies, we get some idea about how consumer care can have an impact on climate change and water consumption. A 'One Day of Washing Around the Globe' survey was conducted to obtain more information about denim washing (consumer care phase) techniques. The emission results for the United States and the United Kingdom/France were almost same in both cases (i.e., conventional and efficient washers). However, to use warm water, we need to use more energy, which creates higher $CO_2$ emissions (in kilograms) than does washing in cold water. The emission results for China were two times higher than those of the United States, and it had a huge impact on environmental effects with respect to climatic changes. Overall, efficient washing techniques may cause less emission potential than conventional washing techniques. The detailed study is described in Fig. 4.9.

Levi Strauss & Co. studied water intake for 1 year's care of denim (Fig. 4.10). Results showed that washing denim once a week consumes a large quantity of water; around 958 L per year for one pair of denim. That is two to four times higher consumption than other techniques such as washing biweekly or monthly. The most important fact from that study is that efficient washing machines consume 30% less water during washing than conventional washing machines with respect to weekly washing.

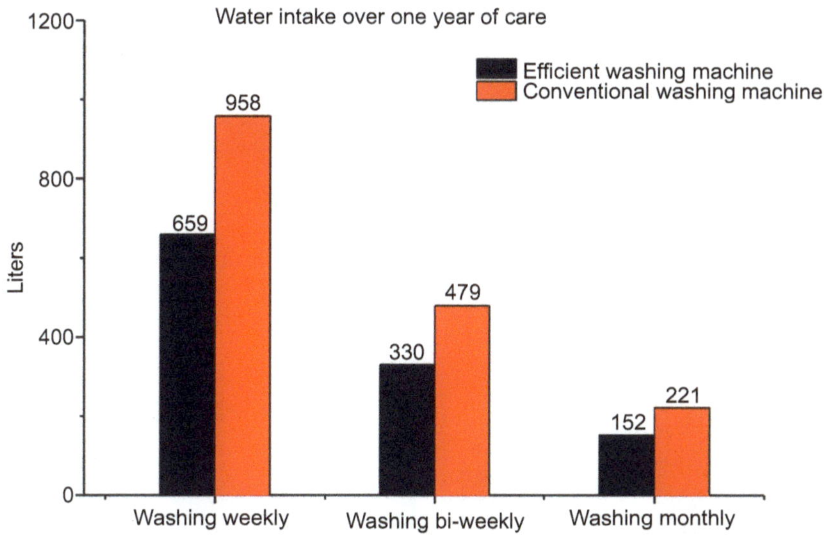


**Figure 4.10** Water consumption of denim washing during consumer care, using different washing techniques [52].
Adapted from The life cycle of a jean, by Levi Strauss & Co., 2015; Copyright 2015 by Levi Strauss & Co.

### *4.5.1.2 Eutrophication*

Over the past century humanity has transformed its water bodies from a clear-water ecosystem into a eutrophic (nutrient-rich) environment. Today it is highly contaminated with nutrients that cause eutrophication, such as algal blooms, red tides, dead zones and a range of related serious problems for the ecosystem and the people living nearby and earning a living from it. It has a significant impact on the aquatic ecosystem, damaging drinking water sources, fisheries and recreational bodies of water. There is no doubt that it damages human health and the entire ecosystem. Eutrophication has been happening silently all over the world and it is a big environmental issue. In the case of denim production, the fibre cultivation stage has a huge impact on the environment, and a 37% share. The second largest eutrophication impact on the environment is consumer care, with a share of 16%. Apart from this, sundries and packing, fabric production and transportation have high impacts to the environment due to the eutrophication process. The cradle to grave eutrophication process of denim, by phase ($PO_4$ emission in grams and percent share by various processes) can be seen in Fig. 4.11.

### *4.5.1.3 Land occupation*

Humans are the major force of change on the earth, transforming the land for agriculture that provides food, shelter, production systems and infrastructure for development and such. However land transformation can affect the planet's physical, chemical and biological systems and directly affect ecosystems. Therefore land is the basis for most biological and human activities on earth. The transformation rate has drastically increased

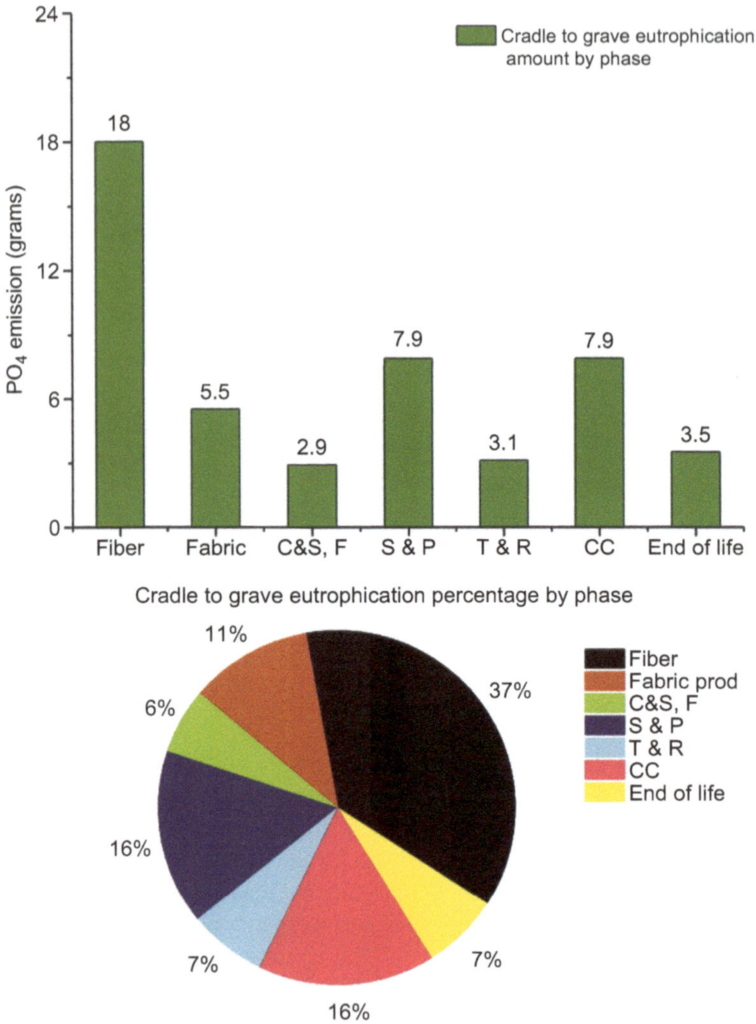


**Figure 4.11** Cradle-to-grave eutrophication amount of denim by phase ($PO_4$ emissions in grams and percent share by various processes). *C & S, F*, cut and sew, finishing; *CC*, consumer care; *Fabric Prod*, yarn formation, fabric formation, colouration; *Fiber*, fibre cultivation to harvesting; *S & P*, sundries and packing; *T & R*, transport logistics and retail [52]. Adapted from The life cycle of a jean, by Levi Strauss & Co., 2015; Copyright 2015 by Levi Strauss & Co.

since the 19th century and has had an impact on the ecosystem, global warming, floods, etc. Owing to the production and agricultural sector, land occupation is unavoidable. Compared to other fibres, cotton consumes a larger quantity of the land for cultivation to production. Its cultivation requires large areas of land as well as production activities because of the number of processes in yarn and fabric formation. To produce one pair jeans, 9.3 $m^2$/year area for cultivation purposes is required, which is 78% of the share.

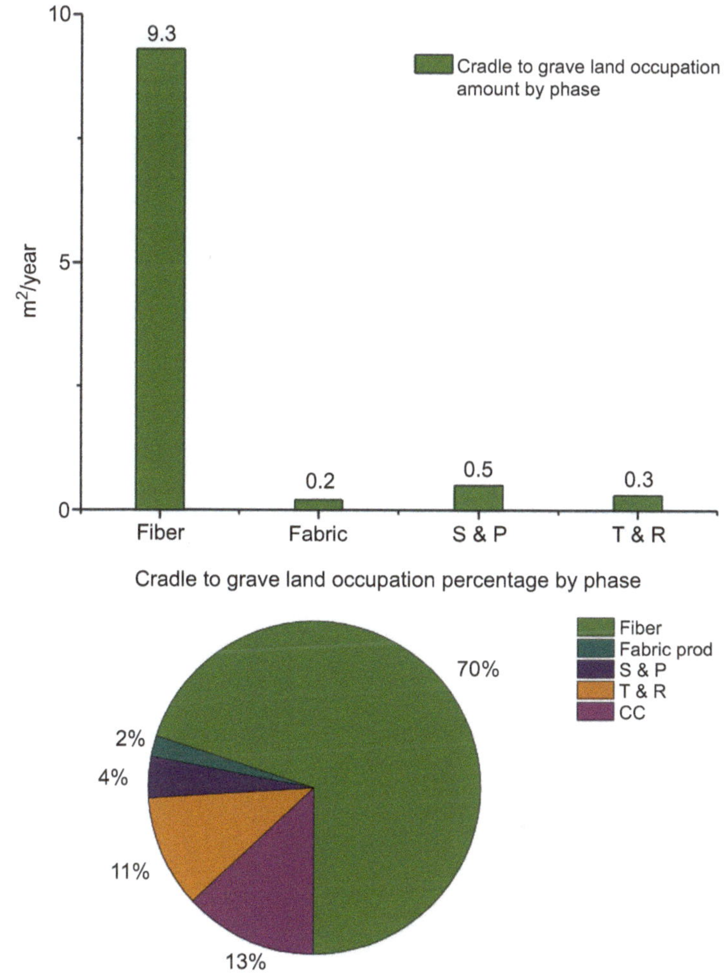


**Figure 4.12** Cradle to grave estimate of land occupation with respect to the production of one pair of denim jeans. *C & S, F*, cut and sew, finishing; *CC*, consumer care; *Fabric Prod*, yarn formation, fabric formation, colouration; *Fiber*, fibre cultivation to harvesting; *S & P*, sundries and packing; *T & R*, transport logistics and retail [52].
Adapted from The life cycle of a jean, by Levi Strauss & Co., 2015; Copyright 2015 by Levi Strauss & Co.

Fibre production has consumed the highest land occupation and entails consumer care, transportation, logistics and retail. The brief results are shown in Fig. 4.12.

### *4.5.1.4 Abiotic depletion*

These are the measures of the gradual loss of natural resources such as minerals, fossil fuels and metals. Natural resources are resources that exist without the actions of humankind. However, natural resources contain renewable and nonrenewable resources.

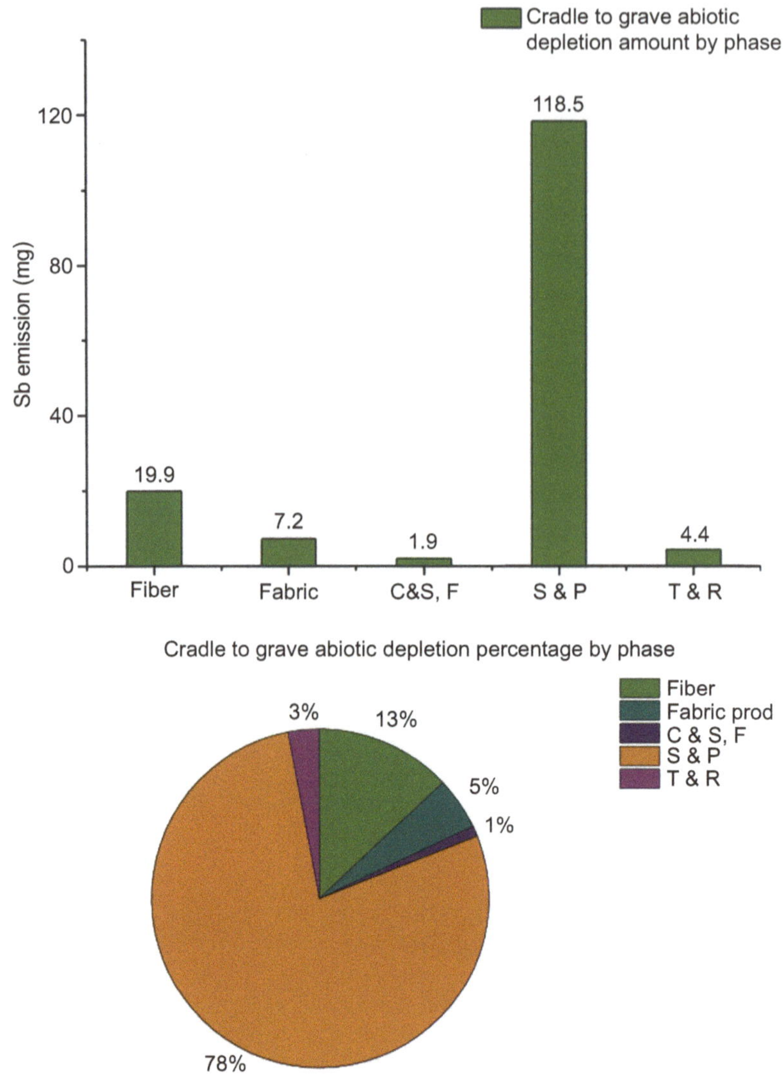


**Figure 4.13** Abiotic depletion cradle to gate to produce one pair of denim jeans. *C & S, F*, cut and sew, finishing; *CC*, consumer care; *Fabric Prod*, yarn formation, fabric formation, colouration; *Fiber*, fibre cultivation to harvesting; *S & P*, sundries and packing; *T & R*, transport logistics and retail [52].
Adapted from The life cycle of a jean, by Levi Strauss & Co., 2015; Copyright 2015 by Levi Strauss & Co.

The result of abiotic depletion with respect to producing one pair of denim jeans is reported in Fig. 4.13. Over the entire life cycle, the sundries and packaging phase of the production process is 118.5 mg and accounts for 78% of the cradle to gate portion of the denim life cycle. Other phases require less; of that, fibre production produces 19.9 mg Sb emission. Next, fabric production, cutting, sewing, finishing and transportation has a small impact on abiotic depletion.

## 4.6 Interpretation and conclusion of results

*Climate change*: In the production of one pair jeans, the fabric production and consumer care stages have the highest impact on climate change owing to the number of processes which use resources.

*Water consumption*: This result mainly reflects litres of water used and not reused, and that fibre production and consumer care require large quantities of water; we need a solution for this.

*Eutrophication*: As to eutrophication in the producing one pair of denim jeans, fibre cultivation has the highest impact. This is because of the use of fertilizer and pesticides, which damage water bodies, causing nutrient-rich runoff water. They generally affect the oxygen content in water, altering aquatic systems.

*Land occupation*: Results shows that the highest land occupation required to produce denim jeans is for fibre cultivation, which obviously requires a huge area. Apart from cultivation, land occupation may occur in the forest to cut trees to produce energy. This practice is still carried out in many developing countries.

*Abiotic depletion*: It is considered to be one of the largest environmental issues caused by textile production. This is the result of the use of natural resources to convert fibre to garments. During denim production, sundries consumed the highest resources with the greatest impact on abiotic depletion.

Materials, production and other findings:

- Fabric assembly, which includes yarn spinning, dyeing, weaving, and fabric finishing, had notable contributions related to their impact on climate change and nonrenewable energy consumption.
- Life-cycle stages that had a minimal contribution to impact include fabric transport, product transport, packaging, production waste, distribution, retail, and end-of-life waste.

The entire life cycle of one pair of Levi's 501 denim jeans is equal to:

- climate change: 33.4 kg $CO_2$ emissions; this is equal to a US car driven 111 km, or 246 h watching TV on a plasma big screen
- water consumption: 3781 L, or 3 days of a household's total water needs in the United States
- eutrophication: 48.9 g $PO_4$ emissions; this is around the total amount of phosphorus found in 1700 tomatoes
- land occupation: 12 $m^2$/year of land, equal to seven people standing in a relaxed manner

Consumers will be given information about the environmental impact of denim production by labeling, which is another potential output of a technique such as LCA. Labeling such as eco-labels can identify better environmental policies used in products. Customer do not have much knowledge about relevant aspects of denim production and how it has been optimized. It is not necessary for them to be flooded with figures and scientific details. They need a clear message that allows them to choose products according to the information given. In practice there are no such customs followed in eco-labeling, which is confusing and leads to misunderstanding. Eco-labeling needs to be based on established methods such as LCAs and linked to best available technology criteria and sustainability standards such as ISO 260000.

## 4.7 Future trends

In the future, the same methodology can be used to find LCAs for other denim garments which may be made from lyocell and other sustainable fibres [54]. It would provide ideas which would make it possible to compare the environmental impacts of similar products with respect to the different fibres. LCA research, which is available with statistical data, is needed yearly to account for changes in fibre production, a company's sustainable initiative, the availability of suppliers and factors that contribute harm to human health, the ecosystem, diversity, and resource availability.

### *4.7.1 Software for life-cycle assessment*

A lot of software is available in the marketplace. The most widely circulated LCA software includes:

- Boustead Consulting Database and Software
- ECO-it: Eco-Indicator Tool for environmentally friendly design: PRé Consultants
- ECOPRO: Sinum Corporate Environmental Management
- EDIP: Environmental design of industrial products: Danish Environmental Protection Agency
- EIOLCA: Economic Input-Output LCA at Carnegie Mellon University
- GaBi 3 – (Ganzheitliche Bilanzierung): University of Stuttgart (IKP)/PE Product Engineering
- IDEMAT: Delft University Clean Technology Institute Interduct Environmental Product Development
- KCL-ECO 3.0: KCL LCA software
- LCAiT: CIT EkoLogik (Chalmers Industriteknik)
- LCNetBase: LCA using traceable US data: Sylvatica
- SimaPro 5.0 for Windows – PRé Consultants
- SPOLD: Society for the Promotion of Life-Cycle Assessment Development
- TEAM (Tools for Environmental Analysis and Management): Ecobalance, Inc.
- Umberto: Advanced software tool for Life Cycle Assessment: Institut für Umweltinformatik

## Acknowledgements

We would like to express our thanks to Levi Strauss & Co. for the useful data, which helped us to make this chapter. The primary author is grateful to the Republic of India for providing the best life and knowledge, and also thanks the Technical University of Liberec and the Ministry of Education, Youth and Sports of the Czech Republic for the doctoral student scholarship.

## References


[1] Hethorn J, Ulasewicz C. Sustainable fashion: why now?: A conversation exploring issues, practices, and possibilities. Fairchild Books Inc.; 2008.

[2] Wakankar DM. 5-Regulations relating to the use of textile dyes and chemicals A2-Gulrajani, M.L. BT – advances in the dyeing and finishing of technical textiles. In: Woodhead publ. ser. text.. Woodhead Publishing; 2013. p. 105–32. http://dx.doi.org/10.1533/9780857097613.1.105.

[3] Levasseur A, Cavalett O, Fuglestvedt JS, Gasser T, Johansson DJA, ørgensen SVJ, Raugei M, Reisinger A, Schivley G, Strømman A, Tanaka K, Cherubini F. Enhancing life cycle impact assessment from climate science: review of recent findings and recommendations for application to LCA. Ecol Indic 2016;71:163–74. http://dx.doi.org/10.1016/j.ecolind.2016.06.049.

[4] Driving demand for denim jeans, cott. Inc.. Supply Chain Insights; 2013. http://www.cottoninc.com/corporate/Market-Data/SupplyChainInsights/Driving-Demand-For-Denim-Jeans/Driving-Demand-for-Denim-Jeans.pdf.

[5] Koszewska M. 15-Life cycle assessment and the environmental and social labels in the textile and clothing industry A2-Muthu, Subramanian Senthilkannan BT – handbook of life cycle assessment (LCA) of textiles and clothing. In: Woodhead publ. ser. text. Woodhead Publishing; 2015. p. 325–44. http://dx.doi.org/10.1016/B978-0-08-100169-1.00015-0.

[6] Muthu SS. Environmental impacts of the use phase of the clothing life cycle. In: Muthu SS, editor. Handb. life cycle assess. text. cloth.. Woodhead Publishing; 2015. p. 93–102. http://dx.doi.org/10.1016/B978-0-08-100169-1.00005-8.

[7] Eryuruk SH. 7-Life cycle assessment method for environmental impact evaluation and certification systems for textiles and clothing. In: Muthu SS, editor. Handb. life cycle assess. text. cloth.. Woodhead Publishing; 2015. p. 125–48. http://dx.doi.org/10.1016/B978-0-08-100169-1.00007-1.

[8] Henry BK, Russell SJ, Ledgard SF, Gollnow S, Wiedemann SG, Nebel B, Maslen D, Swan P. LCA of wool textiles and clothing. In: Muthu SS, editor. Handb. Life cycle assess. text. cloth.. Woodhead Publishing; 2015. p. 217–54. http://dx.doi.org/10.1016/B978-0-08-100169-1.00010-1.

[9] van der Werf HMG, Turunen L. The environmental impacts of the production of hemp and flax textile yarn. Ind Crops Prod 2008;27:1–10. http://dx.doi.org/10.1016/j.indcrop.2007.05.003.

[10] Søndergård B, Hansen OE, Holm J. Ecological modernisation and institutional transformations in the Danish textile industry. J Clean Prod 2004;12:337–52. http://dx.doi.org/10.1016/S0959-6526(03)00049-0.

[11] Nieminen E, Linke M, Tobler M, Vander Beke B. EU COST Action 628: life cycle assessment (LCA) of textile products, eco-efficiency and definition of best available technology (BAT) of textile processing. J Clean Prod 2007;15:1259–70. http://dx.doi.org/10.1016/j.jclepro.2006.07.011.

[12] Eryuruk SH. 7-Life cycle assessment method for environmental impact evaluation and certification systems for textiles and clothing A2-Muthu, Subramanian Senthilkannan BT – handbook of life cycle assessment (LCA) of textiles and clothing. In: Woodhead Publ. Ser. Text.. Woodhead Publishing; 2015. p. 125–48. http://dx.doi.org/10.1016/B978-0-08-100169-1.00007-1.

[13] Hackett T. A comparative life cycle assessment of denim jeans and a cotton T-shirt: the production of fast fashion essential items from cradle to gate. College of Agriculture at the University of Kentucky; 2015. http://uknowledge.uky.edu/cgi/viewcontent.cgi?article=1008&context=mat_etds.
[14] Downey L. A short history of denim. Levi Strauss Co.; 2014. http://www.levistrauss.com/wp-content/uploads/2014/01/A-Short-History-of-Denim2.pdf.
[15] Zhang Y, Liu X, Xiao R, Yuan Z. Life cycle assessment of cotton T-shirts in China. Int J Life Cycle Assess 2015;20:994–1004. http://dx.doi.org/10.1007/s11367-015-0889-4.
[16] Yafa S. Cotton: the biography of a revolutionary fiber, first. USA: Penguin Books; 2006.
[17] Allen W. Fact sheet on U.S.A. cotton subsidies and cotton production. Org Consum Assoc 2004. https://www.organicconsumers.org/old_articles/clothes/224subsidies.php.
[18] Mancini F, Van Bruggen AHC, Jiggins JLS, Ambatipudi AC, Murphy H. Acute pesticide poisoning among female and male cotton growers in India. Int J Occup Environ Health 2005;11:221–32. http://dx.doi.org/10.1179/107735205800246064.
[19] Esteve-Turrillas FA, de la Guardia M. Environmental impact of recover cotton in textile industry. Resour Conserv Recycl 2017;116:107–15. http://dx.doi.org/10.1016/j.resconrec.2016.09.034.
[20] Khan M, Damalas CA. Farmers' knowledge about common pests and pesticide safety in conventional cotton production in Pakistan. Crop Prot 2015;77:45–51. http://dx.doi.org/10.1016/j.cropro.2015.07.014.
[21] Liu EM, Huang J. Risk preferences and pesticide use by cotton farmers in China. J Dev Econ 2013;103:202–15. http://dx.doi.org/10.1016/j.jdeveco.2012.12.005.
[22] Maumbe BM, Swinton SM. Hidden health costs of pesticide use in Zimbabwe's smallholder cotton growers. Soc Sci Med 2003;57:1559–71. http://dx.doi.org/10.1016/S0277-9536(03)00016-9.
[23] Oosterhuis DM, Tom Cothren J, editors. Flowering and fruiting in cotton, the cotton foundation. 2012. https://www.cotton.org/foundation/fandfcontents.cfm.
[24] Fertilizer use-cotton, USDA. (n.d.). http://www.ers.usda.gov/data-products/fertilizer-use-and-price.aspx#26737.
[25] Kalliala EM, Nousiainen P. Life cycle assessment environmental profile of cotton and polyester. AUTEX Res J 1999:8–20.
[26] Patra AK, Pattanayak AK. 16-Novel varieties of denim fabrics A2-Paul, Roshan BT – Denim. In: Woodhead publ. ser. text. Woodhead Publishing; 2015. p. 483–506. http://dx.doi.org/10.1016/B978-0-85709-843-6.00016-0.
[27] McLoughlin J, Hayes S, Paul R. 2-Cotton fibre for denim manufacture BT – Denim. In: Woodhead publ. ser. text. Woodhead Publishing; 2015. p. 15–36. http://dx.doi.org/10.1016/B978-0-85709-843-6.00002-0.
[28] Pardo Martínez CI. Energy use and energy efficiency development in the German and Colombian textile industries. Energy sustain Dev 2010;14:94–103. http://dx.doi.org/10.1016/j.esd.2010.02.001.
[29] Kane C. Environmental and health hazards in spinning industry and their control. Indian J Fiber Text Res 2001;26:39–43. http://nopr.niscair.res.in/bitstream/123456789/24911/1/IJFTR26(1-2)39-43.pdf.
[30] Kalliala E, Talvenmaa P. Environmental profile of textile wet processing in Finland. J Clean Prod 2000;8:143–54. http://dx.doi.org/10.1016/S0959-6526(99)00313-3.
[31] Reddy N, Chen L, Zhang Y, Yang Y. Reducing environmental pollution of the textile industry using keratin as alternative sizing agent to poly(vinyl alcohol). J Clean Prod 2014;65:561–7. http://dx.doi.org/10.1016/j.jclepro.2013.09.046.

[32] Ozturk E, Koseoglu H, Karaboyaci M, Yigit NO, Yetis U, Kitis M. Sustainable textile production: cleaner production assessment/eco-efficiency analysis study in a textile mill. J Clean Prod 2016;138:248–63. http://dx.doi.org/10.1016/j.jclepro.2016.02.071.
[33] Hasanbeigi A, Price L. A technical review of emerging technologies for energy and water efficiency and pollution reduction in the textile industry. J Clean Prod 2015;95:30–44. http://dx.doi.org/10.1016/j.jclepro.2015.02.079.
[34] Periyasamy Aravin Prince. Bio-processing in textiles. Text Mag 2007;48.
[35] Teli MD. 8-Environmental textiles: testing and certification. In: Wang L, editor. Perform. test. text. Woodhead Publishing; 2016. p. 177–92. http://dx.doi.org/10.1016/B978-0-08-100570-5.00009-8.
[36] Jena B, Das BP, Khandual A, Sahu S, Behera L. Ecofriendly processing of textiles. Mater Today Proc 2015;2:1776–91. http://dx.doi.org/10.1016/j.matpr.2015.07.020.
[37] Easton JR. 6-Key sustainability issues in textile dyeing*. In: Blackburn RS, editor. Sustain. text. Woodhead Publishing; 2009. p. 139–54. http://dx.doi.org/10.1533/9781845696948.1.139.
[38] Mohanty RP, Prakash A. Green supply chain management practices in India: a confirmatory empirical study. Prod Manuf Res 2014;2:438–56. http://dx.doi.org/10.1080/21693277.2014.921127.
[39] Sarkis J. A strategic decision framework for green supply chain management. J Clean Prod 2003;11:397–409. http://dx.doi.org/10.1016/S0959-6526(02)00062-8.
[40] Seuring S. Integrated chain management and supply chain management comparative analysis and illustrative cases. J Clean Prod 2004;12:1059–71. http://dx.doi.org/10.1016/j.jclepro.2004.02.006.
[41] Aziz TNAT, Jaafar HS, Tajuddin RM. Green supply chain: awareness of logistics industry in Malaysia. Procedia – Soc Behav Sci 2016;219:121–5. http://dx.doi.org/10.1016/j.sbspro.2016.04.052.
[42] Levi Strauss & Co. 8 bottles one jean; 2013. http://explore.levi.com/news/sustainability/introducing-levis-wasteless-8-bottles-1-jean/.
[43] ISO 14040. Environmental management: life cycle assessment. Principles and Framework; 2006.
[44] ISO 14041. Environmental management: life cycle assessment, goal and scope definition and inventory analysis. 1998.
[45] ISO 14042. Environmental management: life cycle assessment, life cycle impact assessment. 2000.
[46] ISO 14043. Environmental management: life cycle assessment, life cycle interpretation. 2000.
[47] Rebitzer G, Ekvall T, Frischknecht R, Hunkeler D, Norris G, Rydberg T, Schmidt W-P, Suh S, Weidema BP, Pennington DW. Life cycle assessment: part 1: framework, goal and scope definition, inventory analysis, and applications. Environ Int 2004;30:701–20. http://dx.doi.org/10.1016/j.envint.2003.11.005.
[48] Stylianou KS, Heller MC, Fulgoni VL, Ernstoff AS, Keoleian GA, Jolliet O. A life cycle assessment framework combining nutritional and environmental health impacts of diet: a case study on milk. Int J Life Cycle Assess 2016;21:734–46. http://dx.doi.org/10.1007/s11367-015-0961-0.
[49] Ali S. Life cycle assessment, submitt. as coursework phys. 240. Stanford Univ. Fall 2010; 2010. http://large.stanford.edu/courses/2010/ph240/ali2/.
[50] Skone TJ. What is life cycle interpretation? Environ Prog 2000;19:92–100. http://dx.doi.org/10.1002/ep.670190207.

[51] Schrott W, Paul R. 20-Environmental impacts of denim manufacture BT – Denim. In: Woodhead publ. ser. text.. Woodhead Publishing; 2015. p. 563–80. http://dx.doi.org/10.1016/B978-0-85709-843-6.00020-2.

[52] Levi Strauss & CO. The life cycle of a jean. 2015. p. 16–50. http://levistrauss.com/wp-content/uploads/2015/03/Full-LCA-Results-Deck-FINAL.pdf.

[53] Lamnatou C, Chemisana D. Evaluation of photovoltaic-green and other roofing systems by means of ReCiPe and multiple life cycle–based environmental indicators. Build Environ 2015;93(2):376–84. http://dx.doi.org/10.1016/j.buildenv.2015.06.031.

[54] Periyasamy AP, Dhurai B, Thangamani K. Salt-free dyeing – a new method of dyeing on lyocell/cotton blended fabrics with reactive dyes. Autex Res J 2011;11.

# Water footprint of denim industry

5

*H. Pal[1], K.N. Chatterjee[2], D. Sharma[3]*
[1]BPS Mahila Vishwavidyalaya, Sonipat, Haryana, India; [2]The Technological Institute of Textile & Sciences, Bhiwani, India; [3]Amity University Haryana, Gurgaon, India

## 5.1 Introduction

The textile industry has a prominent role in the global economy [1] but it also has an adverse influence on the environment. The denim industry, the products of which are highly used in clothing, is an important subsector of the textile industry. It is at the top of the pyramid of water-consuming textiles, with a total volume of 2900 gallons (approximately 11,000 L) consumed per pair of jeans [2–4]. In addition, dyeing, finishing and washing processes are major sources of water consumption as well as water pollution. Industrial wastewater contains large amounts of dyestuff and alkaline chemicals [5]. Considering the global jeans production estimated to be more than 3.5 billion units [6,7], the extent of environmental damage caused by this industry is quite evident.

The denim industry must undergo a revolutionary change to sustain the world in the production of apparel. Innovations in the way dyestuffs, fibres, weaving fabrics, etc. are manufactured must be made to reduce the water footprint of denim. 'The water footprint is a measure of humanity's appropriation of fresh water in volumes of water consumed and/or polluted' [8]. Consumers also need to know that it is not only the monetary cost that is involved in manufacturing a pair of jeans; a huge environmental cost is associated with them [9].

## 5.2 Water consumption with denim

Apparel production depends heavily on water availability, from growing cotton to manufacturing, and to consumer care at home [10]. Denim has become a universal material in apparel. It is used in almost every type of garment, from jeans and jackets to skirts, dresses and even evening wear [11]. Some of the studies done on the denim industry have a difference of opinion regarding the production of denim jeans. As per the findings, it has been observed that the production of jeans ranges from 3.5 to 5.0 billion pairs per year. According to Luiken et al., the production of jeans is estimated to be more than 3.5 billion pairs. If an average pair of jeans weighs 600 g, the total textile consumption of jeans is above 2.1 million metric tons a year. Considering water consumption at 11,000 L per pair of jeans and production to be 3.5 billion pairs, water resource consumption to the extent of 38.5 trillion L or 38.5 billion $m^3$ will be required. However Garcia mentioned that annual global jeans production is estimated

**Sustainability in Denim. http://dx.doi.org/10.1016/B978-0-08-102043-2.00005-8**
Copyright © 2017 Elsevier Ltd. All rights reserved.

to be 5000 million units; and considering that the average amount of water required to finish one pair is 70 L, this means that 350 million $m^3$ water is consumed in jeans manufacturing. This is twice the water supply needs of all inhabitants of one of the most populated cities of Europe, Madrid (Spain). So this is why denim finishing using less water is becoming a priority for many brands and retailers [6,7].

The main ingredient in denim is cotton, which in itself is a water-intensive crop and is responsible for the high water footprint of denim. During the life cycle of a pair of jeans, approximately 70% of water consumed is used solely for cotton agriculture [10]. It can even take more than 20,000 L of water to produce 1 kg of cotton, which is equivalent to a single T-shirt and a pair of jeans. In addition, use of fertilizers and pesticides with large quantities of water for cotton production has affected large-scale ecosystems, which has affected the health and well-being of people living nearby. The amount of water footprint to produce a pair of jeans depends on the location, irrigation technology, farmers' awareness to conserve water, raw materials content and production technology involved. Dyes and fabric treatments involve the use of chemicals, requiring extensive water consumption and wastewater treatment. Thus the entire denim industry from cotton irrigation to manufacturing is responsible for a high water footprint [4,11].

## 5.3 Concept of water footprint

The concept of a water footprint, introduced in 2002 by Arjen Y. Hoekstra, is an excellent indicator to measure water use. A water footprint refers to the total amount of freshwater used directly and indirectly by an individual, community or country over a period of time. It is a geographic indicator which shows the amount of water used or polluted by a group at a particular location. Its knowledge helps an organization to study water use patterns and gauge the eco-friendly nature of its products to make suitable improvements. It indicates water consumption and water contamination during industrial processes that rendered water unfit for future use [12].

A water footprint is important in the sense of understanding the consequences of the human appropriation of freshwater resources by analyzing how much water is needed for human use compared with how much is available at which place and at what time [14,15].

### *5.3.1 Freshwater*

Freshwater is one of the most precious resources we have, and it is becoming increasingly scarce [10]. There is a need to assess how much freshwater is available and its human appropriation over a certain period. The water footprint basically expresses the human appropriation of freshwater in terms of volume. Comparing humanity's water footprint with actual freshwater availability is part of a water footprint sustainability assessment [13]. Today, one in 10 people in the world do not have access to clean water and one in every three people do not have access to a proper toilet. The water table level is gradually decreasing at a faster rate than it can be replenished [10]. The unsustainable use of freshwater resources by humans is manifested around the world. Human activities are being affected by the overexploitation of water resources. Freshwater scarcity has become an important issue on the environmental agendas of nations, governments, companies, decision makers and the public as well as across the media. A water crisis would have a great impact not only for human societies but also for natural ecosystems. Thus there is a dire need to conserve the most precious water resource and protect vital ecosystems for future generations by reducing water footprints [14,15].

### 5.3.2 Water footprint components [13]

There are three components of a water footprint for both direct and indirect usage calculations: green, blue and grey. Whereas the first two account for total consumption, the last measures the amount of freshwater polluted. Fig. 5.1 shows the components of a water footprint. It also indicates that the unconsumed part of water withdrawals (the return flow) is not part of the water footprint.

#### 5.3.2.1 Green water footprint

The green water footprint refers to the consumption of green water, which further refers to rainwater that does not run off or recharge groundwater but is stored in the soil or temporarily stays on top of the soil or vegetation. A green water footprint is an indicator of the human use of evaporative flow from the land surface, mostly for growing crops or a production forest. A green water footprint is the volume of rainwater consumed during the production process.

#### 5.3.2.2 Blue water footprint

The blue water footprint refers to the volume of surface and groundwater consumption along the supply chain of a product or service. 'Consumption' refers to the losses that occur when water evaporates, returns to another catchment area or the sea or is

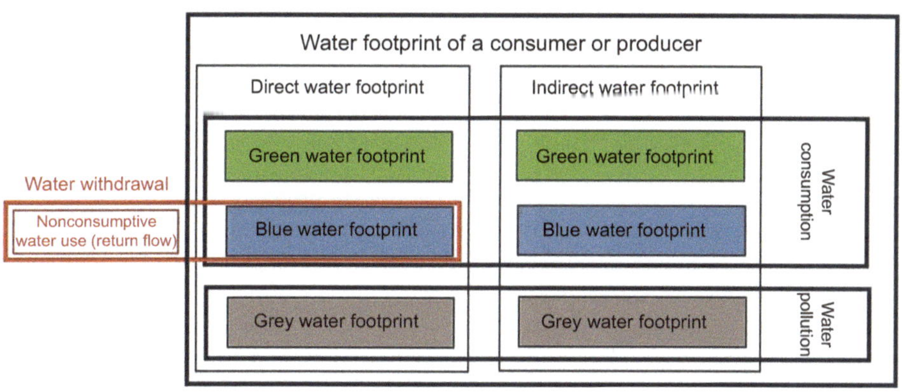


**Figure 5.1** Components of water footprints.
Courtesy Hoekstra AY, Chapagain AK, Aldaya MM, Mekonnen MM. The water footprint assessment manual: setting the global standard. London: Earthscan Publishing; 2011.

incorporated into a product. It also includes water that does not return in the same period (i.e. withdrawn in a scarce period and returned in a wet period).

#### *5.3.2.3 Grey water footprint*

The grey water footprint of a product is an indicator of freshwater pollution that can be associated with the production of a product over its full supply chain. It is defined as the volume of freshwater that is required to assimilate the load of pollutants based on existing ambient water quality standards. It is calculated as the volume of water that is required to dilute pollutants to such an extent that the quality of the water remains above agreed water quality standards.

### *5.3.3 Water footprint of a denim product*

The water footprint of a denim product refers to the total volume of freshwater consumed directly and indirectly to produce that particular denim product, measured over the full supply chain. It is a multidimensional indicator, showing water consumption volumes by source and polluted volumes by type of pollution [13]. The standard units of measurement are cubic metres of water per ton of production, or litres per kilogram, or gallons per pound. The water footprint of a pair of jeans refers to the sum total of the water footprint of each step, or process, required to manufacture it. To manufacture a pair of jeans requires cotton to be grown, fibres to be ginned and spun, and the fabric to undergo weaving, sewing and wet processing ultimately to have the finished product. Each step has a direct water footprint and an indirect water footprint. The direct water footprint of one process becomes the indirect water footprint of the subsequent one. In this way, the full amount of water consumed or polluted is taken into account in the product water footprint. Calculating the water footprint of the denim industry as a whole, which includes the entire product range made from the denim, will indicate the pressure it has put on freshwater resources [14].

## 5.4 Standards and measurement

Global standards as developed and maintained by the Water Footprint Network[1] are mainly applied for 'water footprint assessment'. These include set of definitions and in-depth procedures to calculate the water footprint for processes, products, end users, organizations and countries. These standards also cover methods for a water footprint sustainability assessment and a library of water footprint response options [13].

Water footprints can be measured for a wide variety of entities by summing the water footprint involved during the various processes involved in the complete supply chain or life cycle. By adding the water footprints of products produced or

[1] Water Footprint Network: International learning community providing a platform for sustainability, equitability and efficient water use.

consumed, the water footprint of a company, an industrial sector, an end user or a group of end users can be assessed. The total water footprint of a particular geographical zone is calculated by adding the water footprints of all activities within that zone [15].

Bottom-up and top-down approaches are used to assess the water footprint. The bottom-up approach is used to calculate all types of water footprint accounts, i.e., of organizations, sectors, nations and regions. The water footprint of a nation can be calculated by multiplying all commodities and services used by residents of that nation by the respective water requirements for those commodities and services. The bottom-up approach is generic and precise, but it is data-intensive, especially for large entities. A top-down approach is used to calculate the water footprint of sectors, provinces, nations and regions. It uses trade data and water footprints of traded goods to determine virtual water balances of countries. The water footprint of people living in a region, country or river basin is the total use of water resources in the area under consideration plus the gross virtual water import into the area minus the gross virtual water export. The virtual water import is the volume of water used in other countries to make goods and services imported to and consumed within the country considered. The virtual water export is the volume of water used domestically to make export products which are consumed elsewhere. A bottom-up approach is precise but data-intensive and depends on the quality of consumption data. The top-down approach does not require consumption data, but it requires trade data and therefore is vulnerable to the quality of those data. The top-down approach was used in all early national water footprint studies; but recent studies tend to use the bottom-up approach [15].

Water footprints provide spatiotemporally explicit information on water appropriation for various human activities. All components of water footprints depend on location and time. The same value of a blue water footprint resulting from some activity or process may appear to be less significant in the wet season but of great concern in a dry season. Also a water footprint in one catchment area cannot be compensated for by offsetting activities to reduce the water footprint in another catchment area [16].

Water footprints are also measured as per International Organization for Standarization14046 standards, which is based on a life-cycle assessment (LCA) [17]. LCA is a systematic, phased and standard approach that identifies the impact of water use in certain products, consumers, companies, nations, etc., which can help reduce the amount of water used. It assesses the environmental aspects and potential impacts associated with a product, process or service. 'Life cycle' refers to the major activities in the course of the product's life-span from its manufacture, use, and maintenance, to its final disposal, including the raw material acquisition required to manufacture the product. It specifically looks at the damage to three areas of protection: human health, ecosystem quality, and resources. In addition, regional assessments are equally necessary as the impact of water use depends on its location [18]. LCA also provides information for consumers about the environmental impact of production and through eco-labels that identifies products that have better environmental profiles [19].

## 5.5 Water footprint studies of denim

An assessment of the water footprint for the denim industry accounts for all activities from cotton production to garment manufacturing, transportation and use until disposal. In 2007, Levi Strauss & Co. conducted an LCA study to assess the environmental impact of a pair of Levi's jeans from cotton seed to landfill. This study led to the idea of developing a life cycle–based product environmental impact assessment method (E-valuate) based on primary data which could work as actionable as well as dynamic in nature. The objectives of the assessment approach were to provide designers and developers the information to produce more sustainable products and to provide a scientific method to support any claims of environmental improvement of products [20].

Current LCA studies identify the washing of jeans, both during the production process and by consumers during use, as having the greatest environmental impact. In 2010, Levi's launched its 'Water<Less' jeans collection and claimed to reduce water consumption in the manufacturing process by an average of 28%, and up to 96% for some products in the line. 'Low-impact denim' by Jack & Jones, H2Ø from Springfield and the conscious denim collection by H&M are a few more examples of the efforts to reduce water consumption in the jeans manufacturing supply chain [6].

Jack & Jones carried out an in-depth LCA which has allowed them to define the average water consumption for their five best-selling jeans, which accounts for 40% of their total production. LCA research concluded that the average water consumption of finishing a pair of Jack & Jones jeans is 36 L per garment. By introducing new technologies and innovative thinking in their production, they were able to save more than 40% water and additionally, important savings on energy consumption. The company defines jeans as 'low-impact denim' when they are finished with less than 22 L of water. Levi Strauss & Co. also published the results of an LCA it conducted. Jeanologia, which developed environmental impact measuring (EIM) tool, classify a low-impact process which consumes less than 35 L of water per garment [6].

Huge amounts of pollutants are released during different processing stages of denim, especially dyeing, finishing/washing and rinsing. This wastewater, if released without proper treatment, is harmful to the environment. In Table 5.1, the characterization of typical denim plant wastewater, as well as indigo and sulphur dyeing wastewater, is summarized.

Wastewater released from denim production contains high levels of salts, acids or alkali chemicals, surfactants, dye colours and a high pH concentration. Indigo, sulphur, reactive and vat dyes are the major dye types used in combination with a huge amount of water. Wastewater also contains trace metals. The composition of wastewater from textile processes varies greatly daily and hourly depending on the dyestuff, fabric and chemicals used. Dye consumption is gradually increasing each year owing to increasing demand for the production of denim. If necessary steps are not taken by the denim industry, polluted wastewater can cause serious environmental problems [5].

**Table 5.1** **Characterization of denim dyeing process wastewater**

| Parameter | Indigo dyeing wastewater | Sulphur dyeing wastewater | Denim plant wastewater |
|---|---|---|---|
| Chemical oxygen demand (mg/L) | 750–950 | 2500–3500 | 1500–3100 |
| Colour (Pt-Co) | 5500–7000 | 24,000–34,000 | 1147–3547 |
| Total suspended solids (mg/L) | 50–300 | 100–340 | 150–300 |
| Total dissolved solids (mg/L) | – | – | 4000–8000 |
| pH | 10.5–11.5 | 10–11 | 9–13 |
| Conductivity (mS/cm) | – | – | 6–12 |
| Total phosphorous (mg/L) | – | – | 0–3 |
| Total nitrogen (mg/L) | – | – | 17–23 |

Reproduced from Uzal N. Effluent treatment in denim and jeans manufacture. In: Denim: manufacture, finishing and applications. Woodhead Publications; 2015. pp. 541–61.

## 5.6 Trends in denim industry

A growing awareness for water and environmental protection has translated into consistent efforts to reduce the adverse ecological impact of the production processes of denim. Worldwide consumers are also demanding goods with eco-friendly credentials. The denim industry is responding to this trend by modifying in-house practices and encouraging academic research [21]. Sustainable trends and developments in various spheres of denim industry, from the cultivation of cotton to recyclability in producing denim products, are mentioned subsequently.

### *5.6.1 Trends in cotton production*

Sustainability will hold the key to the future, and it is high time for cotton to become a sustainable fibre: in cultivation and denim manufacturing, as well as in the subsequent processing of garments [11].

Reducing chemical hazards and water consumption is an obvious way to make products more sustainable. New technologies such as drip irrigation, weather stations, underground moisture sensors and smart phone apps are helping cotton farmers use less water from field to gin, with enhanced profitability [9].

To reduce the water footprint, there is a growing trend toward cultivating and producing organic cotton for manufacturing denim. Organic cotton is cotton grown from nongenetically modified plants, that is to be grown without the use of synthetic agricultural chemicals such as fertilizers or pesticides. Its production also promotes and enhances biodiversity and biological cycles [22]. Many retailers, including the top brands, support organic cotton as a sustainable alternative. Consumers are also showing their willingness to pay a premium for denim garments made of certified organic cotton fibre. Although the production of organic cotton is expanding rapidly, conventional cotton still accounts for about 99.9% of the total world output. However, in the future it is probable that this trend is likely to be led by demand from consumers for an organic product [11].

Naturally coloured cotton is also another area of interest for the denim industry in view of its eco-friendly character. The colour in the cotton is a genetically controlled characteristic. Scientists have been able to develop cotton of different colours such as creamy white, brown, green, blue and pink. Other areas of future trends and developments include naturally coloured and genetically modified cotton [11].

### 5.6.2 Trends in novel denim varieties and blends

The cotton supply is volatile and depends on weather conditions. Reasonable prices which depends on cotton's availability and compliance with environmental norms are key concerns for the denim industry. In an attempt to lower the raw material cost and reduce water consumption, textile mills are developing new fibre blends to reduce the amount of cotton in denim. A current trend is substituting cotton with other fibres such as bamboo, viscose and Tencel. For instance, G-Star is using a cotton/viscose/Tencel[2] blend to improve the fabric hand and Nankatan, a Norwegian apparel company, uses 100% Monocel[3] to develop stretch denim [23].

Worldwide research is being conducted to replace cotton in denim with more sustainable fibres. Researchers at the Herriot Watt School of Textiles and Design, in the United Kingdom, developed a jeans using fibre made from sustainable wood using one-fifth of the water needed to manufacture conventional jeans. Cotton is extensively blended with lycra, polyester, lyocell, flax, etc. to develop special types of denim. Although most of the world production of denim jeans is still 100% cotton, the market for stretch denim is one of the fastest-growing segments of jeans manufacture. Cotton blends that use both lycra and polyester, combining both strength and stretch properties, are gaining in popularity [11].

As sustainability is becoming a major focus globally, the use of biodegradable fibres in denim is increasing. Some sustainable varieties include denim made of polylactic acid, soybean and bamboo fibres. Other such variants include organic cotton denim and denim dyed with natural dyes without using metallic mordants [24].

### 5.6.3 Developments and trends in indigo dye technology for denim

To accelerate and succeed the ecological transition in indigo dyeing processes, several actions are implemented to produce sustainable and fashionable denim. These include the use of non-indigo dyes and the development of clean indigo dyeing processes which minimize the adverse environmental impact [25].

Several environmental issues are associated with the conventional process of denim dyeing with indigo [21]. Indigo dye leads to major effluent problems during the dyeing process, and later in the washing of denim garments. Natural indigo has been

[2] TENCEL is the lyocell fibre, extracted from wood.

[3] Monocel is a range of dyed yarns made from bamboo.

completely replaced by synthetic indigo, which seems to be more sustainable, but the biosynthesis of indigo would be really sustainable [24]. Future research and development activities will focus on the microbial synthesis of indigo and environmentally friendly technologies for indigo reduction [21]. Alternative reducing systems such as organic reducing agents, biological reduction, electrochemical reduction and catalytic hydrogenation of indigo have been explored [24].

Technologies are available for the biosynthesis of indigo and the total replacement of sodium hydrosulphite in indigo dyeing, but their commercial acceptability has yet to be established because technologies featuring environmental benefits sell only in combination with cost reduction [21].

ITV Denim Italy collaborated with Levi Strauss & Co. to develop a natural dye developed from red wine. The dye is free from chemicals, is colourfast, and it uses less water than conventional indigo dyeing [23].

A breakthrough dyeing process is Advanced Denim, which operates completely without indigo. It offers a great variety of colours, needs much less water and energy, and produces no effluents [24].

The demand for natural dyes has been increasing in many countries, particularly in the west. This trend is due to the health hazard and environmental pollution issues associated with many synthetic dyes. Therefore, there is revival of interest in natural dyes. An entirely new concept of denim production without using indigo is emerging [21].

### 5.6.4 Trends of reducing water washing of denim garments

Dry treatments or nearly water-free treatments are slowly becoming a sustainable trend for replacing traditional wet treatments in denim washing, such as laser treatment, ozone treatment and dry ice blasting. Laser treatment is a water-free, colour fading treatment of denim and is an ecological and economical process. Lasers can create local abrasion, fabric breaks and a 'used' look effect with excellent reproducibility and higher productivity. Because it is an automatic system, the chances of human error are eliminated in laser treatment. In ozone treatment, the ozone generated in the equipment can provide a bleaching effect. Commercially available ozone equipment is operated like a washing machine but without much use of water for the colour fading process. In dry ice blasting, the working principle is like sandblasting using abrasive materials. However, in this case, dry ice is used for blasting. The advantage of using dry ice is that after blasting, the residual material is carbon dioxide gas with no secondary contaminants. In coming years, denim and jeans washings may involve only such dry or water-free treatments [26].

To measure the washing effectiveness, the Spanish Company, Jeanologia has developed EIM software that assesses the environmental impact of the garment finishing processes. It is the first of its kind and is specifically designed for the garment finishing industry. The software assesses the environmental impact of garment laundry processes, categorizing the impacts under four individual headings: water consumption, energy consumption, chemical impacts and impact on worker health. Moreover, it allows the user to compare different processes and assess their subsequent results.

It provides valuable information to determine whether process modifications and integrated technologies used for a particular denim processing route have led to the desired objective of water reduction [6].

The software benchmarks the impacts in each category against a predefined environmental threshold. First, it classifies each individual category; second, the entire process into either low, medium or high impact. The results are presented in an easily understood, colour-coded display.

### 5.6.5 *Trends in denim recovery and recycling*

To reduce the water footprint, efforts are made not only in the reduction of water consumption, chemical used and substitution of cotton but also in recycling of the denim products. Innovative approaches are being used for denim recovery and recycling. One such approach is adopted by ISKO[4] and Nudie[5] Jeans Company. Post consumer denim is ground into pulp to be converted into a denim yarn which is strong enough to be woven into a new denim fabrication. Levi Strauss & Co. in one of its product i.e., Waste<Less uses post consumer polyethylene terephthalate products collected from recycling programs that are manufactured into yarn and then woven into a denim fabrication. Mud Jeans, a denim apparel company, offers its customers to pay a monthly fee to rent their product for 1 year. After 12 months, a consumer can return the jeans, exchange them for a new model or keep them. The company deconstructs the returned jeans to create new products [23]. Denim is also triturated and converted into nonwoven felts for insulation in the construction and automotive industries. Nowadays the possibility of manufacturing denim from other waste materials is also becoming important. There is an increasing trend for recycling of denim, and a number of fashionable products are already available in the market with a high content of recycled denim fibres [24].

## 5.7 Challenges

Despite all of the innovations happening on the technological front to reduce the water footprint, the denim industry still has to adopt it effectively. Continuous rises in demand for denim are resulting in huge amounts of water pollution owing to dyeing, washing and finishing processes. Because of the complex nature of effluents containing a high amount of chemicals, suspended solids, dispersing agents and trace metals, it is not easy to treat wastewater to meet the desired standards. A limited work has been reported on the treatment of denim industry wastewater for reuse in production plants. Alternate technologies must be investigated and assessed for wastewater treatment to be used effectively [24]. Even other sustainable alternatives such as the use of organic denim are facing difficulties in dyeing technology. Dyeing of organic denim is challenging for dyers because it is time-consuming and requires skilled workers. This

[4] ISKO is a Turkish textile manufacturer and distributor of denim.
[5] Nudie Jeans is a Swedish clothing brand.

leads to high processing costs, which prevents organic denim from currently being more popular [25]. Recycling of denim waste is a challenging business and has numerous constraints related to the proper use of discarded denim apparel. The current state of denim recycling is merely mechanical recycling and there is a need to introduce improved recycling technologies [7]. The denim industry still has to adopt sustainable technologies in an effort to reduce its water footprint.

## 5.8 Conclusion

The overall impact of the denim industry on the water footprint is significant especially in areas such as cotton growing, dyeing and washing. Although modern research has focused on developing technology to reduce the overall impact, there is still a long way to go. The denim industry needs to adopt sustainable approaches such as the use of organic cotton, a reduction in dyes and chemicals, and the production of less complex wastewater. Other environmentally friendly alternatives include electrochemical reduction, minimal application technologies and reduced water washing. There is a need to spread awareness among consumers to look for denim which has been manufactured with the highest environmental standards, maintaining sustainability throughout its life cycle. Even product labeling must communicate clearly to consumers grading regarding the sustainability of a product, from low to high. Apart from technology upgradation and consumer awareness, strict regulations from governments on reducing the blue and grey water footprint components can have a crucial role in the optimal use and allocation of scarce freshwater resources. Water footprint–specific policies in isolation are not sufficient; rather, policies that address poverty, land use, trade, pollution, agriculture, food security and population should be considered together to meet these reduction goals.

## Sources of further information


1. Chico D, Aldaya M, Garrido A. A water footprint assessment of a pair of jeans: the influence of agricultural policies on the sustainability of consumer products. J Clean Prod 2013.
2. Paul R. Denim: manufacture, finishing and applications. Woodhead Publishing; 2015.
3. http://www.alternet.org/environment/levis-ceo-want-conserve-water-dont-wash-your-jeans.
4. http://www.engineerlive.com/content/new-concept-sustainable-denim-production.
5. http://gizmodo.com/5920267/it-takes-a-hell-of-a-lot-more-water-than-you-think-to-make-jeans-burgers-pizza-and-other-stuff.
6. http://levistrauss.com/sustainability/planet/.
7. http://news.asiaone.com/News/Latest+News/Science+and+Tech/Story/A1Story20120706-357560.html.
8. http://phys.org/news/2012-06-sustainable-denim-green-jeans.html.
9. https://www.rwlwater.com/jeans-maker-reduces-water-footprint/.
10. http://www.treehugger.com/clean-technology/how-many-gallons-of-water-does-it-take-to-make.html.

11. https://www.usfashionindustry.com/news/off-the-cuff-newsletter/2144-transforming-the-denim-industry.
12. http://waterfootprint.org.
13. http://waterprint.net/jeans.html.
14. https://en.wikipedia.org/wiki/Water_use.

## References

[1] http://www.texprocil.org/cdn/gallery/11/4d3dda720b6f748a85f72a24b08a21c1.pdf.
[2] A new concept for sustainable denim production. Available from: http://www.engineerlive.com/content/new-concept-sustainable-denim-production; 2013.
[3] National Geographic. A special issue on "Water: our thirsty world". 2010.
[4] Waterprint (online). 2016. Available: http://waterprint.net/jeans.html.
[5] Uzal N. Effluent treatment in denim and jeans manufacture. In: Denim: manufacture, finishing and applications. Woodhead Publications; 2015. p. 541–61.
[6] Garcia B. Reduced water washing of denim garments. In: Denim: manufacture, finishing and applications. Woodhead Publications; 2015. p. 405–23.
[7] Luiken A, Bouwhuis G. Recovery and recycling of denim waste. In: Denim: manufacture, finishing and applications. Woodhead Publications; 2015. p. 527–40.
[8] Water footprint network (online). 2016. Available: http://waterfootprint.org/en/water-footprint/what-is-water-footprint/.
[9] Transforming the denim industry (online). 2015. Available: https://www.usfashionindustry.com/news/off-the-cuff-newsletter/2144-transforming-the-denim-industry.
[10] Levi Strauss & Co. Water (online) 2016. Available from: http://levistrauss.com/sustainability/planet/.
[11] McLoughlin J, Hayes S, Paul R. Cotton fibre for denim manufacture. In: Denim: manufacture, finishing and applications. Woodhead Publications; 2015. p. 15–36.
[12] Water footprint concept (online). 2015. Available: http://www.gktoday.in/blog/water-footprint-concept-of-blue-water-greenwater-grey-water/.
[13] Hoekstra AY, Chapagain AK, Aldaya MM, Mekonnen MM. The water footprint assessment manual: setting the global standard. London: Earthscan Publishing; 2011.
[14] Water footprint (online). 2016. Available: http://waterfootprint.org.
[15] Ercin AE, Hoekstra AY. Carbon and water footprints: concepts, methodologies and policy responses. United Nations World Water Assessment Programme, UNESCO; 2012. ISBN: 978-92-3-001095-9.
[16] Hoekstra AY. The water footprint of modern consumer society. Earthscan; 2013.
[17] ISO (International Organization for Standarization). ISO/CD 14046: life cycle assessment – water footprint – requirements and guidelines. Geneva: ISO; 2012.
[18] Water use (online). 2016. Available: https://en.wikipedia.org/wiki/Water_use.
[19] Schrott W, Paul R. Environmental impacts of denim manufacture. In: Denim: manufacture, finishing and applications. Woodhead publications; 2015. p. 541–61.
[20] Levi Strauss & Co. Levi Strauss & Co. life cycle approach to examine the environmental performance of its products (online) 2011. Available from: http://lsco.s3.amazonaws.com/wp-content/uploads/2011/01/LSCO-Life-Cycle-Approach-to-Examine-the-Environmental-Performance-of-its-Products.pdf.
[21] Chavan RB. Indigo dye and reduction techniques. In: Denim: manufacture, finishing and applications. Woodhead Publications; 2015. p. 37–67.

[22] Organic cotton (online). 2016. Available from: https://en.m.wikipedia.org/wiki/Organic_cotton.
[23] Regan C. Role of denim and jeans in the fashion industry. In: Denim: manufacture, finishing and applications. Woodhead Publications; 2015. p. 191–217.
[24] Paul R. Denim and jeans: an overview. In: Denim: manufacture, finishing and applications. Woodhead Publications; 2015. p. 1–11.
[25] Meksi N, Mhenni MF. Indigo dyeing technology for denim yarns. In: Denim: manufacture, finishing and applications. Woodhead Publications; 2015. p. 69–105.
[26] Kan CW. Washing techniques for denim jeans. In: Denim: manufacture, finishing and applications. Woodhead Publications; 2015. p. 313–56.

# Carbon footprint in denim manufacturing

6

*T. Karthik, R. Murugan*
PSG College of Technology, Coimbatore, India

## 6.1 Introduction

'Eco-chic', 'environmentally conscious', 'ethical consumerism', 'sustainable fashion' and 'clothing with a conscience' are fashion buzzwords. Even more familiar, 'natural', 'organic' and 'green' are adjectives intended to shape consumers' minds and purchases regarding more sustainable choices. Sustainability is a growing phenomenon in the world of fashion. Globalization has provided the necessary impetus for the apparel industry to produce and deliver fashion at a rapid rate [1–3]. Also, stringent enforcement of environmental laws by regulatory authorities shows that there is growing recognition regarding moving toward a more sustainable model in the textile and clothing industry.

Climate change, also known as global warming, is resulting in large-scale irreversible changes globally. A major promoter of this is the emission of greenhouse gases (GHGs), which are caused by the production and consumption of fuels and other manufacturing materials. The greenhouse effect is responsible for keeping the earth warm. High concentrations of GHGs affect the earth's equilibrium, putting it out of balance, which will lead to extreme variations in weather. Unchecked burning of fossil fuels continuously releases carbon dioxide ($CO_2$) into the atmosphere [2]. The greenhouse effect is caused by GHGs that trap heat in the earth's atmosphere. Uncontrolled global warming could influence global eco-regions. A rise in temperature globally will influence ecosystems, which could lead to several species being forced out of their territory because of changing conditions while others flourish [4–10]. Apart from these major drawbacks, minor effects of global warming are a rise in sea level, changes in climatic conditions and lessened snow cover, which may influence the ecosystem as well as human activities. Burning of fossil fuels to generate energy could increase the carbon footprint into the atmosphere.

The carbon cycle is an important cycle of the earth, and human activity is significantly influencing the various physical and chemical activities responsible for the earth's climate [11]. Atmospheric concentrations of $CO_2$ have been increasing for decades, creating profound implications for future climate change, the productivity of agriculture and the function of ecosystems. Much research has demonstrated that global warming mainly results from the increasing emissions of GHGs as a result of human activities. The major contributing factor to global warming is $CO_2$, which accounts for approximately 80% of emissions from developed economies. To keep

**Sustainability in Denim. http://dx.doi.org/10.1016/B978-0-08-102043-2.00006-X**
Copyright © 2017 Elsevier Ltd. All rights reserved.

track of one's GHG emissions is to be aware of one's carbon footprint. Carbon footprints are a measure of the impact that human activity has on the environment in terms of the quantity of $CO_2$ equivalents; it is measured in kilograms or tons.

The carbon footprint assessment (CFP) is an important approach for the control and management of GHG emissions. The method is based on the life-cycle concept and is used to assess GHGs released during the manufacture of goods and services, either throughout their entire life cycle or within specified temporal or spatial boundaries. The carbon footprint is the total amount of $CO_2$ and other GHGs released during the full life cycle of a process or product. It is calculated for the period of 1 year and is expressed in terms of the amount of $CO_2$ or its equivalent of other GHGs released [12]. Other kinds of GHGs which could be released are ozone and methane [13,14]. These GHGs are also considered for the evaluation of the carbon footprint but they are normally converted into equivalent $CO_2$ (i.e., the amount of $CO_2$ that could lead to a similar influence on global warming).

The major idea behind CFP originates from life-cycle analysis (LCA). As per the International Organization for Standardization (ISO) 14040 series, an LCA involves a methodical set of measures for compiling and investigating the inputs and outputs of products and energy and the related environmental impacts directly related to the functioning of a product through its entire life cycle, and within a given set of boundary conditions [15]. When accounting for $CO_2$ (or other GHG) emissions along the life cycle of a product, the term 'carbon footprint' is used. The idea behind 'footprint' comes from an ecological footprint accounting system, in which 'ecological footprint' is used as an indicator to measure the consumption of resources and waste assimilation necessities based on the economy or human population by means of a productive land area [16].

Denim is one the most popular types of clothing in the world; millions of pairs are made and sold in many different kinds of styles and colors. Having a pair of jeans has become the most personal thing a person can wear because it is said that it molds to the body. The reasons why jeans have become a neutral garment is that they can be worn in different ways, their composition is simple and they have changed with time. Denim production is a major subsector in the textile industry. Denim products are some of the most highly used in textile clothing, with continuous fashion and consumer preference, especially by young people. A survey by Cotton, Inc. showed that the global denim jeans market was projected to grow by 8%, from \$55 billion in 2015 to \$59 billion by 2021, with Latin America and Asia expected to lead the increase.

Denim manufacturing and other textile-based industries continue to be a source of specific environmental pollution because of considerable amounts of mineral substances that are used to process textile raw materials and semifinished products. The life cycle of a pair of jeans is divided into two categories: the manufacturing stage and the use of the jeans until the end of its cycle. The production stage is responsible for approximately 59% of the climate change impact, whereas the use of the jeans and end-of-life process accounts for the remaining 41%. This causes jeans to be responsible for about 2 kg of $CO_2$ per functional unit.

The first part of this chapter provides the basis for the carbon footprint and its estimation in the denim manufacturing process. Furthermore, the carbon footprint in various stages of denim manufacture, such as cotton production, denim yarn manufacture, fabric production, garment manufacture, transportation and distribution, are dealt with in detail. In the second part of the chapter, the carbon footprint of denim processing, such as dyeing and washing, and the carbon footprint during the consumer user phase are elaborated upon. In addition, strategies to reduce the denim fabric carbon footprint, such as 3R concept and other energy efficient technologies and processes, are briefly discussed.

## 6.2 Carbon footprint standards and methods of calculation

The major objective behind the application of standards is to ensure easy application in practice [17]. ISO standards for LCA, which has been in existence for about 3 decades, have had a vital role in shaping carbon footprint standards: ISO 14040 for LCA and ISO 14064 for GHGs. Yet another standard, ISO 14067, is under development [18]. Another development is the PAS 2050, by the British Standards Institute, which covers a wide range of diverse products based on the LCA method. It is the most detailed and comprehensive product carbon footprint methodology available [17,19]. However, one of the main issues in the PAS 2050 methodology is the varying results arising owing to the freedom given to use any database. For example, one product within a specific system boundary can have two different values depending on the databases used for calculations, because the standard does not require the use of specific databases. In the absence of primary data, different values are often used from various databases [20].

Other standards in wide use are the GHG Protocol of the World resources Institute and the World Business Council for Sustainable Development and their Product Life Cycle Accounting and Reporting Standard [18], and the Basic Guideline of the Carbon Footprint of Products of the Japanese Industrial Standards Committee.

Methodologically, there are three ways to carry out a life cycle–based CFP analysis, depending on the scale of application:

1. a process- or product-based analysis (or bottom-up) approach which accounts for the environmental impacts of individual products from cradle to grave or only on-site. In this method, mostly first-order, and some second-order impacts are considered [21]. This also depends on the question of how to select boundaries. If used to make carbon footprint comparisons of various products, it is important to mention the boundaries used because it can cause confusion or errors. Furthermore, if applied within a larger context such as a sector (industry, household or government), estimates are based on assumptions and therefore could be far from reality;
2. an environmental input–output analysis (or top-down) approach can be used in a broader context. Input–output analysis in economic accounts provides a picture of all economic activities at the sector level. A detailed understanding of micro-level impacts is hidden or limited owing to aggregation of data and homogeneity of the system required by this approach [20,22];
3. combining the strengths of these two approaches, a hybrid approach is proposed in which the process-based and input–output methodologies are integrated. However, hybrid modeling is rarely practiced because of the high methodological and data sophistication required [18,23].

## 6.3 Carbon footprint in textile and clothing sector

Sustainable fashion has been around since the early 1600s, as a way of life rather than a choice. Before the Industrial Revolution, people conserved resources because of the high costs of labour involved in the conversion which are basically meant for the wealthy people [24–26]. Raw materials (textiles) for apparel came from nature and required a long time to be transformed into fabrics. The labour required to produce a high-quality product contributed to the high cost of fabrics and apparel. Hence, quality fabrics were worn by wealthy people as a way to express their wealth in society. By the end of the 18th century, because of the Industrial Revolution, several types of machinery were introduced to spin yarn and weave fabrics, and the textile industry flourished. Synthetic dyes, manufactured fibres, ready-to-wear clothes, fashion magazines, etc., were products of the Industrial Revolution that caused the textile sector to mature [27].

The textile and clothing industry has created a lot of material wealth for humans but it also consumes a lot of energy, which causes massive emissions of $CO_2$ and other GHGs. To protect the environment for human survival, the industry should develop a low-carbon way to achieve sustainable development [28,29]. Because of the long and complicated nature of the supply chain, the textile and clothing sector is responsible for creating a significant carbon footprint and is a major source of GHGs. An environmental impact occurs at each stage in the life cycle of garments, from the cultivation of fibres or the production of synthetic fibres to the various manufacturing sequences involved in converting fibres to fabric and the dyeing and finishing of garments for delivery to consumers. Disposal of products after use and the discharge of effluents also cause significant impacts [30]. Cheap clothing has become a disposable product and most European consumers recycle it, donate it or dispose of it; the remainder goes to landfills. Regardless of the economic downfall globally, increasingly more garments, home textile materials and other textiles are purchased by the consumers annually.

The problem is more severe in the developing world, where textile production contributes to a larger percentage of the gross domestic product and processing facilities are out of date. The carbon footprint of a textile product varies depending on the fibre used and the method of processing. The carbon footprint of any product can be examined by the 'level of embedded energy' within the product. This is the net amount of energy consumed at each stage of the production process. Despite stringent environmental laws and regulations, the textile industry cannot meet these requirements satisfactorily. Several reasons lie behind the huge impact of textiles on the environment:

- The vast majority of fibres produced are synthetic and petrochemical-based fibres such as nylon and polyester consume a huge amount of energy to manufacture. In addition, chemicals used during the manufacture of these materials end up as toxins polluting the air, soil and water.
- Traditional methods of cultivating cotton contribute heavily to global warming and are detrimental to the environment. Cotton cultivation necessitates enormous quantities of water and energy and the intensive use of chemical fertilizers and pesticides.
- Bleaching and coloring of fabrics also warrants an enormous quantity of water, dyes, chemicals and energy. Around 1 million tons of chemicals and dyes is used every year to color fabrics.
- The wet finishing process uses a huge amount of water and energy.

The textile industry is a huge industry and a massive polluter. It uses two things to a great extent: water and chemicals. Wet processing of textiles involves desizing, scouring, mercerizing, dyeing, printing, finishing, etc., and requires a lot of chemicals to be applied on fibres or yarns for fabrics. Water is required at almost all stages of fabric processing. It is used as a solvent to dissolve chemicals and for washing between the various sequences of wet processing. The bleaching operation releases an organo-chlorine compound which is harmful to the environment. At least 2000 types of chemicals are employed by this sector [29,31].

## 6.4 Denim

Among various textile products, denim has received wider acceptance compared with other fabrics. It is worn by various age groups and classes and both genders. Denim fabric is a hard and durable 3/1 warp-faced twill cotton fabric with indigo-dyed warp yarn and white color filling yarns, with a weight of 14.5 ounces/square yard. As long ago as the 17th century, denim was used to make trousers, upholstery and awnings [32].

### 6.4.1 Definition of denim and jeans

- The Textile Institute defines denim in *Textile Terms and Definition* thus: 'Denim is conventionally a 3/1 warp faced twill fabric made from yarn dyed warp and un-dyed weft yarn and Jean is a 2/1 warp faced twill fabric used chiefly for overalls or casual wear'.
- Based on the *Encyclopedia of Textile Finishing*, edited by Professor Dr Hans-Karl Rouette, published by Springer, 'Denim is a firm 2/1 or 3/1 cotton twill-weave fabric (work wear twill) with a dyed blue warp and raw white weft. The warp is sometimes dyed during sizing'.
- 'Jeans originally an American term derived from Genes (cloth from Genoa). The term was coined in the USA to describe tight-fitting trousers made of cotton twill, usually woven with a dark blue dyed warp and undyed weft, stitched on seams and pockets and having copper rivets here and there. As a rule, blue jeans are dyed with indigo and thus have limited color fastness to rubbing and washing'.
- Conventionally, denim means sturdy, indigo-dyed cotton fabric woven with a dyed warp yarn and natural fill yarn available in various weights and qualities that is used to make pants, skirts, jackets, shirts and shorts, etc.; jeans are pants made from denim.

### 6.4.2 Main features of denim fabrics

Denim fabrics are generally woven from coarse yarn and are dyed with indigo dyes in warp. They are hard wearing, high-density fabrics that are high gram per square meter. Denim is popular for long-term use and is a fashion fabric for casual wear. General features of the denim fabric are as follows [33]:

- cotton or cotton polyester blended, durable and heavy-weight twill-weave
- warp-faced twill fabric and colored warp yarns dominate on the face and the white filling yarns on the back side of fabric
- colored warp yarn and white filling yarn
- left-hand twill with a blue (indigo) warp, white weft for use in apparel in a variety of weights

- weight of the denim ranges from 203.46 to 474.74 $g/m^2$ or more in a 2/1 or 3/1 interlacing pattern
- may be napped or printed, made with spandex or other stretch yarns, or otherwise modified for fashion

### 6.4.3 Raw materials of denim fabrics

To manufacture a better denim fabric, processing conditions have to be optimal as far as the raw material is concerned. For raw cotton, the ring and rotor spun yarns made from it should have the following quality specifications [34]:

- minimum staple length, 2.7 cm
- short fibres (less than 0.5 inches) percentage, below 40%
- fibre fineness, 4.0–4.5 µg/inch
- yarn characteristics intended for use in denim manufacturing should at least conform to the Uster 25% standard
- common count range of denim warp yarns is 50–90 Tex and weft yarns 75–120 Tex; finer yarns as fine as 25 Tex in twill or plain weave are repeatedly used in denim shirts
- yarn twist multiplier: 4.5–5.0 for warp yarns and 4.2 for weft yarns
- low yarn hairiness
- better yarn strength and uniformity

### 6.4.4 Classification of denim fabrics

Denim fabric can be classified based on several factors:

1. Based on weight/unit area:
   a. light weight: 4.5–7 oz/square yard
   b. medium weight: 7–10 oz/square yard
   c. heavy weight: 11 and above oz/square yard
2. Based on colour:
   a. indigo vat
   b. sulphur black
   c. pigment
3. Based on slub yarns
   a. warp yarns are slub only
   b. weft yarns are slub only
   c. cross-hatch: both-ways slub, i.e., both warp and weft yarns are slub
4. Based on fabric structure/design:
   a. plain
   b. twill (S-twill, Z-twill, herringbone, or zigzag twill)
   c. right-hand twill, diagonals run toward to the right
   d. left-hand twill, diagonals run toward to the left

### 6.4.5 Manufacturing process of denim

Denim, in particular blue jeans, has the biggest market share in sales of textile products around the globe owing to its popularity in all countries and age groups. The denim production value chain is optimized for mass production and with production facilities to meet the global demand. The higher demand and production of denim could lead to a significant influence on the environment [34]. Hence, technological developments in every step of denim production could provide cost savings to the industry and the environmental impact of denim. In evaluating the environmental impact of denim manufacture, it is vital to take a holistic approach by considering all of the activities involved in manufacturing denim, such as raw materials, manufacturing, transportation, use and disposal.

The denim manufacturing process is similar manufacturing grey fabric up to the weaving process, except that for denim fabric the warp is dyed at the sizing stage compared with dyeing at the finishing stage in normal fabric production [35]. Fig. 6.1 shows the entire textile value chain in the manufacture of denim fabric, from cotton fibre to consumer use. The process flowchart of denim fabric manufacturing is shown in Fig. 6.2, which involves spinning, warp preparation, fabric manufacturing and finishing processes.

### 6.4.5.1 *Spinning*

The first process in denim manufacturing is the spinning of yarn in either a ring spinning or rotor spinning system. The fibre to yarn conversion is composed of several processes such as opening and cleaning, carding, drawing, roving and spinning. For rotor spinning, the roving process can be eliminated. Opening and cleaning of fibres are carried out in a series of machines in a blow room. Carding individualizes the fibres and removes neps, short fibres and remaining trash, and produces a rope-like form called sliver. The uniformity and parallelization of fibres are improved through a drawing process. In rotor spinning, the drawn sliver is fed into the rotor machine to convert it into yarn. In ring spinning, the drawn

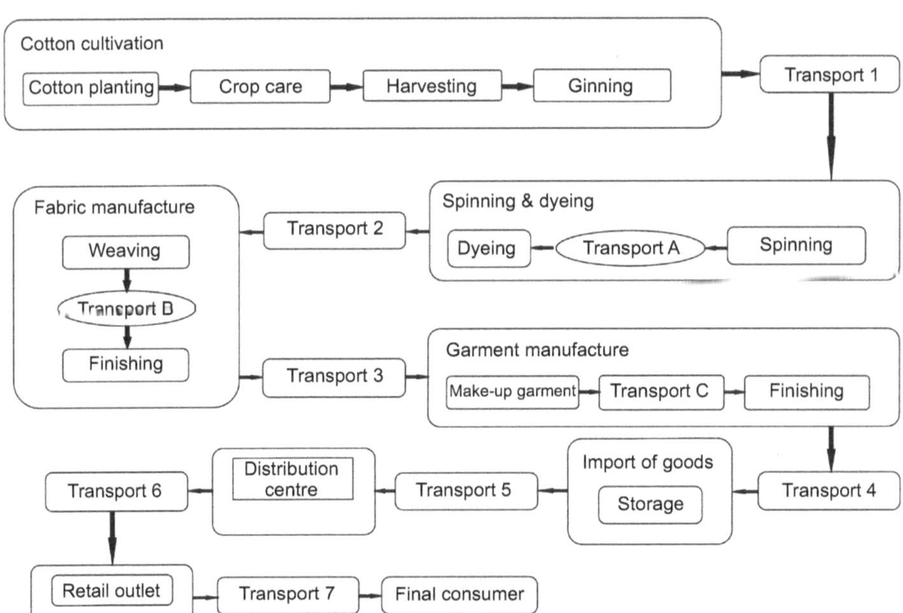


**Figure 6.1** Textile value chain in denim manufacturing from cotton to consumer.

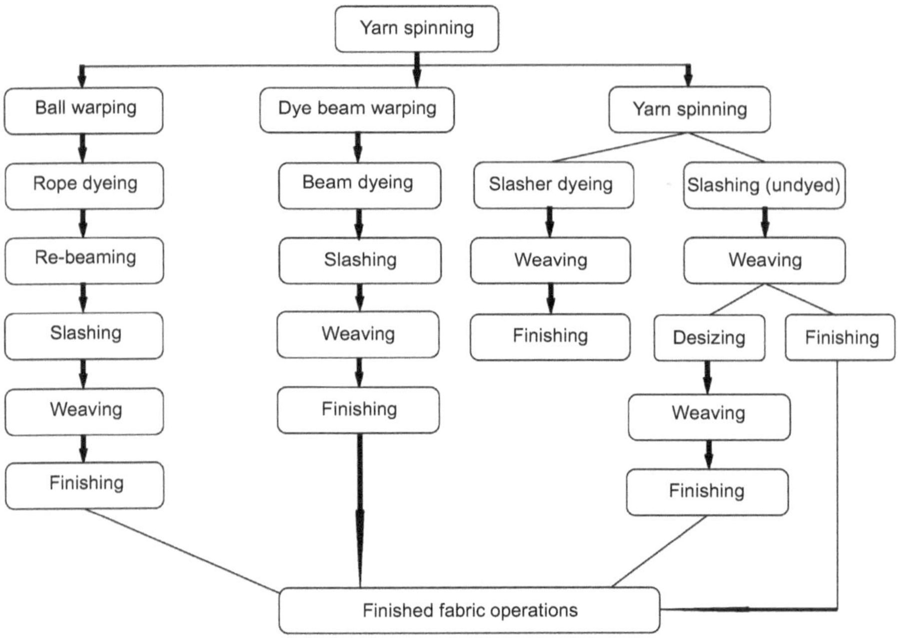


**Figure 6.2** Process flowchart of denim fabric manufacturing.

sliver is further reduced to a roving form in a speed frame and then is converted into yarn in a ring frame [36–38]. Generally, denim fabric is 3/1 warp-faced twill fabric made from an indigo-dyed warp and an undyed weft yarn.

### 6.4.5.2 Warp preparation: dyeing and sizing processes

Dyeing and sizing of warp yarns are carried out in two methods:

- Yarn from several beams is combined to produce a warp sheet and then is dyed and sized on the same machine.
- A rope is made by combining about 350–400 yarns and about 12–14 ropes are passed near each other through the continuous dyeing machine. After dyeing, ropes are dried on drying cylinders and then a worker's beam is prepared. After that, sizing is carried out by a conventional method.

Tin denim manufacturing, dyeing and sizing processes can be classified into three categories:

- continuous indigo-rope dyeing and sizing
- continuous indigo dyeing and sizing
- indigo–back beam dyeing and sizing

#### Continuous indigo-rope dyeing and sizing

Yarn from ring cops is converted into bigger packages as cone or cheese in cone winding and is positioned on the ball warper on which 350–400 yarns are formed into a rope and are cross-wound into a ball shape according to the length of the warp beam. During this process, lease rods are inserted at frequent intervals as required for further processing on a long chain beamer. According to the size of the rope dyeing plant, 12–24 ropes can be dyed at a time [39–41]. The ropes are then opened on the long chain beamer through a tension roll and expansion comb, wound onto a back beam and then sized.

#### Continuous indigo dyeing and sizing

In this method, instead of ropes, back beams are processed on the dyeing/sizing machine. The warp yarn is indigo-dyed, oxidized, dried and sized in one continuous process. Although this method is less cumbersome compared with rope dyeing, the risk of individual thread breakage is greater than for rope dyeing [41,42].

#### Indigo–back beam dyeing and sizing

In this method, indigo dyeing and sizing are done in two different stages. In the first stage, back beams are indigo-dyed, oxidized, dried and wound onto a batch roll; in the second stage, the batch roll is sized, dried and wound onto a weaver's beam [43].

### 6.4.5.3 Weaving

The weaving process can be carried out on conventional weaving systems such as automatic looms or shuttle-less looms such as air jet, rapier, and projectile weaving machines. The slow speed of the conventional shuttle loom results in less production and extreme wear and tear to machines. Generally, denim fabrics are woven in shuttle-less weaving machines.

### 6.4.5.4 Finishing

The woven fabric wound in a roll form is inspected for defects on fabric inspection machines for immediate corrective actions. The denim fabric then goes through a range of finishing processes such as singeing, brushing, washing and drying. Singeing and brushing removes impurities from the fabric's surface and aids in making the fabric's surface even. The finished denim fabric is then sent for making-up, where the fabric is cut into the required width as per the size [44,45]. The finished denim fabric is then inspected for defects such as uneven dyeing, bleaching and dyeing defects, oil stains, patches, etc., and then is categorized according to the quality.

### 6.4.5.5 Garment manufacturing

Garment manufacturing is a process in which the two-dimensional fabric is transformed into a three-dimensional garment. Garment manufacturing involves several operations such as pattern making, spreading, cutting, sewing, pressing, inspecting and packaging. Accessories and trims are selected based on requirements and style aspects of the garment. Generally, the output from this process is the final product of the manufacturing sector, excluding some special effects conveyed on it owing to the requirements of functions or design concepts.

## 6.5 Carbon footprint and its estimation

The concept of a carbon footprint revolves around transportation and heating issues. The fabric necessitates a large $CO_2$ footprint. According to US information, the textile industry is the fifth largest contributor to the carbon footprint among many countries. The textile industry is large and produces a large amount of GHG. Textile production is estimated to produce around 60 billion kilogram of fabric. The total quantity of water and energy required to manufacture a fabric includes 6–9 trillion litres of water and 1074 billion KWh of electricity. The Bureau of International Recycling states that collecting 1 kg of used cloth could reduce around 3.6 kg of $CO_2$ emissions. Recycling textile products instead of disposing of them could reduce 6000 L of water, 200 g of pesticides and 300 g of fertilizer [46].

To evaluate the carbon footprint in fabric, the 'embodied energy' in the fabric has to be considered, i.e., the total energy used at each step of the process needed to create the particular fabric. To determine the embodied energy in the fabric, the energy used in two separate fabric production steps is determined [47]:

- The first step is to find out from what type of fibre the fabric is made, because that gives an idea about the energy needed to make the fibres used in the yarn. The type of fibres influences the carbon footprint of the fabric to an extent.
- Next, the energy needed to convert the yarn into fabric has to be taken into account. Once any material becomes a 'yarn' or 'filament', the amount of energy and conversion process to weave that yarn into a textile is consistent irrespective of the type of fibre.

### 6.5.1 Factors influencing the carbon footprint of the fabric

- **Use phase**: This segment of the life cycle of a product influences GHG emissions to a greater extent. The largest contributor to a garment's life-cycle GHG footprint is laundering. Garments requiring the largest energy during washing, drying and ironing needs higher energy input; hence laundering itself accounts for 40%–80% of total life-cycle GHG emissions. Machine drying consumes the highest energy and produces higher GHG emissions in garments. On the other hand, hand washing and dry cleaning require lower energy, leading to lower GHG emissions.
- **Raw material**: The type of fibre has a vital role in GHG emissions. For example, synthetic fibres produce comparatively high GHG emission owing to higher energy requirements in fibre production compared with natural fibres. On the other hand, plant fibres such as cotton and linen have lower GHG emissions. Linen requires a relatively low amount of pesticides, fertilizers and irrigation compared with cotton, leading to much less GHG emissions. Animal fibres, especially wool, emits higher GHG owing to methane emissions from sheep. The type of washing, laundering and ironing process also depends on the type of fibres. For example, wool needs to be dry-cleaned or hand washed rather than machine washed and linen needs more ironing.
  - ***Sourcing and manufacturing locations***: GHG emissions vary depending on the sourcing and manufacturing locations: for example, as a result of differing energy sources.
  - ***Dyeing***: The dyeing process has little impact on the emissions of GHG.
  - ***Garment makeup***: Garment assembly has a considerable impact on GHG emissions.
  - ***Packaging***: According to many studies, packaging has a low GHG impact because packaging is often composed of recycled materials. Packaging generally depends on the type of garment.

- ***Transportation***: The transportation of goods has a small portion of a garment's total carbon footprint. However, it depends on the distance products are transported and whether transportation is ship-based or air-based. GHG emissions could increase significantly if air transport is used in any stage of product manufacturing.
- ***Garment lifespan***: The short life of clothing could increase total GHG emissions in ways not considered in a typical garment LCA. For example, a garment which is laundered only once before being discarded will have lower GHG emissions compared with that of a garment used and laundered more than 100 times.
- ***Garment disposal***: GHG emissions resulting from the disposal of clothing are meager as per the LCA and are caused by the emission of a small quantity of methane during the decomposition of natural fibres. Certain disposal options reduce GHG emissions, however. The incineration of natural fibres in a waste-to-energy plant may shift the use of petroleum-based raw materials.

The carbon footprint in jeans manufacturing was determined by various denim manufacturing companies such as Levi Strauss & Co., Blue Dot and Crystal Denim to ensure that their jeans are sustainable. Levi Strauss & Co., one of the world's leading denim manufacturers, started a self-analysis program in 2006. They analysed the company's internal and external environmental impacts throughout their product line. In their LCA, the life cycle of jeans from 'cradle to grave' was reported as a collection of seven independent systems. Analysis started in the cotton production sector and then covered step-by-step denim material production such as fabric production, garment manufacturing, transportation and distribution, consumer use and recycling. Furthermore, the company listed the impact of the waste stream from the product manufacturing process in landfills. In their report, they mentioned that making one pair of denim jean required almost 920 gallons of water and 400 MJ of energy and expelled 32 kg of $CO_2$. They also pointed out that the major quantity of water consumed (68%; 2565 L) was in fibre production. The percentage of water consumption is comparatively high: in all phases of garment production likely in fabric manufacturing, 6% (236 L); during sewing and finishing, 1% (34 L); in washing and packing, 2% (77 L); and at the customer's end, 23% (860 L). Through their carbon reduction programme, the Crystal group, a Hong Kong–based denim manufacturing group, reduced the carbon intensity 2.21–1.77 kg $CO_2$ per piece. Similarly Mud jeans went through several steps to reduce their carbon footprint in jeans manufacturing and reduced the carbon footprint of blue jeans from 23.45 to 5.13 kg $CO_2$e by streamlining the process and raw materials.

## 6.6 Carbon footprint in each stages of denim manufacturing

Levi Strauss & Co. classified the LCA of denim into seven different independent systems [48]:

1. cotton production
2. fabric production
3. garment manufacturing
4. transportation and distribution

5. consumer use
6. recycling, which then goes back to Step 2
7. waste stream in a landfill

The carbon footprint in the life cycle of denim with respect to the quantity and percentage is shown in Figs 6.3 and 6.4, respectively. The figures show that the consumer care phase (use phase) has the highest impact on climate change (driven by the high use of nonrenewable energy), followed by fabric manufacturing and transportation.

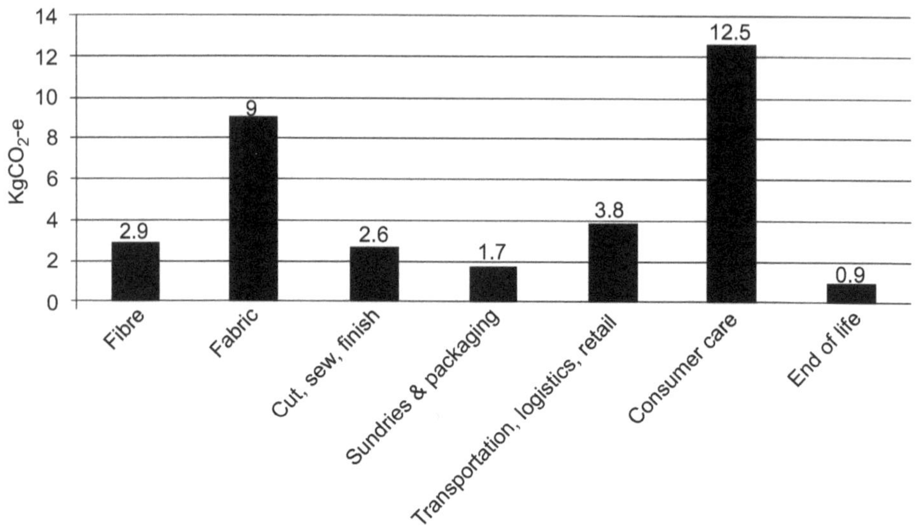


**Figure 6.3** Climate change impact in the life cycle of denim jeans, in quantity. Levi Strauss & Co. Life cycle of a jean, reproduced with permission.

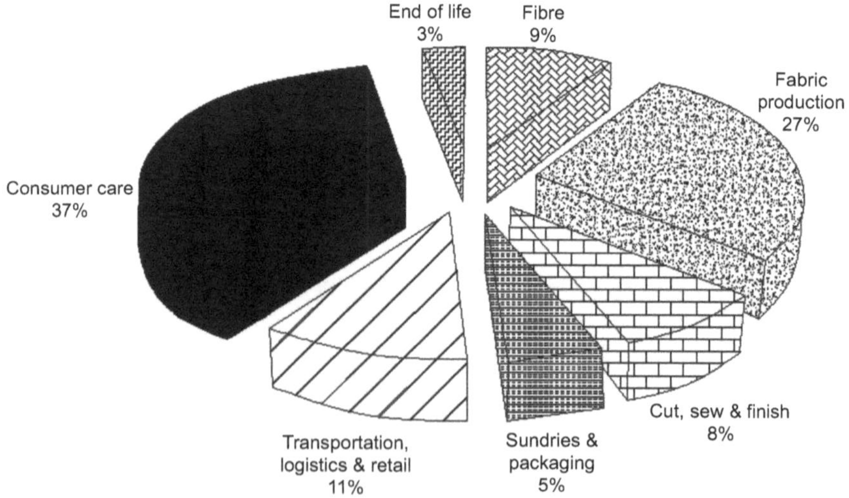


**Figure 6.4** Climate change impact in life cycle of denim jeans, in percentage. Levi Strauss & Co. Life cycle of a jean, reproduced with permission.

### 6.6.1 Cotton cultivation and harvesting

The cultivation of natural fibres involves several factors such as field preparation, planting operations such as irrigation, weed control, pesticide, and the type of harvesting, which needs to be considered to reduce the carbon footprint. Cotton cultivation and harvesting are indispensable for denim manufacturing, although they are resource intensive and involve political and social aspects [49].

Cotton is considered one of the world's major crops; it contributes to about 3%–5% of the world's cultivatable surfaces. Cotton cultivation and harvesting are usually known for their unsustainable nature, owing to the use of pesticides, fertilizers and water. Cotton production can be both a contributor to and a 'victim' of climate change; it contributes around 0.3%–1% of total global GHG emissions. Current cotton-growing methods and practices are not considered sustainable because they damage water, soil and allied ecosystems, and contribute to high social costs and are a threat to regional economies, depending on cotton farming and associated textile industries.

To manufacture a pair of jeans, approximately 1.5 pounds of cotton is essential, which in turn requires about 1500 gallons of water. To cultivate and harvest the cotton, numerous stages of transport are necessary, and based on the number of steps each supplier can take in one facility, the quantity of carbon emissions will change. By taking into consideration the use of insecticides in cultivation, cotton growers use 25% of chemicals applied worldwide, including highly toxic organophosphates. Machines such as pesticide sprayers and farm equipment, which run on oil, require about a pound to harvest sufficient cotton to manufacture a single pair of jeans. Not only are the environmental impacts from these chemicals apparent within the industry, the social impact is compelling and sometimes is overlooked. To prepare the soil, harvest the cotton and transport cotton to the factories, a large amount of energy is used. In total energy consumption is around 60 MJ/kg [50,51].

The social impact of high-priced pesticides is horrible in India. In 2003 alone, more than 17,000 farmers committed suicide. Most were from cotton-growing areas where costly chemicals and genetically modified cotton seeds left them deeply in debt. The use of genetically modified cotton seeds could be an alternative method to conventional pesticide spraying, which is also revealed to be a different threat to social and environmental sustainability. Although India's share of $CO_2$ emissions is only 3.11%, the embodied energy of organic cotton from India is higher than conventionally produced cotton from the United States because the yields are much less in India, requiring more land to cultivate the same quantity of cotton. The energy distributions in cotton farming were studied by Nigam et al. and are shown in Fig. 6.5.

The figure shows that fertilizer consumes a major portion of the energy input along with diesel used for farming operations and electricity for irrigation. Fig. 6.6 shows GHG emissions from cotton farming. The study reveals that the cotton farming stage has 2.1076 $CO_2$eq/kg cotton yarn without allocation and 0.8430 $CO_2$eq/kg cotton yarn on a mass basis and 1.8336 $CO_2$eq/kg cotton yarn on an economy basis. From the figure can be seen that the major contribution is from the use of synthetic fertilizers. As much as 60%–70% of seed is available from seed cotton during ginning.

The workforce required to process 1 kg of cotton is calculated to be 0.002 worker h/kg. The electrical energy required to process 1 kg of yarn is 0.5863 MJ/kg. Based on the electrical energy consumed for ginning, the GHG emissions are calculated to be 0.1310 $CO_2$ eq/kg cotton yarn [52].

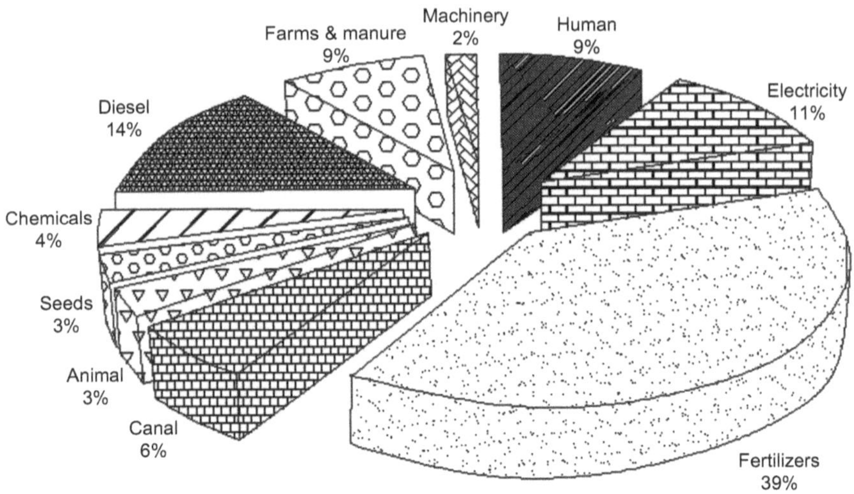


**Figure 6.5** Energy distribution in cotton farming.
CIAT, reproduced with permission. Clothing Industry Training Authority (CIAT), Apparel supply chain–carbon assessment about jeans producing. http://www.cita.org.hk/wp-content/uploads/2013/10/Fty_rpt_Eng_20130912.pdf.

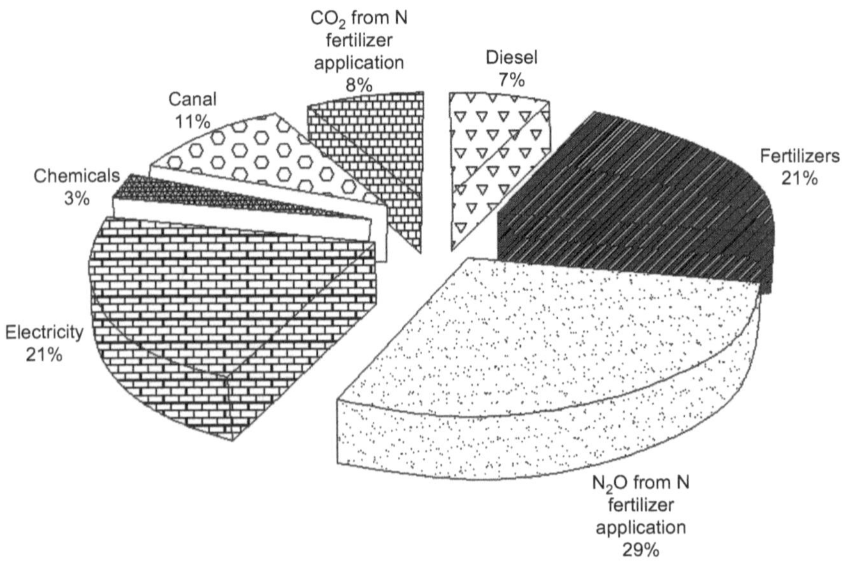


**Figure 6.6** Greenhouse gas emissions from cotton farming.
CIAT, reproduced with permission. Clothing Industry Training Authority (CIAT), Apparel supply chain–carbon assessment about jeans producing. http://www.cita.org.hk/wp-content/uploads/2013/10/Fty_rpt_Eng_20130912.pdf.

## 6.6.2 Cotton spinning

The value chain for cotton yarn production involves the following processes:

1. cotton cultivation and the harvesting phase, including fuel or energy-intensive material inputs such as fertilizers, herbicides, seeds, diesel fuel and electricity for irrigation, machinery and labour used for agriculture.
2. transportation to move agricultural products from the farm to the ginning site
3. ginning process, which consumes electricity
4. transportation to spinning industry
5. conversion of fibre to yarn, which uses mainly electricity and water for humidification
6. packaging

A process map for the manufacturing of carded ring yarn is shown in Fig. 6.7. Spinning starts with the blow room in which opening, cleaning, blending and dust removal are performed using beaters and grid bars. The cleaned tufts are then transferred to the carding machine through pipelines called a chute feed system. The fibres are then individualized, neps and short fibres are removed, remaining trash is cleaned and sliver is formed in carding. The drawing process improves the evenness and parallelization of fibres in sliver through doubling and drafting. Then the sliver is thinned to the roving stage in a speed frame, twisted and wound on a roving bobbin for the ring frame, where it is further thinned into yarn. In the case of rotor yarn, the slivers are directly converted into yarn.

The GHG emissions distribution study conducted by the Clothing Industry Training Authority in the cotton ring spinning industry is shown in Figs 6.8 and 6.9. The ring spinning process and ring frame machine contribute to a larger extent to GHG emissions and electricity consumption, followed by the winding process and machine. Incorporating energy-efficient processes and materials into these processing stages would offer a potential opportunity for carbon reduction [53]. Auxiliary production facilities used in the production process, such as air compressors, lighting, electric fans, humidification systems, and automatic waste extraction, have a vital role in energy consumption and hence a potential opportunity to reduce carbon emissions.

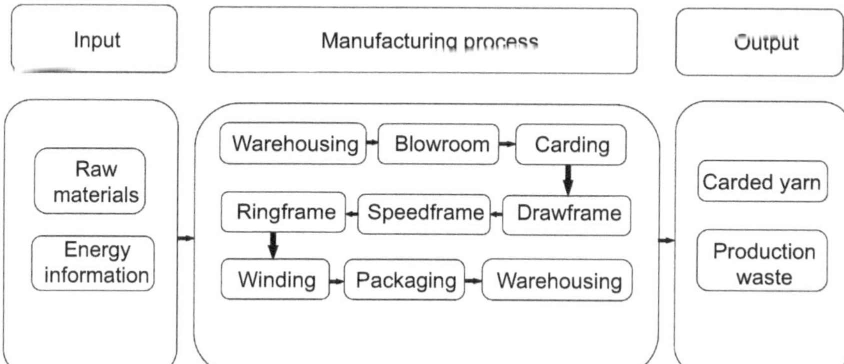


**Figure 6.7** Process flowchart of carded yarn.

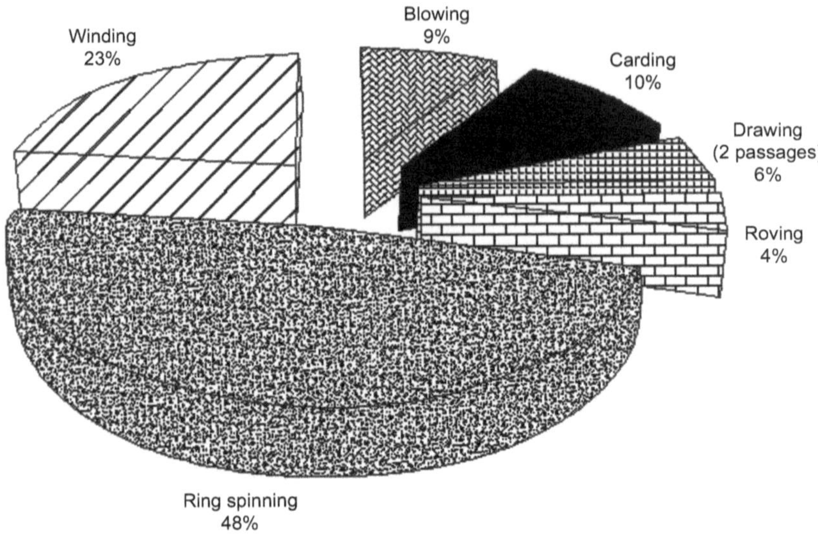


**Figure 6.8** Greenhouse gas emission distribution in spinning, by process.
CIAT, reproduced with permission. Clothing Industry Training Authority (CIAT), Apparel supply chain–carbon assessment about jeans producing. http://www.cita.org.hk/wp-content/uploads/2013/10/Fty_rpt_Eng_20130912.pdf.

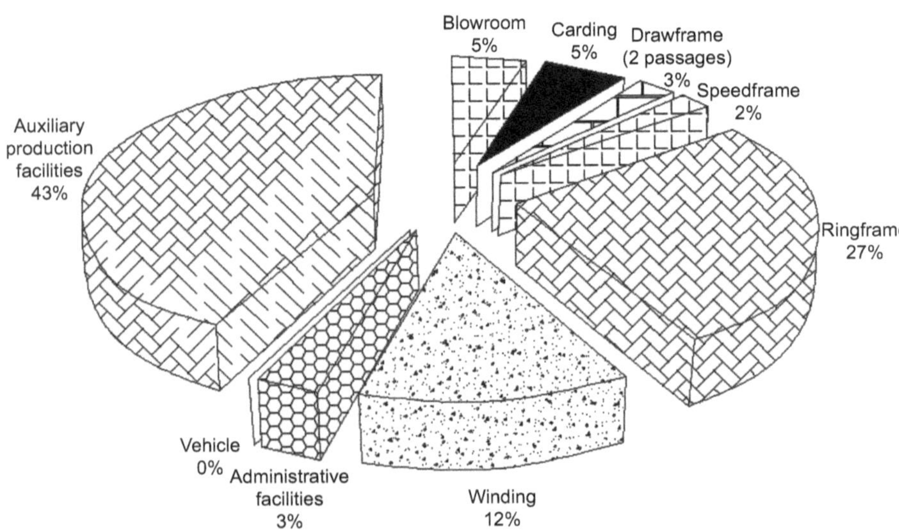


**Figure 6.9** Greenhouse gas emission distribution in spinning, by emission source.

### 6.6.3 Dyeing and sizing

Rope dyeing is most commonly used in denim manufacturing because of the better uniformity achieved in this method compared with slasher dyeing. It produces a ring-dyeing effect in yarn, i.e., colour on the sheath of the cotton yarn without colour (white) on the

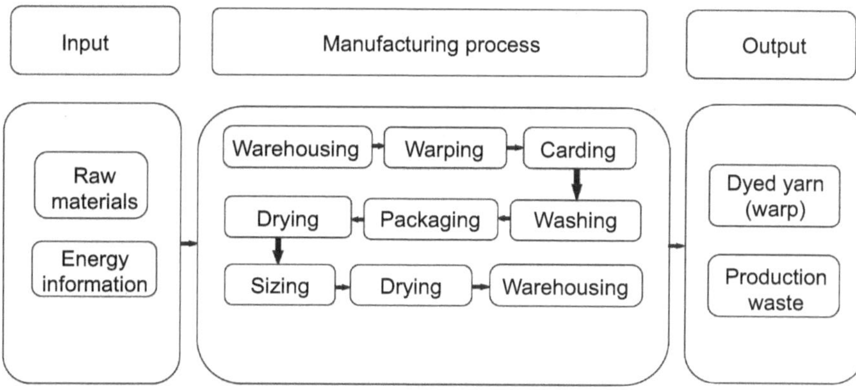


**Figure 6.10** Process map of yarn dyeing.

core of the yarn, because it forms a coating on the outer layers of the yarn and fibre during the dyeing process. The most commonly used dye for denim is indigo [54]. This type of dye is known as a vat dye; it is insoluble in water and has less affinity toward cellulose fibres. Hence, indigo dyes are converted into leuco form, which forms a coating on the outer layer of yarn. The indigo is then made insoluble through oxidation by exposing the yarn to air. During sizing, a size add-on of 8%–12% is applied over the warp thread to improve the strength and frictional resistance of the warp yarn [55,56]. A process map for dyeing and sizing for denim manufacturing is shown in Fig. 6.10.

An average pair of jeans has around three-quarters of a pound of chemicals in it. Although natural indigo dyes can be produced, most indigo dyes are synthetically produced. The distribution of GHG emissions in cotton dyeing and sizing, process by process, and emission sources during the denim process are shown in Figs 6.11 and 6.12, respectively. Fig. 6.11 shows that the dyeing and sizing processes themselves consume around 95% of total energy. Analysing the energy source for each process and the behaviour of machines during the process may provide a possibility of reducing the carbon footprint in these processes [57]. Fig. 6.12 shows that steam boilers are major consumers of energy, about 80%, followed by dyeing and sizing machines.

Any kind of dye is basically a polluting process. Textile processing units need a nearby water source to break down dyes and to wash and treat the fabrics. Synthetic indigo dyes are normally sulphur based and have serious detrimental effects for both the people that work with them and the natural environments where they are discharged [58]. In addition to sulphur-based dyes, sodium hydroxide, hydrosulphate and formaldehyde are used to manufacture synthetic dyes. Even if the color is removed from wastewater, this does not necessarily mean that the water is clean: Sulphur dye residues tend to remain in wastewater even after treatment. Use of natural indigo dyes could minimize the carbon footprint level.

### 6.6.4 Denim fabric weaving and finishing

A process map of denim fabric weaving and finishing is shown in Fig. 6.13.

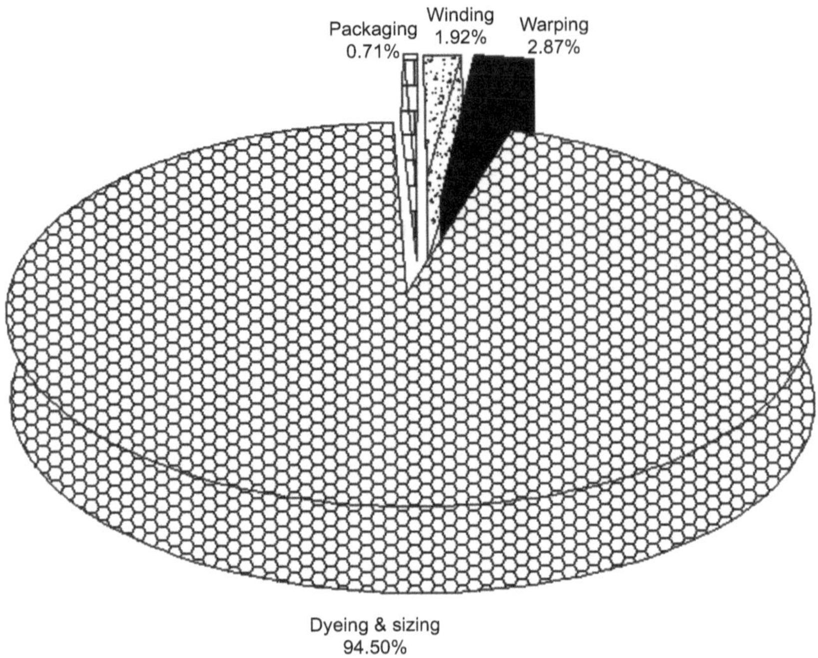


**Figure 6.11** Greenhouse gas emissions in cotton dyeing, by process.
CIAT, reproduced with permission. Clothing Industry Training Authority (CIAT), Apparel supply chain–carbon assessment about jeans producing. http://www.cita.org.hk/wp-content/uploads/2013/10/Fty_rpt_Eng_20130912.pdf.

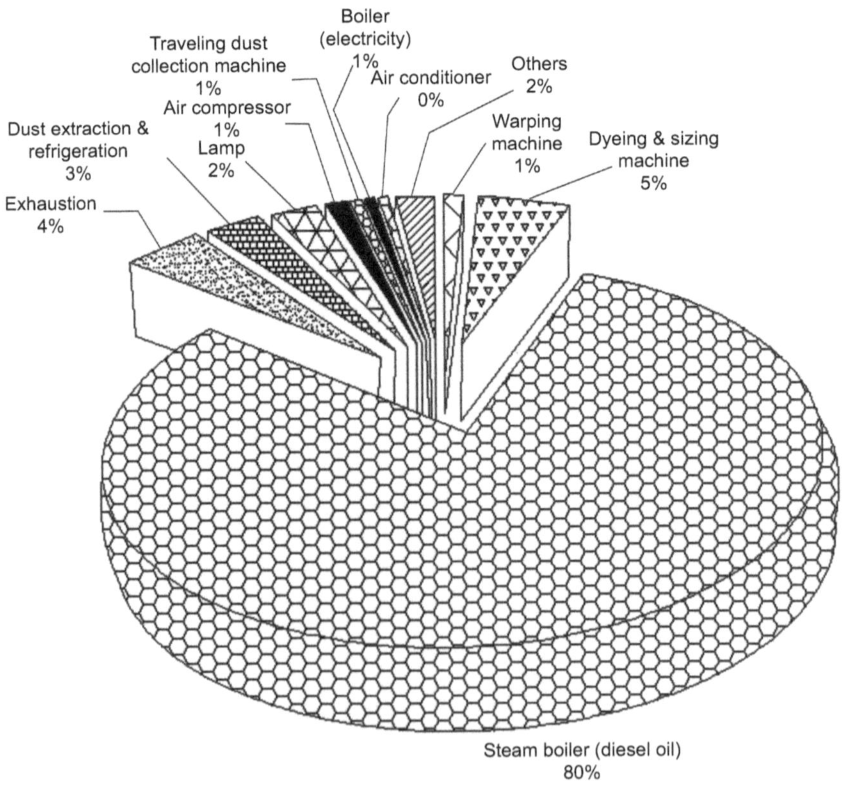


**Figure 6.12** Greenhouse gas emission distribution in cotton dyeing, by emission source.

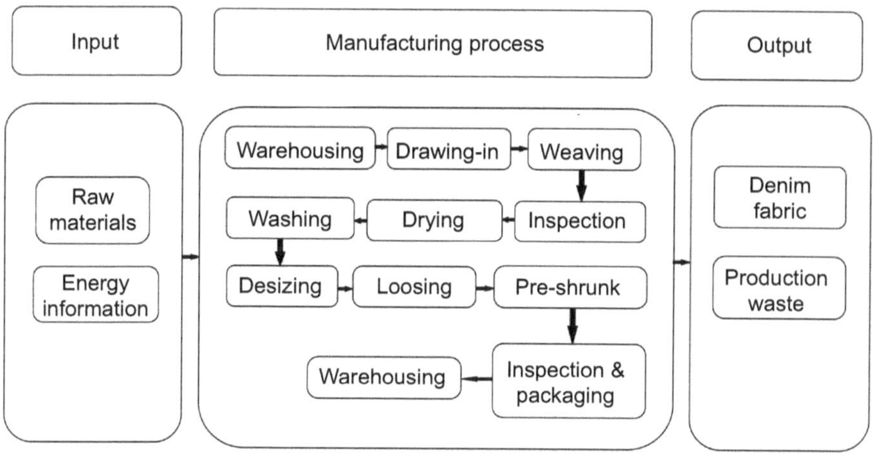


**Figure 6.13** Process map of denim fabric weaving and finishing.

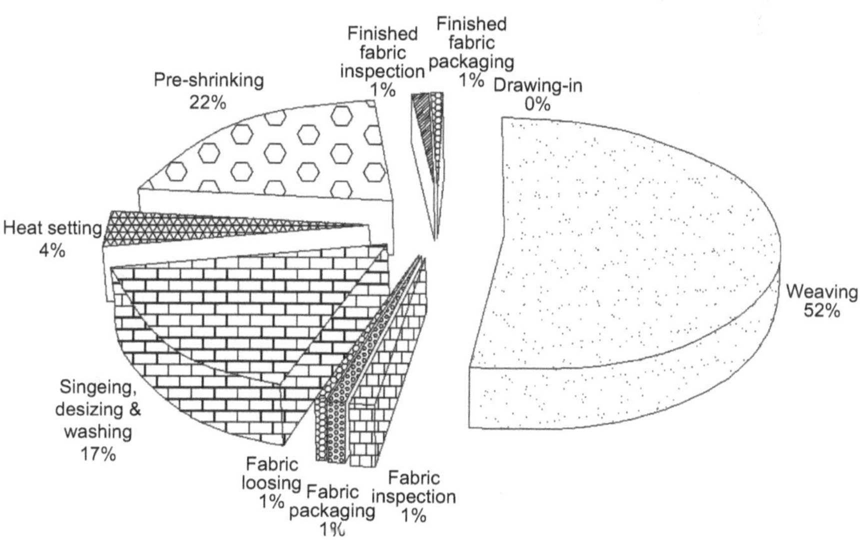


**Figure 6.14** Greenhouse gas emissions in fabric weaving and finishing processes.
CIAT, reproduced with permission. Clothing Industry Training Authority (CIAT), Apparel supply chain–carbon assessment about jeans producing. http://www.cita.org.hk/wp-content/uploads/2013/10/Fty_rpt_Eng_20130912.pdf.

The distribution of GHG emissions in denim fabric weaving and finishing, by process and emission source during the denim process, is shown in Figs 6.14 and 6.15, respectively. Fig. 6.14 shows that the weaving process alone contributes to about 50% of GHG emissions in the entire process. Fabric finishing processes such as preshrinking, singeing, desizing, washing and heat setting contribute around 40% of GHG emissions. Hence, weaving and finishing processes have to be thoroughly analyzed, including auxiliary production facilities, to reduce the carbon footprint [57,59].

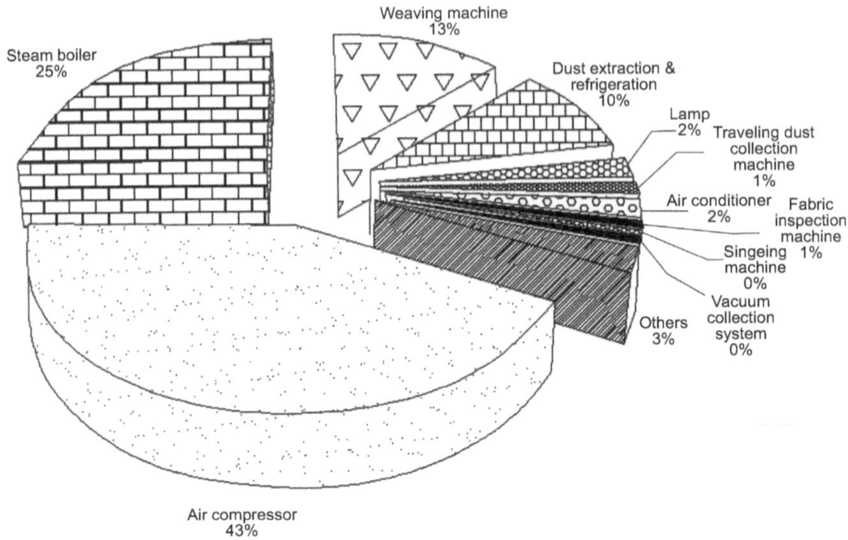


**Figure 6.15** Greenhouse gas emission distribution in weaving and finishing, by emission source.
CIAT, reproduced with permission. Clothing Industry Training Authority (CIAT), Apparel supply chain–carbon assessment about jeans producing. http://www.cita.org.hk/wp-content/uploads/2013/10/Fty_rpt_Eng_20130912.pdf.

Fig. 6.15 shows that air compressors contribute to about 40% of total GHG emissions, followed by steam boilers (light oil) and weaving machines. Compressed air is an essential motive power in many processes of textile manufacturing. Air jet weaving needs a huge amount of compressed air. Because of the higher GHG emissions of air compressors (40%), they are considered an important issue for carbon footprint reduction [60]. The second major contribution is from steam boilers (20%) which are used to provide steam for finishing processes such as desizing, drying and preshrinking. Weaving machines combined with air compressors contribute the most GHG emissions in this sector.

### 6.6.5 Garment manufacturing

Garment manufacturing (otherwise known as the makeup process) in when two-dimensional denim fabrics are converted into three-dimensional garments. It involves pattern making, cutting, assembling and sewing, and finishing processes. The entire process map in denim garment manufacturing is shown in Fig. 6.16.

The distribution of GHG emissions in garment manufacturing, by process and emission source, is shown in Figs 6.17 and 6.18, respectively. Fig. 6.17 shows that the final ironing of garments before packing contributes about 58% of GHG emissions compared with other processes. The main garment manufacturing processes such as cutting and spreading account for about 17% of total GHG emissions. Ironing, which uses electric power and to remove creases and wrinkles, consumes a huge amount of energy [61–63]. The low production rate of the cutting process leads to higher GHG

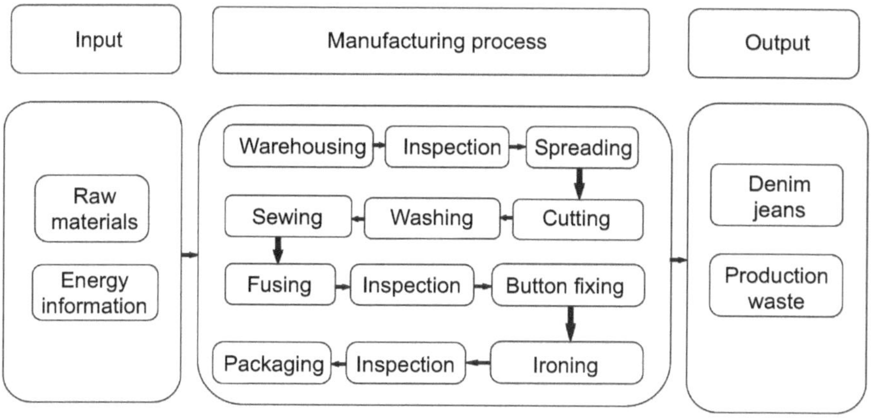


**Figure 6.16** Process map in denim jeans manufacturing.

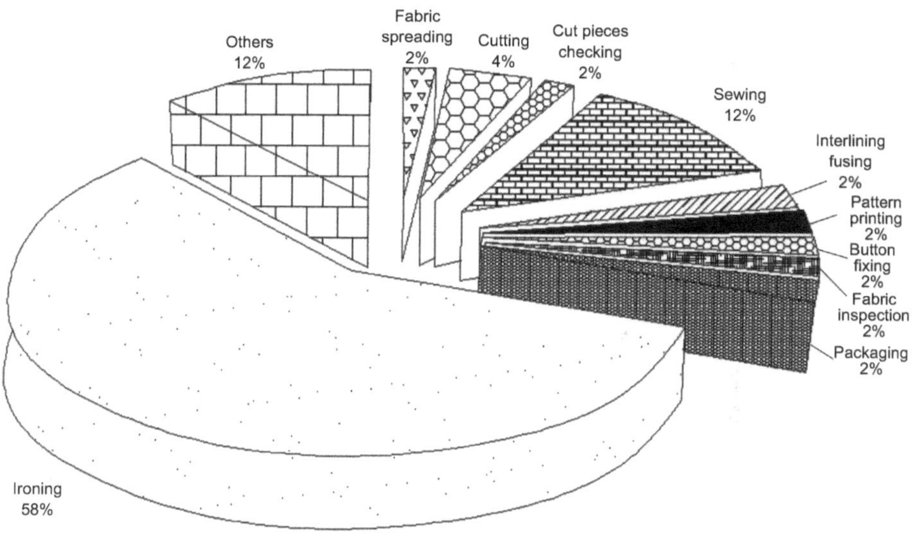


**Figure 6.17** Greenhouse gas emissions in denim garment manufacturing process.

emissions. Fig. 6.18 shows that the steamer, lighting, sewing machine, exhaust fan, electric iron, air conditioner and fusing machines are major GHG emission sources and have to analyzed to reduce the carbon footprint.

### 6.6.6 Denim washing

To give jeans distinctive visual effects, a unique washing process is after the garment or fabric stage, such as a standard wash, stonewash, enzyme wash, chemical wash, bleach wash or destruction wash. The jeans-washing effect design is an important step of jeans development because it attracts consumers [64]. The washed products are sent back to the garment manufacturing factory for final inspection and packaging. This is

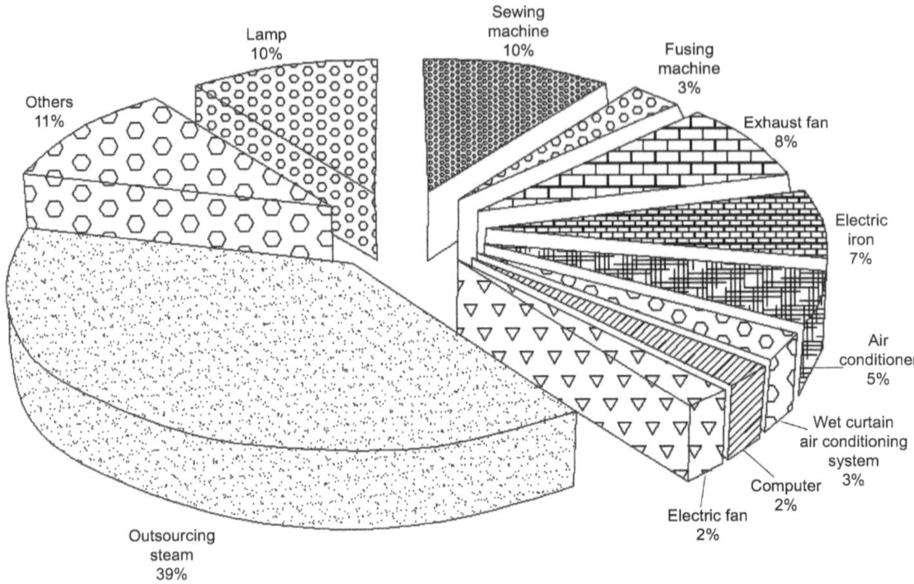


**Figure 6.18** Greenhouse gas emission distribution in denim garment manufacturing, by emission source.

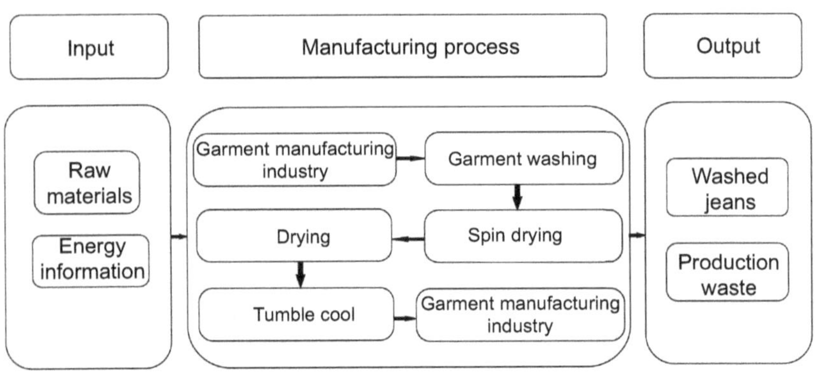


**Figure 6.19** Process map in denim washing.

the end of the assessment of the carbon footprint for jean in the manufacturing sector. The entire process map in denim washing is shown in Fig. 6.19.

The distribution of GHG emission in garment washing, by process and emission source, is shown in Figs 6.20 and 6.21, respectively. Fig. 6.20 shows that only four simple processes are involved in denim washing, of which the main washing process contributes more than 60% of total GHG emissions [65]. The second major contributor for GHG is a drying process. These two processes comprise the majority of GHG emissions with a total share of about 80%, which should be considered for opportunities to reduce carbon.

Fig. 6.21 shows that outsourcing steam holds the biggest share of the total GHG emissions which are the essential elements for washing processes by providing

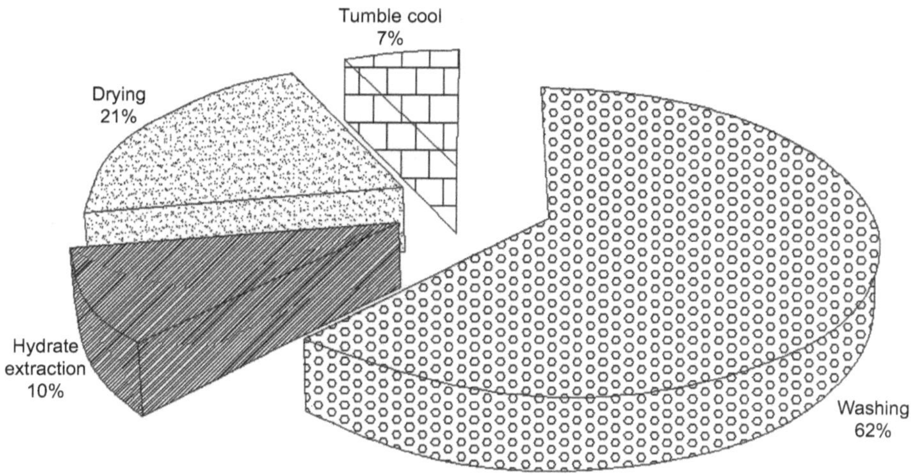


**Figure 6.20** Greenhouse gas emissions in denim washing process.

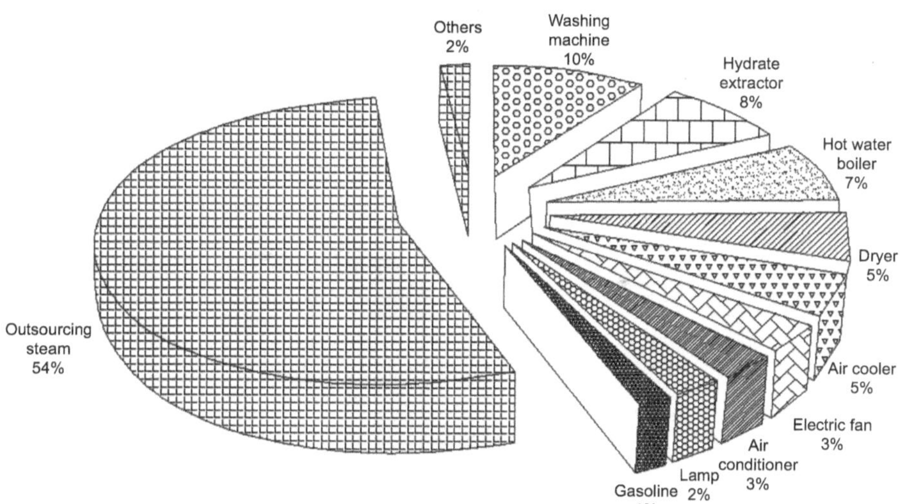


**Figure 6.21** Greenhouse gas emission distribution in denim washing, by emission source.

steaming to enhance the washing effects. The electricity used to run washing machines contributes about 10% of GHG emissions.

## 6.6.7 Carbon footprint of denim in consumer use phase

The highest energy use and GHG emissions in the life cycle of denim come from the consumer use phase of jeans. Although overall water consumption in the jeans life cycle is accounted for at the cotton-growing stage, at around 68%, consumer care comes in at 23%. Based on these figures, it seems that the consumer use phase may have the greatest impact on the environment than any other phase. Fig. 6.22 shows

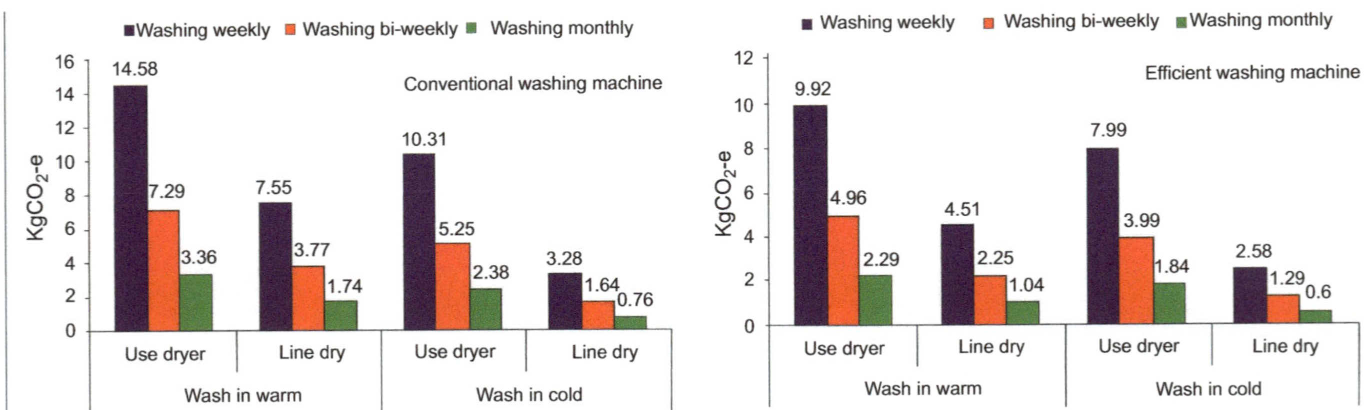


**Figure 6.22** Impact of jeans on climate change from 1 year of consumer care. Levi Strauss & Co. Life cycle of a jean, reproduced with permission.

that the overall impact on climate change is higher in the consumer use phase (37%), followed by fabric production (27%) [60].

The impact of a change in climate owing to consumers washing and drying jeans varies significantly based on the method of washing, washing frequency, and washing equipment. Fig. 6.22 shows the influence of the frequency of washing jeans, type of washing machine and washing conditions on the carbon footprint of jeans during the consumer use phase. From Fig. 6.22 it is evident that washing jeans with energy-efficient washing machines accounts for about a 30% lower carbon footprint compared with conventional washing machines; a further 30% reduction in carbon footprint was noticed by washing jeans in cold water compared with warm water. Similarly, washing jeans biweekly reduces the carbon footprint around 50% compared with washing them weekly, and a further reduction of 50% by washing them monthly [48]. Washing denim garments after they are worn 10 times instead of twice reduces energy use, the impact on climate change and water intake by up to 80%.

Fig. 6.23 shows the influence of washing machines and washing conditions on the carbon footprint on 1 day of washing jeans over various countries. Fig. 6.23 demonstrates that washing jeans in an energy-efficient washing machine and in cold water reduces carbon emissions compared with washing jeans in a conventional washing machine and in warm water [66,67]. Furthermore, the figure shows that consumers in China are leading the pack when it comes to laundering jeans, because they mostly wash in cold water and air-dry. American consumers have the highest water intake and use of nonrenewable energy. Consumers in the United Kingdom and France mostly air-dry their jeans but they use more hot water than do American or Chinese consumers. Consumers in the United States, United Kingdom and France wash their jeans more frequently than do consumers in China.

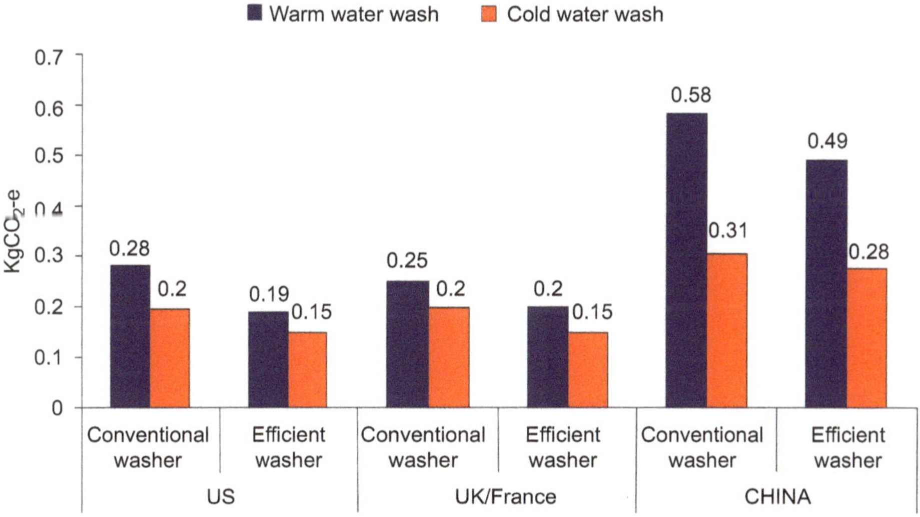


**Figure 6.23** Global impact on climate change of 1 day of washing jeans.
Levi Strauss & Co. Life cycle of a jean, reproduced with permission.

## 6.7 Strategies to reduce carbon footprint in denim manufacturing

A global paradigm shift toward cleaner and greener processes and products is in progress. A systematic approach with continuous improvement can minimize the carbon footprint of textiles. Textile industries have realized the benefits of increasing energy efficiency and thus reducing the costs of fuel and electricity. The following activities should be considered to reduce total life-cycle GHG emissions [68]:

- *Product design for the environment*: The decision regarding most significant life-cycle GHG emissions has to be done in the design stage of the product, so consideration of GHG and other environmental impacts should be incorporated into design decisions.
- *Care requirements*: Product care requirements and labels should be considered and consumers have to be educated so that garments are not washed more than necessary.
- *Enhancing product durability*: More hard-wearing products minimize GHG emissions substantially by spreading the 'sunk costs' of manufacturing emissions over a longer product life.
- *Review raw materials and engage with suppliers*: Consider greater use of plant-based fibres where garment specifications permit, engaging with suppliers to reduce emissions and shifting toward organic fibres to reduce GHG types.
- *Transport*: If air freight is used at any point in the supply chain, consider options to reduce it.

Environmental sustainability can be realized by considering the full life cycle of clothing, from the design and materials sourcing process onward to disposal. Total energy used in each stage of the process to produce the fabric is known as embodied energy and is the sum total of the energy required to produce the fibre and yarn as well as to weave those yarns into fabric. Beyond fibre production, dyeing and finishing are the highest consumers of water and energy in the textile processes and have the highest opportunities for energy and water savings and efficiency enhancements [69]. Once the magnitude of a carbon footprint is known for a product, a stratagem can be developed to reduce it by developing new technologies, an improved process and product management, changed Green Public or Private Procurement, consumption strategies, carbon capture, etc. Strategies that can be used to reduce the carbon footprint are:

- adopting the 3R concept (see subsequent discussion)
- developing standards and labels
- educating consumers

### 6.7.1 *Adoption of 3R concept*

#### 6.7.1.1 *Reduce*

The carbon footprint in textiles can be reduced by processes and technologies which use relatively fewer raw materials and consume less energy, slowing down climate change.

## Use of organic and wood-based fibres

To produce a single pair of jeans, about 3000 gallons of water is required, excluding chemicals, dyes, the energy required for drying and the workforce needed for production. Hence, there is a need to design sustainable practices and technology in the product supply chain. The battle is on to produce technology and processes that will lessen the impact of denim production on the environment as well as on the health of the workers making the items. To start with, more environmentally friendly fibres and dyes are needed which can use less water and safer chemicals [70].

By substituting traditional fibres with organic fibres and not using petrochemical-based fertilizers and pesticides during production, we can ensure a few GHGs are emitted. Some green fibres are bamboo, linen, soya bean, abaca, hemp and polyester recycled from used clothing. Besides having a low carbon footprint, natural fibres can be decomposed by microorganisms. Thus the fixed $CO_2$ in fibre is released and the cycle is closed. Also, natural fibres can sequester carbon, in which $CO_2$ from the atmosphere is absorbed by plants through photosynthesis and stored as carbon in biomass and soils [71].

Cultivation and organic cotton are inevitable for making sustainable denim. However, cultivation has to meet some standards that limit the use of toxic and agrochemicals. Hence, a more fastidious approach has to be practiced in organic cotton farming at all stages, including enhancing soil fertility, preharvest defoliation, and cotton picking. Growing organic cotton not only enhances natural ecosystems; it also increases human participation and a healthy lifestyle [24].

Replacing organic fibres for conventionally grown fibres is little better because it needs less energy for production and lower costs, and produces less $CO_2$. A study reported that organic fibres release 4.3% less GHG than conventional fibres and use only around 63% of energy for farming than conventional fibre farming. The important aspect is in handling organic matter; because soil organic matter is based on carbon, increasing the level of organic matters directly relates to carbon sequestration. Organic farming also supports environmental and social aspects and eliminates the use of synthetic fertilizers, pesticides which cause some serious problems. Because no pesticides are applied to organic cotton, organic cotton yields are as much as 50% lower than conventionally produced cotton, which necessitates more land to produce organic jeans. Organic agriculture is regarded as a climate change tool that can be used to battle global warming [72,73].

There is buzz in sustainable denim manufacturing about the use of denim from wood-based fibres. One eco-friendly fibre which can be used at a sustainable level to produce jeans is Tencel, made from wood pulp cellulose. The modal fibre is $CO_2$ neutral and can be combined with Tencel fibres to produce denim with stretch and performance. Some manufacturers use bamboo fibre, which is fast-growing and needs less water to cultivate. It does not require fertilizers and pesticides to manufacture sustainable jeans.

## Reduction of water and chemicals in denim processing

Dye which can be employed in denim processing should have the property of reducing water use. Miguel Sanchez, global head of Lenzing product line for special dyes, says the firm is developing a dye that will reduce water usage by up to 92%. Most water consumed in colouring is in the washing-off stage (93%). Additional water is then required to dry the yarn through steam (1%). In total, around 100,000L of water is used in 10,000m of yarn. Ideally water should go back to where it came from and close the loop. We really need to treat this water and send it back to nature [74]. The use of water for dyeing can be reduced by:

- using dyestuffs with 100% fixation capacity
- using 100% biodegradable dyestuffs and chemicals and ensuring the highest fixation possible in the application process
- minimizing washing-off
- recover and recycling everything
- replacing dye with another type of abundant solubilizing agent

Denim can be dyed using natural indigo to make it more sustainable. Besides the low yield of natural indigo production in plants, there is a problem because of the insolubility of indigo in water. Hence, the extraction of indigo from the natural plant requires a huge amount of alkali which contains a small amount of heavy metals. After the dyeing process most of the heavy metal can be seen in the blue jeans. The denim industry currently uses prereduced indigo crystals for dyeing, or a less sustainable powder indigo which has to be reduced at the mill using alkaline sodium hydrosulphite, the by-product of which is water. However the process poses a danger to human health because manual labour involved and there is exposure to waste [56].

The manufacturing and purification processes involved in producing synthetic indigo have a vital role in using of amounts of reducing agents and the quality of indigo dyes. The use of aniline as a starting material for the synthesis of indigo would not affect the environment because the quantity of it would be below critical limits. Based on the eco-efficiency analysis developed by BASF, it was shown that the total eco-balance of the production of blue jeans based on prereduced indigo and using an electrochemical dyeing (ECD) process would be optimal.

## Energy-efficient processes

It is high time for the textile industry to implement energy-efficient and environmentally sound technologies to save costs and reduce the carbon footprint [56]:

- The ECD could replace the chemical reducing agent hydrosulphite completely with electrons for the dyeing of denim fabrics. This technique minimizes wastewater pollution considerably and enhances the ecological situation in the dye house. However, because of a labeling issue, the technology was not supported by retailers. This is the case with sulphur dyeing, especially sulphur black, the most important dyestuff for black denim. Currently, in sulphur dyeing process, the environmental friendly glucose is used as the best option as it causes huge chemical oxygen demand and Chemischen Sauerstoffbedarfs values in the effluent.
- After the completion of the yarn-dyeing and weaving process, the denim fabric is treated for special effects before going for making-up. The processing technologies which require a minimal amount of water, energy and chemicals in the process are needed for the eco-friendly processing of jeans. Innovative machines apply finishes to fabrics using foam, which conserves water.
- An ozone bleaching system can save up to 40% of energy and water and reduce cotton loss by 50%. For cleaner denim products, peroxide, oxygen and ozone bleach in garment finishing creates less environmental impact.

- Revolutionary air technology for dyeing requires only 25% of water compared with the conventional system and reduces energy and chemical consumption.
- Digital printing, using ink from the dyes, wastes neither fabric nor ink, does not use harmful salts and significantly reduces the environmental impact.
- Color-fast finish is a single-stage process for textiles and can reduce the processing time and $CO_2$ emissions.
- Industrial enzymes, which are basically proteins, replace strong chemicals used to remove impurities from the fibre or fabric; they reduce energy costs and water consumption and improve the feel of the fabric.

#### Effective laundry management

Effective laundering management in the context of sustainability means eliminating certain detergents and surfactants such as alkyl phenol and ethoxylates, and other chemicals such as phosphates and chlorine. Peracetic acid is considered a good alternative to chlorine bleach because it is less hazardous and more biodegradable.

#### Others

Setting up establishments that use more natural light for illumination, upgrading existing lighting systems with energy-efficient ones and installing heat recovery plants are all ways to reduce the carbon footprint.

### 6.7.1.2 Reuse

Effluents of chemically treated textiles are normally discharged in water. Treatment of wastewater obtained from chemically treated textiles is a must. Use of chrome mordant dyeing and limiting the emission of copper, chromium and nickel into water reduces impurities in dyes and pigments. Using dyeing carriers with high chlorine content should be avoided. During bleaching, an alternative agent can be used that is less hazardous or completely harmless. One challenge in the denim industry is the decolouring of wastewater. The concept involved in decolouring is zero liquid discharge. An electrochemical device is added to the water use area to treat the spent water before sludge is separated. Because no chemicals are used in the treatment, the process requires low energy with higher efficiency and reduced costs. DyStar introduced LoopTEC technology for wash water reuse and indigo recycling for denim dyeing; the company claimed 95% less water consumption and 95% less caustic demand, 95% less effluent volume and 75% energy savings.

### 6.7.1.3 Recycle

In denim manufacturing, the concept of reusing waste and raw materials is dominant in the production chain. Denim waste is being used for other application segments such as fleece and insulation material and to make denim bags. Denim recycling begins with using worn-out blue jeans as raw material. The pure cotton blue jeans are made free of all metal parts and the labels are shredded. After deinking, the shredded denim can be used in the paper industry. Also, the recycled cotton can be mixed with fresh cotton as a raw material in the production chain [69].

'Real' denim recycling employing used-up denim material as a raw material in the denim manufacturing was evaluated from only a few producers such as Orta Anadolu in Turkey. The denim sector is huge enough to attain remarkable technical and economical process if a number of denim brand manufacturers agree to the mutual concept. To implement this concept, only the absence of nonrecyclable raw materials is necessary. The denim is removed from other metal components and labels from the jeans by shredding. The shredded denim containing recycled cotton can be used as a raw material by mixing it with virgin cotton to manufacture denim. The recycled denim, called 'R-Jeans', could create an attractive marketing concept, especially for the growing sector of environmentally conscious customers.

Recycled cotton materials from used blue jeans are used as a part of carpet backing and sound absorption in Ford cars. Postconsumer cotton used in a car is obtained from approximately two pair of jeans. Another way to personalize old jeans is to recycle them into sandals to make soft, springy shoes; it makes a perfect accessory for both eco-warriors and funky fashionistas.

## 6.7.2 Development of standards and labels

### 6.7.2.1 The global recycling standard

This standard was developed to verify claims related to recycled goods. The percentage of recycling varies at 95%–100%, 70%–95% and 30%, respectively, for classifying them into gold, silver and bronze levels. Scientific certification systems have put forth criteria to be followed by 'recycled' materials. In addition to strict environmental processing, raw material specifications and social responsibility are included to ensure the health and safety of workers in accordance with International Labour Organization norms. The carbon trust, based in the United Kingdom, has developed the world's first carbon label for clothing in association with continental clothing. This label provides the carbon footprint of garments right from raw materials and manufacturing, to use and disposal [24].

### 6.7.2.2 Global Organic Textile Standard

Internationally, the Global Organic Textile Standard (GOTS) is recognized as a tool for understanding environmentally sound production systems and social accountability in the textile sector. It covers the entire supply chain from raw material production to the distribution of all natural fibres. It actually means the use of certified organic fibres and the prohibition of an entire list of synthetic chemicals for use in the processes. Fabric that meets the norms of GOTS standards is more than just fabric; it offers a promise to keep the environment (air and water) pure and the soil renewed. Future generations will not be detrimentally affected by providing fabrics of GOTS standards, because of the reduced carbon footprint. The minimum requirement for certification from GOTS is that a product must be 67% recyclable or biodegradable.

### 6.7.3 Educate consumers

Consumer education regarding the huge carbon footprint in the textile value chain will inspire consumers to change their habits and attitudes. That is, consumers should minimize their purchase of fabrics that are blends of synthetic fibres and shift their focus toward 'organic fabrics'. In addition, they should be on the lookout for stamps from GOTS, SMART and Cradle to Cradle certifying agencies, paying attention to the carbon footprint of the fabrics purchased. On the other hand, retailers are also seeking sustainability requirements along their supply chains to maintain and expand their customer base [56,75].

Low-carbon manufacturing programmes and carbon accounting in factories, carbon footprint calculation projects, and benchmarking energy consumption across the textile and apparel supply chains are a few strategic measures required to reduce the carbon footprint of textile and apparel industry.

## 6.8 Conclusion

The textile industry across the globe is continuously taking steps toward reducing the carbon footprint to meet challenges for a more sustainable future. Awareness of environmental issues among consumers who insist on textile products that comply with environmental standards is also increasing. These are healthy trends that will continue to drive the industry toward offering eco-friendly products.

As a rule, in the future, innovative as well as eco-friendly processes and products should be economically attractive if plans related to environmental protection initiatives and the treatment of wastewater are carried out realistically. Most significant part in the ecological assessment and comparison of denim manufacturing concept is analyzing its features for the complete denim manufacturing process. Therefore only a total process comparison can offer a real comparison. The total sustainable denim production approach was highlight by DyStar, in its econfidence denim concept. The econfidence concept in the production of sustainable denim involves: ECD on slasher and rope dyeing ranges, use of minimal application processes in the denim manufacturing value chain, electrochemical bleaching in garment dyeing/finishing machines (laundry machines), combination processes and recycling concepts in denim laundries and garment finishing mills, and eco-efficiency evaluation of denim processes.

Sustainable denim manufacturing in the near future will be a process involving all steps of the complete denim manufacturing value chain. The optimal solution comes from balancing the use of recycled materials, advanced and energy-efficient machinery and equipment for manufacturing and use of optimized process parameters in manufacturing. In this process of development it would be helpful if retailers (global jeans brands) implement a global jeans recycling concept and use eco-labeling on best available technology parameters. A reduction in the global warming potential of denim manufacturing and production can come from using minimal resources, water, chemicals and energy as well as a minimal carbon footprint.

## References


[1] Brundtland GH, Khalid M. Our common future, report of the world commission on environment and development. UK: Oxford University Press; 1987.

[2] Gardetti MA, Torres AL. Sustainability in fashion and textiles: values, design, production and consumption. UK: Greenleaf Publishing; 2013.

[3] Gardetti MA, Torres AL. Entrepreneurship, innovation and luxury: the any Savoirs des Peuple case. J Corp Citizsh 2013;52:55–75.

[4] Karthik T, Gopalakrishnan D. Impact of textiles in environmental issues and environmental legislation, Part – I. Text Rev 2012;7(3):14–21.

[5] Karthik T, Gopalakrishnan D. Impact of textiles in environmental issues and environmental legislation, Part – II. Text Rev 2012;7(4):16–21.

[6] Karthik T, Gopalakrishnan D. Impact of textiles on environmental issues, Part – I. Asian Dyer 2012;9:52–8.

[7] Karthik T, Gopalakrishnan D. Impact of textiles on environmental issues, Part – II. Asian Dyer 2013;9:45–51.

[8] Karthik T, Gopalakrishnan D. Impact of textiles on environmental issues, Part – III. Asian Dyer 2013;9:40–3.

[9] Karthik T, Gopalakrishnan D. Impact of textiles on environmental issues, Part – IV. Asian Dyer 2013;9:43–6.

[10] Karthik T, Gopalakrishnan D. Eco-friendly fibres of the future. Asian Text J 2012;21:67–71.

[11] Prakash C, Maruthavanan T, Parvathi C. Environmental impacts of textile industries. Indian Text J 2009;23:22–6.

[12] Defra (Department for Environment, Food and Rural Affairs). Sustainable clothing roadmap briefing note. 2008. December 2007 (updated March 2008) http://www.defra.gov.uk/environment/business/products/roadmaps/clothing/documents/clothing-briefing-Dec07.pdf.

[13] Gopalakrishnan D, Karthik T. Eco-friendly fibres – textile contribution to minimize global warming. Text Rev 2012;7(9):13–5.

[14] Fletcher K. Sustainable fashion and textiles. London: Earthscan; 2008.

[15] Climate change and the textile industry. 2014. https://oecotextiles.wordpress.com/2014/10/15/climate-change-and-the-textile-industry/.

[16] Anonymous. Carbon Footprint. http://www.carbonfootprint.com/carbonfootprint.html.

[17] Weidema BP, Thrane M, Christensen P, Schmidt J, Løkke S. Carbon footprint a catalyst for life cycle assessment? J Ind Ecol 2008;12(1):3–6.

[18] ErtugErcin A, Hoekstra AY. Carbon and water footprints. 2012. http://waterfootprint.org/media/downloads/Ercin-Hoekstra-2012-Carbon-and-Water-Footprints_1.PDF.

[19] Business for Social Responsibility. Apparel industry life cycle carbon mapping. 2009. https://www.bsr.org/reports/BSR_Apparel_Supply_Chain_Carbon_Report.pdf.

[20] Bolwig S, Gibbon P, Jones S. The economics of smallholder organic contract farming in tropical Africa. World Dev 2009;37:1094–104.

[21] Lenzen M. Errors in conventional and input output based life-cycle inventories. J Ind Ecol 2001;4:127–48.

[22] Vivek D. Carbon footprint of textiles. 2009. http://www.domain-b.com/environment/0090403_carbon_footprint.html.

[23] Muthu SSK, Li Y, Yan Hu J, Ze L. Carbon footprint reduction in the textile process chain: recycling of textile materials. Inst Text Cloth Fibers Polym 2012;13:1065–70.

[24] Mehta R, Goyal C. Role of carbon footprint in textile and apparel industry. 2015. http://www.textilevaluechain.com/index.php/article/technical/item/235-role-of-carbon-footprint-in-textile-and-apparel-industry.

[25] Zamani B, Svanstrom M, Peters G, Rydberg T. A carbon footprint of textile recycling-a case study in Sweden. J Ind Ecol 2014;19(4):1–12.
[26] Chaudhary H. Assessment of carbon footprint in textile sector – a case study based on garment manufacturing unit. Pak Text J 2011:44–5.
[27] Muthu SS, Li Y, Hu JY. Carbon footprint reduction in the textile process chain: recycling of textile materials. Fibre Polym 2012;13:1065–70.
[28] Kuby M, He C, Trapido LB. The changing structure of energy supply, demand, and CO emissions in China. Ann Assoc Am Geogr 2011;101:795–805.
[29] Choudhury AKR. Green chemistry and the textile industry. Text Prog 2013;45:3–143.
[30] Hasanbeigi A. Energy-efficiency improvement opportunities for the textile industry, LBNL-3970E. Berkeley (CA): Ernest Orlando Lawrence Berkeley National Laboratory; 2010.
[31] Bevilacqua M, Ciarapica FE, Giacchetta G, Marchetti B. A carbon footprint analysis in the textile supply chain. Int J Sustain Eng 2011;4:24–36.
[32] Downey LA. Short history of denim. San Francisco (CA, USA): Levi Strauss & Co.; 2007.
[33] Paul R, Pardeshi PD. Everything you ever wanted to know about denim. Text Horiz 2003:12–6.
[34] Paul R. Denim manufacture, finishing and applications. Cambridge (UK): Woodhead Publishing; 2015.
[35] Sullivan J. Jeans: a cultural history of an American icon. New York (NY): Penguin Group, Inc.; 2006.
[36] Adkins J. The evolution of jeans: American history 501. Mother Earth News 1990;22:60–3.
[37] Anon. Technical bulletin: denim fabric manufacturing. NC, USA: Cotton Inc.; 2004.
[38] Iqbal S, Kolhatkar AW. Denim fabrics: ring vs rotor spun yarns. Indian Text J 2009;120(1):28–32.
[39] Paul R, Malanker JV. Denim: the blue magic material. Asian Text J 1997;6(12):32–6.
[40] Rathod A, Kolhatkar A. Physical properties of denim fabrics, Part 2. Man-Made Text India 2013;41(9):305–8.
[41] Chakraborty JN, Chavan RB. Dyeing of denim with indigo. Indian J Fibre Text Res 2004;29:100–9.
[42] Chong CL, Poon LW, Poon WC. Improving indigo dyeing colour yield. Text Asia 1995;26:55–7.
[43] Cotton Incorporated. Denim fabric manufacturing. Tech Bull 2004, vol. 1010. North Carolina, USA.
[44] Etters JN. Advances in indigo dyeing: implications for the dyer, apparel manufacturer and environment. Text Chem Colour 1995;27:17–22.
[45] Paul R, Naik SR. Denim series – part-XI. The technology of ring dyeing – an economically viable, environment friendly development. Text Dyer Print 1997;30(8):16–22.
[46] McGill M. Carbon footprint analysis of textile reuse and recycling. 2009. https://workspace.imperial.ac.uk/environmentalpolicy/public/McGill09_textiles_Exec%20Summary.pdf.
[47] Zamani B. Carbon footprint and energy use of textile recycling techniques. 2011. http://publications.lib.chalmers.se/records/fulltext/146872.pdf.
[48] Levi Strauss & Co. Life cycle of a jean. 2010. http://www.levistrauss.com/sustainability/product/life-cycle-jean.
[49] Ajootian A. Saving natural color–how an organic cotton breeder changed the world. Org Consum Assoc 2010. http://www.organicconsumers.org/clothes/color090804.cfm.
[50] Better cotton initiative: why better cotton. Better cotton initiative: home. 2010. http://www.bettercotton.org/index/18/why_better_cotton.html.
[51] Chapagain A, Hoekstra A, Savenije H, Gautam R. The water footprint of cotton consumption: an assessment of the impact of worldwide consumption of cotton products on the water resources in the cotton producing countries. Ecol Econ 2006;60:186–203.

[52] Nigam M, Mandade P, Chanana B, Sethi S. Energy consumption and carbon footprint of cotton yarn production in textile industry. Int Arch Appl Sci Technol 2016;7(1):6–12.
[53] Fitzgerald B. How textile manufacturers are making denim more sustainable. http://www.lesouk.co/articles/material-inspiration/how-textile-manufacturers-are-making-denim-more-sustainable.
[54] American Chemical Society. Sustainable denim manufacturing process creates 'green' jeans. 2012. http://phys.org/news/2012-06-sustainable-denim-green-jeans.html.
[55] Brands S. Patagonia out to change the 'Filthy Business' of denim. 2015. http://www.sustainablebrands.com/news_and_views/products_design/sustainable_brands/patagonia_out_change_filthy_business_denim.
[56] Schrott W. Sustainable denim – eco-labeling and environmentally friendly denim production. 2011. http://www.denimsandjeans.com/environment/sustainable-denim-ways-to-environmentally-friendly-denim-production/3563.
[57] Clothing Industry Training Authority (CIAT). Apparel supply chain – carbon assessment about jeans producing. http://www.cita.org.hk/wp-content/uploads/2013/10/Fty_rpt_Eng_20130912.pdf.
[58] Tyler Michael Cole. Sustainability is the new premium jeans. 2016. https://baumwoll-boerse.de/wp-content/uploads/2016/03/SI-Cole-Sustainability.pdf.
[59] DuFault A. Sustainable denim round up. EcoSalon – conscious culture and fashion. 2010. http://ecosalon.com/sustainable-denim-round-up-17brands-we-love.
[60] Levi Strauss & Co. A product life cycle approach to sustainability. 2010. http://www.levistrauss.com/sites/default/files/librarydocument/2010/4/Product_Lifecyle_Assessment.pdf.
[61] Wambuguh D, Chianelli RR. Indigo dye waste recovery from blue denim textile effluent: a by-product synergy approach. New J Chem 2008;32(12):2189–94.
[62] Levi Strauss & Co. Carbon Disclosure project CDP 2009(CDP7). 2009. http://www.levistrauss.com/wp-content/uploads/2014/01/Carbon-Disclosure-Project-CDP-2009-Info-Request.pdf.
[63] U.S. Department of Energy. Emissions of greenhouse gases in the United States 2009. 2011. http://www.eia.gov/environment/emissions/ghg_report/pdf/0573(2009).pdf.
[64] Pishgar-Komleh SH, Sefeedpari P, Ghahderijani M. Exploring energy consumption and $CO_2$ emission of cotton production in Iran. J Renew Sustain Energy 2012;4(3):1–15.
[65] Ajibola VO. Recovery of indigo from a textile effluent in Nigeria. J Sci Ind Res 2001;60:735–7.
[66] Balan DSL, Monteiro RTR. Decolonization of textile indigo dye by ligninolytic fungi. J Biotechnol 2001;89:141–5.
[67] Sule AD, Bardhan M. Recycling of textile waste for environment protection – an overview of some practical cases in the textile industry. Indian J Fibre Text Res 2001;26:223–32.
[68] Vautier M, Guillard C, Herrmann J. Photocatalytic degradation of dyes in water: case study of indigo and of indigo carmine. J Catal 2001;201:46–59.
[69] Dystar. Wash water reuse and indigo recycling on denim dyeing ranges. 2015. http://www.dystar.com/wp-content/uploads/2015/11/Indigo_water_resuse_looptec_6_print_Nomarks.pdf.
[70] Michaud JC, Farrant L, Jan O, Kjær B, Bakas I. Environmental benefits of recycling—2010 update. 2010.
[71] Wang Y. Fiber and textile waste utilization. Waste Biomass Valoriz 2010;1:135–43.
[72] Carbon credit. http://www.investopedia.com/terms/c/carbon_credit.asp#axzz1sqTy0KQF.

[73] Apparel Export Promotion Council. Emerging issues in apparel trade – sustainable development and carbon neutrality report. 2012. http://aepcindia.com/app/webroot/files/pdf/publication_13606_1402390268.pdf.
[74] Roos G. UK launches first carbon footprint label for retail clothing. 2009. http://www.environmentalleader.com/2009/03/27/uk-launches-first-carbon-footprint-label-for-retail-clothing.
[75] ICIS Chemical Business. Market emerging for green textile chemicals. 2009. http://www.icis.com/resources/news/2009/08/31/9243747/market-emerging-for-green-textile-chemicals.

## Further reading

[1] Joy A, Sherry JF, Venkatesh A, Wang J, Chan R. Fast fashion, sustainability, and the ethical appeal and luxury brands. Fash Theory 2012;16(3):273–96.
[2] Vijayaraaghavan NN, Karthik T. Multi-component fiber technology for medical and other filtration applications. Synth Fibres 2004;33(1):5–8.
[3] Camp S, Clark G, Duane L, Haight A. Life cycle analysis and sustainability report: Levi Strauss & Co. Jeans. http://www.uvm.edu/~shali/Levi.pdf.

# Denim processing and health hazards

7

*A.P. Periyasamy, J. Militky*
Technical University Liberec, Liberec, Czech Republic

## 7.1 Introduction

A large number of industries use dyes and pigments to colour their products. The textile industry, including the denim industry, is the largest consumer of such dyes and pigments. This industry requires a huge volume of water of high purity and generates an equally large volume of wastewaters. Estimates indicate that 1.1 million tons of synthetic dyes are manufactured every year and approximately 17% of synthetic dyes used every year are lost during manufacture and processing operations. If these dyes are present in effluent in a small amount, i.e., 0.1 mg/L, it is noticeable and colour that has persistent nature flows far away in rivers, canals and other water bodies. Humanity is facing a major challenge to provide clean water to a vast majority of the population around the world. With sources of fresh water insufficiently available and with increasing pressures of population and industrialization, reuse of wastewater has become an absolute necessity. The increase in legal pressure and restrictions is forcing industries to accept responsibility for the treatment of waste to minimize pollution. Current commercial textile pollution prevention practices include material substitution, process modification, inventory control, better management techniques, recovery and reuse. The textile industry has developed a relatively comprehensive approach to pollution prevention by source reduction for several types of wastes. Further efforts are required to deal with important environmental issues such as colour residues in textile dyeing and printing wastewater [1–3]. When the environment is polluted, that influences our health and sometimes it modulated the genetic makeup and psychological factors. The majority of the denim process belongs to the handling of harmful chemicals [4–6]. There are potential health and hazardous issues associated with this industry. The aim of this chapter is to study health and environmental issues with respect to harmful denim processes such as sandblasting, PP spray, chlorine bleaching, hand scraping. It also provides the alternatives and practical limitations to such harmful process on denim processing.

## 7.2 Environmental aspects of denim processing

Denim processing houses are the largest consumers of water and chemicals as well as producers of wastewater. The sources of pollution are the natural impurities extracted from the fibres and chemicals used in to process fibres. The main factors affecting the quality and quantity of the effluents generated include unit operations comprising the

Sustainability in Denim. http://dx.doi.org/10.1016/B978-0-08-102043-2.00007-1
Copyright © 2017 Elsevier Ltd. All rights reserved.

overall denim process and the extent of in-house conservation measures in practice. In the denim industry, natural indigo and synthetic indigo were used to colour the warp sheet. The total dye consumption of the textile industry alone is in excess of $10^7$ kg/year worldwide; of that denim has a vital role because of its highest market potential. Even if it is estimated that a maximum of 90% of dye goes onto the fabric, approximately $10^6$ kg/year of dyes would be discharged into waste streams by the textile industry. It is estimated that 280,000 tons of textile dye are discharged into industrial effluents every year worldwide [7,8]. The chemical reagents used are diverse in chemical composition, ranging from simple organic and inorganic compounds to various polymers and complex synthesized organic products [9]. However, the textile industry (including denim) has a major impact on most of the nation's economy, but also the environmental quality of life of many communities [10–12]. The overview is shown in Fig. 7.1, which provides a clear idea of the consumption and emissions that cause ecological concerns.

### 7.2.1 Air pollution

Energy and power are required for domestic, transportation and industrial operations. The burning of fossil fuels leads to the addition of pollutant gases such as carbon monoxide, carbon dioxide, nitrogen, sulphur and carbon particles and adds to the air temperature. Industrial emissions and chlorinated products also add potential contaminants to the atmosphere. The end results are issues such as acid rain, global warming, air toxicity, climatic changes and depletion of the ozone layer. Air pollution caused by the denim industry is also a major cause of concern. Boilers, power generators and transportation of goods (from raw materials to finished jeans) produce pollutants that are released into the air. The impact of air pollution can cause severe health problems to living organisms (including humans) owing to the release of toxic gas into the atmosphere. Therefore it is imperative to reduce air pollution emitted by the denim industry. The area where textile factories are located has contaminated the air and is

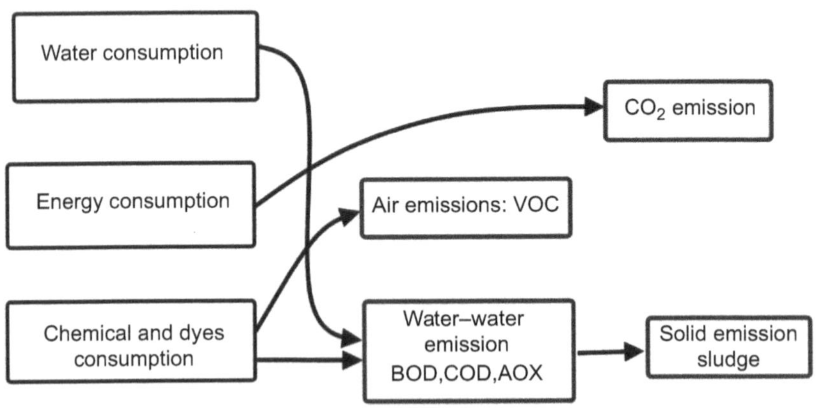


**Figure 7.1** Scheme for pollution route in the denim industry. *AOX*, adsorbable organic halogen; *BOD*, biological oxygen demand; *COD*, chemical oxygen demand; *VOC*, volatile organic compound.

one of the most serious threats to the environment as well as to health [13]. It generates fumes and called volatile organic compounds (VOC) that can be varied, countless and spread everywhere. Mostly VOC is not acutely toxic but it has chronic effects owing to its low concentration, and symptoms develop slowly [14–16]. Especially dry finishing on denim can generate a large amount of VOCs. The concentration of VOCs may vary from 10 to 350 mg/m$^3$ carbon [17,18]. It has endangered the life of humans and various other species on earth. Global warming is a direct result of pollutants released by such industries. VOCs also cause harmful disease and health issues to people who are exposed to the pollutants in the long term [19,20].

### 7.2.2 Water pollution

Overconsumption of water for domestic, agricultural and industrial needs ends in the contamination of water sources, surface and groundwater depletion, and degradation. Water, which is a major abiotic element for the support of life is rendered useless for consumption by way of organic, mineral, toxic, bacterial and physical pollution. Water pollution is the alteration of physical, chemical and biological properties of water in a water body, which renders it unsuitable for consumption. Water pollution results in water becoming nonpotable, rendering it unsafe for human or animal health, industry, pisciculture and agriculture or recreation. This includes its influence on land, thermal pollution and marine pollution [3].

Generally, the textile industry produces 40–65 L of wastewater per 1 kg of products [21]. The large volumes of wastewater generated also contain a wide variety of hazardous and toxic chemicals used throughout processing. Among the various processes, colouration is the major source of pollution in denim; however, it contains dyes, dyeing additives and other chemicals. Obviously, dyes contain heavy metals which are highly toxic. These pollutants are toxic to aquatic life and include various toxic elements such as salts, surfactants, ionic metals and their complexes, formaldehyde, toxic organic chemicals, biocides and toxic anions, detergents, emulsifiers and dispersants [22]. Therefore, wastewater from the textile industry is highly polluted and dangerous, especially when it becomes mixed with other chemicals and is disposed of untreated. Characteristics of wastewater from this industry are a high pH, biological oxygen demand and chemical oxygen demand; and high concentrations of total dissolved solids, suspended solids, chlorides, sulphates and phenols, as listed in Tables 7.1 and 7.2. Most of these chemicals and dyes are not readily biodegradable and cause severe health and environmental issues [9,22].

#### 7.2.2.1 Impact of water pollution on environment

Water pollution is the biggest threat for living organisms, including humans. It has a vital role in overall changes on the environment [24–29]:

- decrease in dissolved oxygen level
- increase in biological oxygen demand
- eutrophication of lakes
- detrimental to fish life and water use for drinking, bathing, industry and recreation

## Table 7.1 Various effluents from the textile industry produced and released [23]

| | |
|---|---|
| Acids | Lowers pH, causes corrosion, liberates $Na_2S$, destroy microorganisms, destroys self-purification system of water |
| Alkalis | Increase pH, liberates hydrogen ions, destroy microorganisms, destroys self-purification of water |
| Dissolved solids | Chlorides, sulphates, nitrates, bicarbonates of various inorganic metals; plants and vegetable kingdom affected by osmotic changes, affects biological organisms, BOD |
| Suspended matter | Insoluble wastes, destroy self-purification of streams, reduces photosynthetic activity of plants, chokes gills of fish, organic solids undergo purification giving rise to solids buoyed by gas, leads to floating masses, bad odor results |
| Dyes | Change colour of water, high BOD, high COD, leads to cancer, disabilities and disorders |
| Oils | Absorption of $O_2$ from air into water affected, destroys animals, plant life in water, high BOD |
| Synthetic detergents | Lowers biodegradability, adds to toxicity, accelerates chain reactions, increases floating masses and affects natural aeration |
| Toxic metals | Self-purification of streams destroyed, causes skin diseases, disabilities and disorders, destroy microorganisms, aquatic plants and animals are killed, compounds containing mercury affect food chain |
| Gaseous pollutants | Free chlorine, hydrogen sulfide, etc. Increases toxicity, adds to oxygen demand, kills microorganisms, destroy animals and plant kingdom |
| Heated effluents | Reduces dissolved oxygen concentration, aquatic life is affected, destroys animal and plant kingdom, affects entire ecosystem because of temperature |
| Radioactive wastes | Ill effects are numerous; induces metabolic changes, disorders, disabilities, genetic damage, chronic diseases, cancer, abnormal birth of humans which are also common to plants and animals |

*BOD*, biological oxygen demand; *COD*, chemical oxygen demand.
Based on R.B. B, P.A. K, R. A, Cotton textile processing: waste generation and effluent treatment. J Cotton Sci 2007;11:141–53

- many of the dyes are carcinogenic in nature, which causes cancer to consumers
- generation of odour problems
- affects air penetration to land and water
- affects photosynthesis of plants
- because of toxicity, increased disorders, disabilities, abnormal births and chronic diseases have surfaced in all life beings
- toxicants are capable of causing injury by way of impairing or killing living organisms, when they exceed threshold concentrations (heavy metals, acids, alkalis, phenols, cyanides, pesticides and pentachlorophenolare examples of toxic contaminants)
- causes biodegradability problems (e.g., detergents)
- report suggests that fertility of women is affected
- leads to chronic diseases, abnormal births and disorders in humans

Table 7.2 **Water pollution during denim manufacturing**

| Origin | Characteristics of pollutants |
|---|---|
| Sizing | Starch, generally coloured, high BOD, suspended solids, poly(vinyl alcohol), odor, softener, oils fats |
| Desizing | Starch, hydrolyzed starch, enzymes, salt, acidic pH |
| Scouring | Alkalis, surfactants, saponified oils, hydrolyzed pectins, proteins, suspended solids, oil, high pH, silicates with high COD, natural colours, TDS |
| Bleaching | Chlorines, hypochlorite, alkali, peroxides, silicates, suspended solids, fatty alcohol |
| Dyeing | Dyes, salt, alkalis, acids, detergents, heavy metals such as chromium, copper, high BOD, COD, TDS |
| Printing | Dyes, alkali, acids, chromium, copper, thickeners polymers, detergents high BOD and COD, waxes, oils fatty alcohol, formaldehyde from binder, urea, surfactants, suspended solids |
| Finishing | Silicones, oxidizing agents, formaldehyde, enzymes, cationic compounds organic and inorganic compounds, suspended and dissolved solids |

*BOD*, biological oxygen demand; *COD*, chemical oxygen demand; *TDS*, total dissolved solids.

Table 7.3 **Water pollution during denim dyeing**

| Dyes class | Type of pollution |
|---|---|
| Indigo | Alkali, oxidizing agents, reducing agents |
| Sulphur | Alkali, oxidizing agents, reducing agents |
| Pigments | Unfixed pigments, dispersing agent, cationic retarders, binders |
| Reactive | Unfixed dyes, salt, alkalis |

In the denim industry, most processing is carried out with the help of water, which acts as a medium. Nevertheless, these processed waters are the biggest source of pollution. This polluted water has a large quantity of contaminants such as dyestuff (indigo/sulphur and another dyestuff), solubilizing chemicals, buffer/pH controller, electrolyte, dyeing assistants and sizing ingredients (Tables 7.3 and 7.4). However, the dyes themselves contain heavy metals and hazardous pollutants. There is a huge market potential for denim and it is not possible to dye with natural indigo (i.e., there is an insufficient production of natural indigo), so researchers have developed synthetic indigo, which contains toxic chemicals such as aniline and *N*-methylaniline residues. Apart from synthetic indigo, denim is often dyed with sulphur, reactive pigment and direct dyestuffs. Most of these dyes contain heavy metals such as chrome, copper and zinc. These synthetic dyestuffs are not biodegradable and are toxic to water-living organisms as well as other organisms. Generally in the colouration process, efficiency has never been achieved more than 85% (of course in the laboratory that may be possible), so remaining colour stays behind in the water. Thus the polluted water

**Table 7.4 Water pollution during denim washing**

| Washing steps | Chemicals |
|---|---|
| Prewash | Enzyme, surfactants, detergents |
| Bleaching | Chlorine, adsorbable organic halogens, detergent |
| Antichlor wash | Sodium/disodium thiosulfate |
| Stone/acid/enzyme wash | Stones, enzymes, acids, pH, biological oxygen demand issues |
| Softener | Fatty acids |

**Figure 7.2** Water streams in Xintang, China.
Image courtesy: Greenpeace.

becomes coloured, which is the biggest threat to biodiversity [30]. Similarly, the sizing agents, auxiliary and chemicals used in the colouration process are not consumed in the respective processes. During rinsing (washing), they may be washed off and stay in the wastewater, which can have a different environmental impact.

In 2010, the Greenpeace organization be published a survey about the Xintang, Guangdong province (the denim capital of the world), located in China (Fig. 7.2) [31]. The Greenpeace organization collected 21 polluted samples from Xintang and Gurao. Of 17 samples they found five heavy metals (cadmium, chromium, mercury, lead and copper). In one sample cadmium exceeded China's national limits 128 times [31]. The World Trade Organization reported that 1.3 billion people from China were drinking contaminated water every day. This practice continues to occur in most denim manufacturing countries including China, India, Bangladesh and Mexico (Fig. 7.3). In 1998, the World Bank Group published a report stating that process wastewater, domestic sewage and contaminated storm water and runoff must meet the maximum limits shown in Table 7.5 before being discharged to surface waters [32].

The wastewater that flows into drains corrodes and clogs sewer pipes. If allowed to flow into drains and rivers, it affects the quality of drinking water in hand pumps,

**Figure 7.3** Coloured streams in Mexico caused by the denim industry.
Taken from https://en.wikipedia.org/wiki/Water_pollution#/media/File:Nrborderborderentry threecolorsmay05-1-.JPG

**Table 7.5 Limits for process wastewater, domestic sewage and contaminated storm water discharged to surface waters, for general application**

| Parameter | Maximum value |
|---|---|
| pH | 6–9 |
| Biological oxygen demand | 50 mg/L |
| Chemical oxygen demand | 250 mg/L |
| Oil and grease | 10 mg/L |
| Total suspended solids | 50 mg/L |
| Heavy metals (total) | 10 mg/L |
| Cyanide (total) | 1 mg/L |
| Chlorine | 0.2 mg/L |
| Sulphide | 1 mg/L |

making it unfit for human consumption. However, the effluent treatment plant produces sludge containing chromium and other heavy metals which are highly toxic and should be treated and disposed of in a secure landfill.

## 7.3 Health hazards of denim processing

### *7.3.1 Spinning and weaving*

Although the spinning industry is an old and traditional industry, not much attention has been paid to environmental and personal hazards caused by this industry. Nevertheless, with increasing awareness toward to environmental concerns, it is important to study and resolve environmental issues and pay attention to workers' health and welfare with respect to this industry [34,35]. As a result, many rules and regulations were created by the standardized organization which helps to solve these issues. As far as the spinning and weaving industry is concerned, there are huge health problems and air pollution is generated.

The health hazards can be classified into two main categories: environmental health and physical health. Within environmental health hazards, fibre dust (e.g., cotton, Tencel, polyethylene terephthalate) and noise can cause severe health risks [36,37]. Fibre dust can fly around, contributing maximum health hazards to workers. Fibre dust is continual and persistent in the spinning and weaving industry. It is a mixture of fibre and fragments from seed coats and organic and inorganic dust; it is also exposed to particles of pesticides and soil [34]. Generally, exposure of fibre dust and other related particles leads to respiratory disorders among workers. The fatal disease byssinosis, commonly known as brown lung, occurs among people working in this industry as a result of excessive exposure to fibre dust. In addition, it creates irritation to the eyes and nose, resulting in allergies. Symptoms of this disease include tightening of the chest, coughing, wheezing and shortness of breath [38,39]. Byssinosis can be found in workers on their first working day (i.e., after a holiday or probably a week off). The level of byssinosis is very severe for the workers who are involving in operating carding, ring spinning, winding, and weaving. The dust can be retained in the lungs, which is very dangerous to the health. Proper dust collecting systems can ensure a dustless environment.

Additional humidification systems are unavoidable and are used to keep the work environment more comfortable. However, in humidification systems there is some risk to health when air is taken from the atmosphere; it contains a lot of contamination, which causes a health risk to workers. It can be minimized through properly working filtration units within the humidification systems.

In 1978, the US Occupational Safety and Health Administration (OSHA) published a new law that was enforced in 1982, which specified the upper limit of dust in the cotton industry at $0.2\,mg/m^3$. However, dust includes other contaminates such as vegetable matter, fibre fragments, fibres, fungi, bacteria, pesticides, nonfibrous matter and other materials during the harvesting and ginning process. This system is still followed globally. OSHA also stated that it is it is compulsory for the textile industry to protect employees from overexposure of dust [40].

Physical hazards are mainly caused by accidents such as fire and machines (e.g., the flyer in the speed frame). However, these can be minimized by proper training with precautions, improved process design, and protective clothes.

Noise is another health risk in the spinning and weaving industries. In the long-term, exposure to high levels of noise may damage the eardrums and cause hearing loss. It also creates other problems such as fatigue, changes in pulse rate, fluctuations in blood pressure and sleep disorders. Generally the lack of efficient maintenance to machinery is a major source of noise pollution. It causes serious health effects but organizations are not concerned because the effects are not immediately visible. A study of 100 employees was conducted in the spinning industry located in Ichalkaranji, India. The outcome of this study showed that 81.2% of employees were at risk for partially damaged hearing and a few needed treatment [41]. This study also extended to the weaving industry: almost 78.4% of employees were affected by hearing problems [41].

## 7.3.2 Sizing

Sizing for denim warp yarn is essential to reduce breakage during weaving operations. On the weaving machine, warp yarns are subjected to several types of action, i.e., cyclic strain, flexing, abrasion at various loom parts and interyarn friction. To achieve this process a lot of ingredients are needed, the most important of which is adhesive [42,43]. However, most of these ingredients are not certified by Global Organic Textile Standard 2.3.5. Typically, the sizing recipe for the denim warp is [43]:

| **For 100 L of liquor** | |
|---|---|
| Modified starch | 8 kg |
| Acrylate size | 4 kg |
| Textile wax | 0.2 kg |
| Size concentration | 8% |
| Sizing temperature | 85°C |
| Size pickup | 9%–10% |

For almost 100 L of water with 12.5 kg of ingredients and a size pick-up of 9%–10%, the remaining size paste will be in the bath; it can be removed during the rinse operation and becomes a pollutant.

## 7.3.3 Dyeing of denim warp

Around 3 billion metres of denim are produced worldwide. About 80% of this is still dyed in classic blue, black-blue or blue-black shade using of indigo and sulphur black dye. The remaining 20% is dyed in different shades using sulphur, indanthrene and another class of dyes [30].

### 7.3.3.1 Methods of dyeing

#### Rope dyeing

In rope dyeing, several ropes of yarn are dyed in a conventional manner; this is a traditional dyeing process. It is still in vogue and on occasion remains the preferred dyeing method.

#### Slasher dyeing

Slasher dyeing has become popular to achieve higher productivity and reproducibility. In this process, a warp sheet of yarn is prewetted, dyed and sized in a single passage through the machine. The latest machine provides additional facilities for mercerizing and quick oxidation in the same passage.

### 7.3.4 Processing of denim

The appearance of denim fabrics is continually being modified to suit fashion trends and cater to the demands of all age groups and their individual tastes. The changes in denim over the past few years have been drastic and dramatic, including the production of ecologically optimized denim fabrics and Green Jeans. Over and above the processing of denim in fabric form, it is possible to achieve several modifications when it is processed in garment form.

Most dyes used for the denim industry are recognized as respiratory sensitizers. They create occupational asthma during the inhalation of these respiratory sensitizers. These dyes also cause allergic symptoms such as a runny or stuffy nose, watery or prickly eyes, wheezing, tightness of the chest and breathlessness if dyes based on benzidine have the potential to cause cancer [4,8,9,22,39]. An alternative process to dyeing is listed in Table 7.6.

### 7.3.5 Softeners

Most denim fabric on the market has a stiff and hard handle that is attained by avoiding the desizing process; handling can be further modified by application softeners. A certain percentage of the denim market is for soft and washed denim. In such cases, the fabric is desized and washed between the singeing and preshrinking steps. Depending on the requirements, traditional indigo-dyed denim can be overdyed with or without desizing. Ecru denim can be dyed in different colours using normal place-dyeing processes. Heat setting after desizing is an important step for Lycra-containing denim fabrics. Mercerizing is also done to obtain flat-finish denim [44].

### 7.3.6 Brushing and singeing

The aim of these operations is to obtain a clean uniform and smooth appearance to the fabric. After brushing the denim fabric clean to remove lint, fluff and loose impurities, it is singed to burn away protruding fibres from the surface, which otherwise impart a fuzzy look. Thus, this process enhances colour and brightness. A mixture of gas and air in the right proportion is supplied to burners to give a blue-coloured flame of the desired length. The flame emerges from a set of burner nozzles and the width of this set is adjusted to suit the width of the fabrics. The fabric is drawn over the flame at a high speed so that the flame impinges on the fabric surface and burns protruding fibres from the surface as well as fibre ends present in the interstices of the warp and weft without damaging the fabric. During the singeing operation, it is necessary to use proper exhaust methods to collect the fumes, or it will create a huge health problem to the singeing operators while they inhale the fumes (Fig. 7.4).

### 7.3.7 Surface abrading (emerizing)

The feel and appearance of denim fabric can be considerably improved or modified by emerizing or sueding. The resulting soft, fluffy flannel effect achieved is extremely pleasant to wear. It is possible to refine the fabric's surface and take advantage of

**Table 7.6 List of possible issues with methods, chemicals and their alternatives for denim, and their practical limitations**

| Existing process | Possible issues | Alternative process | Advantages | Limitation |
|---|---|---|---|---|
| Sandblasting | Potential silicosis, lung cancer, life threat | Ice blast, eco-blast, light stoner by Jeanologia (Spain) | Environmentally friendly<br>Harmless process | High machinery cost<br>Requires qualified employers<br>Finished effects not as good as sandblasting |
| PP spray<br>PP brush | Potential health hazardous (possible damage to respiratory systems), environmental impacts | Light PP spray by Jeanologia (Spain), laser fading | Less pollution load<br>Natural effect<br>Reduced health issues | Laser is not suitable for all fabric (polyester may melt during laser treatment); proper safety precaution is required |
| Hand scraping | Flying lint and dyes particles, potentially health hazardous | | | |
| Whiskering (hand and pattern) | Potentially health hazardous | | | |
| Grinding (large area damage) | Flying lint and dyes particles, potentially health hazardous | Laser fading | | |
| Three-dimensional finishing | Potential skin cancer, environmental impacts | Zero-formaldehyde chemicals, reduced content of formaldehyde chemicals (dimethylol dihydroxyethyleneurea) | Improved worker and consumer safety | High cost for formaldehyde content<br>Low durability on effects |
| Bleaching | Adsorbable organically bound halogens issues, toxic in nature | Ozone bleaching | Enhance industrial productivity<br>Energy efficiency<br>Water consumption<br>Reduction of pollution load | Bleach effect not as good as chlorine bleaching<br>High machinery cost |

*Continued*

Table 7.6 **Continued**

| Existing process | Possible issues | Alternative process | Advantages | Limitation |
|---|---|---|---|---|
| Sulphur dyeing | Change colour of water, high biological oxygen demand, high chemical oxygen demand, lead to many health issues, environmental pollution | Reduction by glucose [74,75]. | Easily mineralized by bacteria<br>Less pollution load<br>Odourless environment | Not yet commercialized owing to less efficiency |
| | | Prereduced dyes | Less effluent load<br>Easy process | High cost<br>Maintaining proper process conditions |
| Indigo dyeing | | Electrochemical reduction [76] | Reduction of use<br>Strong oxidizing/reducing agent<br>In future it might be BAT | Requires special machinery<br>Poor fastness properties<br>High machinery cost |
| | | Prereduced indigo dyes (from BASF and Dystar) | | |
| Sodium hydrosulphite | | Catalytic hydrogenation process (by Dystar and BASF) [12] | Reduces need for hydrosulphite<br>Produce deeper colour | High chemical cost |
| Dyeing | | Coating [12] | Prereduced indigo with eco-friendly binder<br>Alternative for padding (warp dyeing)<br>Low water and energy consumption<br>Less pollution load | Lot-to-lot shade variation<br>Requires special machinery |
| | | Spraying [12] | Spraying on surface<br>Less water and energy consumption<br>Less pollution load | Unevenness effect<br>Requires special machinery |
| | | Foam applications [12] | Prereduced indigo used along with nitrogen as form medium<br>Less water and energy consumption<br>Less pollution load | Lot-to-lot shade variation<br>Requires special machinery |

*BAT*, best available technology; *PP*, potassium permanganate.

**Figure 7.4** Typical singeing machine for denim fabric.
Image courtesy: Prism Textile Machinery Pvt. Ltd, India.

current fashion trends such as adopting a 'peach skin effect'. Emerizing can be carried out for all types of denim. It is more popular on ecru denim. Similarly, during a surface-abrading process such as singeing, a lot of protruding fibres and dust will fly around the machine, which is highly dangerous to the operator if they inhale them. Thus the company must provide proper protective clothing to avoid such health problems.

## 7.3.8 Desizing

The purpose of desizing is to remove size particles from denim. Traditional desizing is performed using detergent and soda ash or hydrochloric acid or oxidative desizing agents, which are associated with many limitations and drawbacks. Because of the uncontrolled and nonspecific reaction, cellulose material is damaged and loses its strength. With the introduction of the enzyme-based desizing process, the limitations and drawbacks of traditional desizing can be overcome. Enzymatic desizing process is performed by α-amylase enzyme [45,46]. Most people believe that enzymes are environmentally friendly and do not creating health hazards, but they can create severe problems. The Enzyme Technical Association published the booklet *Working Safely With Enzymes*, in which they specified that care should be taken to avoid skin contact and inhalation of aerosols. Enzymes can be used safely with no adverse health effects through the use of good work practices, engineering controls, and appropriate personal protective equipment [47]. Prolonged skin contact may create irritation that can increase with time and the concentration of enzymes. Skin irritation is most likely to appear in body areas where perspiration occurs, i.e., hands, armpits, groin and feet, and around tight-fitting clothing areas such as the cuffs, waist, collar and facial areas in contact with face masks. Some enzymes do not cause much irritation to the skin; however, all kinds of enzymes create irritation and severe health problems when they are in contacts with the eyes. Therefore, safe handling practices should be taught to operators as well as whoever is handling enzymes. Also, it is advised that operators use self-protective clothing such protective wear, hand gloves, safety goggles and face masks [48].

Most enzymes are water soluble; therefore, water should be flushed on the skin where it may be exposed to enzymes. Later it may be is necessary to wash the skin again with mild soap and water. If the clothes are exposed or contaminated with enzymes, it is necessary to remove them and shower and change the laundered clothes. The contaminated clothes can be washed with normal procedures. If workers inhale enzymes, take the workers out of the area and consult a physician. In the case of eye contact, rinse the eyes for at least 15–20 min and then consult an ophthalmologist [47,48].

### *7.3.9 Bleaching technology*

After a satisfactory pattern is produced by stone or enzyme washing, the general colour depth of denim is adjusted to match the customer's sample using a bleaching process. The purpose of bleaching indigo-dyed denim fabric is to decolourize the indigo dye to a certain level to obtain a desired shade. In general, sodium hypochlorite (NaOCl) is used as a bleaching agent. It is a harmful chemical and contains chlorine. These chemicals belong to adsorbable organically bound halogens as well as VOCs [49].

Sodium hypochlorite solution is clear and slightly yellow or green with a strong chlorine odour. It is cited in the US Department of Transportation, International Agency for Research of Cancer, and US Environmental Protection Agency as a hazardous substance [50]. Sodium hypochlorite may release chlorine and hypochlorous acid when it acidifies. However, it creates significant damage in low concentrations as well. Therefore, it creates pulmonary complications such as acute respiratory distress syndrome, which usually result from aspiration and often contribute to death. An excess amount of sodium hypochlorite can be used in denim bleaching; it is mandatory to provide safety training to people who handle these chemicals.

After the bleaching process, residual hypochlorite needs to be removed from the denim. If not, it may damage the fibres, particularly those with a low pH (by the absorption of $CO_2$ gas from the atmosphere). Residual hypochlorite is also called as residual chlorine and it can be removed by treatment with reducing agents. This process is also known as antichlor treatment. In denim, sodium metabisulphite or thiosulphate is used as a reducing agent. When it mixes with water, it releases $SO_2$, a pungent, unpleasant-smelling gas that can cause breathing difficulties and harm humans. When it is in direct contact with skin it may cause burns, irritation, itching and redness.

Symptoms of sodium hypochlorite poisoning include:

- burning red eyes
- chest pain
- coma
- coughing (from fumes)
- a gagging sensation
- low blood pressure

- pain in the mouth or throat
- possible burns to the oesophagus
- skin irritation on the exposed area, burns, or blistering
- shock
- slow heartbeat
- stomach or abdominal pain
- vomiting

Limitations to sodium hypochlorite bleaching are that it:

- is harmful to human health
- causes corrosion to stainless steel
- requires antichlor treatment
- is not possible to remove chlorine entire from denim
- can cause severe environmental pollution from chlorine in the effluent

### 7.3.10 Alternatives to sodium hypochlorite bleaching

Based on the use of innovative technology and methods, there is alternative process which is also eco-friendly in the process and not does not cause health concerns. Most such processes are still at the laboratory level and unpopular in the industry, but this scenario can change in the near future owing to awareness about global warming and its potential. Among innovative technologies used in denim are enzymatic bleaching, ozone bleaching and laser engraving. These alternatives may replace sodium hypochlorite but none can provide the same effect. There have been many research articles published in the field of ozone bleaching, but it is not as popular commercially owing to the technical and economical drawbacks. Obviously, it cannot produce the same effect as sodium hypochlorite, and it requires a huge investment. The consumer is not interested in buying denim products with a higher cost and the manufacturer has to reduce process costs, which results in the use of harmful chemicals.

### 7.3.11 Enzymatic bleaching

Laccase is the newest enzyme class to be introduced for denim bleaching and other finishing processes. Laccases are multicopper oxidizers that catalyze the oxidation of indigo and produce bleaching effects. Laccase alone cannot produce bleach effects; it requires a mediator. Therefore, the mediator is a low–molecular weight organic compound which mediates electron transfer from indigo to molecular oxygen. In the presence of an aqueous medium, the enzyme is oxidized and attacks the mediator, converting it into free radicals. The free radicals generated then attack the indigo and convert it into oxidized products. The benefits provided by laccase processing include minimal loss of strength, reduced time, reproducible performance, reduced water consumption and a safe alternative to hypochlorite and other bleaching technologies [51]. Conventional hypochlorite bleaching of denim is cheap, fast and efficient, but it is harsh to the environment and denim. Although it has many benefits, this belongs to the enzyme group and the handling procedure is the same as with other enzymes which have already been explained.

### 7.3.12 Ozone fading

Ozone ($O_3$) is a reactive gas made of three oxygen atoms. Normal oxygen gas ($O_2$) consists of two atoms. It is normally created in the upper atmosphere by the action of sunlight on normal oxygen. In the industry, ozone is formed by electrical charges such as lightning, which turns normal oxygen into ozone. Because ozone is an oxidizing agent it has a bleaching property [52]. Ozone can fade indigo-dyed denim by an oxidation process. Ozone can bleach faster than other chemicals and it helps clean stains in 3 s. The optimum time for ozone bleaching is 15 min whereas conventional bleaching requires 30–50 min. Therefore ozone increases the production per shift. Ozone requires two or three rinsing processes, compared with six to seven washes using traditional methods, so it saves a huge quantity of water. Furthermore, replacing some traditional finishing with ozone reduces effluent, including the sludge pumice stones create.

Although it has several advantages, if it inhaled it will damage or kill living cells. Thus companies should consider safety to be an important factor to save living organisms (including humans). It is therefore mandatory to maintain proper ozone bleaching systems composed of a filtered/refrigerated air compressor, oxygen concentrator, ozone generator, exposure chamber and ozone destruct unit. Specific equipment features affect the safety and reliability of ozone denim processing equipment.

The Canadian government reports that the exposure of 0.1 parts per million (ppm) of ozone causes minor irritation to the eyes, nose, throat and ears; it also causes headaches. They reported that 50 ppm can cause death. For ozone bleaching, industries use 5000–7000 ppm; it is a more highly dangerous process than any other. OSHA standards set the maximum allowable exposure of workers operating ozone technology in the United States at 0.1 ppm per 8 h, time-weighted average, and 0.3 ppm per 15 min. Therefore, it is mandatory to teach safety features, and safety practices should be of paramount importance in selecting, installing, operating, and maintaining ozone finishing equipment in any facility for workers who are operating ozone gas and ozone finishing machinery. For ozone safety processing, there are three types of equipment features: fill mechanisms, door locks and monitor/alarm systems.

*Fill mechanisms*: Whenever a breach occurs, by using a negative fill system it can suck all of the air into the exposure chamber instead of spilling ozone into the workplace. This ozone finishing equipment can absorb ozone gas into the exposure chamber with negative force, which reduces accidents to workers.

*Door locks*: The exposure chamber is allowed to open the door when the ozone falls below 0.2 ppm. The exposure chambers are locked with a UV light-based ozone sensor. These setups can help to protect workers, but there is no assurance that ozone has been evacuated before a worker opens the chamber.

*Monitor and alarm systems*: It is mandatory to keep a dual alarm (audio and visual) and it should be linked to the ozone monitors. It ensures that workers are alert during accidents. All of these systems should be perfectly maintained to ensures workers' safety.

### 7.3.13 Laser fading

Laser fading on denim can be a strong alternative compared with conventional physical and chemical processed used for the aged and worn look (vintage) on denim fabrics [53]. The mechanism of this process thus: The laser beam decomposes the dye on the surface and the resulting dyes turn to vapor and are vented away. Fading can happen only in the place where the laser beams are in contacted with the fabric. Two types of laser are used: solid and gas based. The wavelength for solid-based lasers is 1 and 10 μm for gas-based [54]. The efficiency of fading depends on the wavelength, power density and pulse width of the laser beam [55]. Laser fading is environmentally friendly compared with other fading processes such as sandblasting and acid washing [56].

#### 7.3.13.1 Health and hazards concerned with laser fading process

##### Electric shock

Rather than laser exposure, most people were killed by electrocution as a result of the laser electronics [57]. If some workers are affected in this way, immediately provide first aid and do not allowed them to perform further work with the laser. During maintenance, the power supply should be unplugged from the electrical outlet [54].

##### Potential injuries to the eyes

There is great potential to damage the eyes during the fading process (i.e., laser beams) [58]. Stefan Wyrsch reported that some lasers have the chance to result in severe damage to the retina [59]. In particularly, class 4 can damage the interior tissues of the eyes; sometimes loss of vision may occur. The level of injury may vary by class of laser and duration, according to the American National Standards Institute and OSHA.

##### Injuries to the skin

Skin injuries are less severe than those to the eyes. However, the chances of exposure to skin are higher than those to the eye because of its greater surface area [57]. There should not be a large area of skin exposed during normal laser working conditions. Laser-exposed skin injuries can be divided into two types: (1) thermal and (2) photochemical induced. Direct contact or exposure to lasers can create thermal injuries. However, these will be painful but they are not serious. Proper beam management and avoiding the laser beam will prevent such injuries.

Hazard control methods can be divided into three sections: (1) administrative controls (labels, signs, standard operating procedures, etc.), (2) engineering controls (barriers, blocks, etc.) and (3) protective controls (eyewear, uniform, etc.) [54].

### 7.3.14 Water jet fading technology

A hydro jet creates a type of surface finishing on denim by using water jet nozzles. This treatment can enhance the surface, texture, and durability of denim. The degree of wash-down effects depends on the nozzle's pressure. The results obtained with blue indigo are extremely good. This process requires no source of chemicals and it is pollution free. By using a water recycling system, the technique can be employed in economical and environmentally friendly denim processing.

### 7.3.15 Dry finishing

Denim finishing is a finishing method applied to a garment using new technologies and equipment that enable the desired results to be obtained. To finishing denim fabrics, a range of treatment methods is used. They all aim at new possible effects in the fabric's appearance, such as a mill wash or rinse wash, stone wash, moon wash, sand wash, bleach, overdyed look, damaged look, scrubbed look, sandblast, PP spray, whisker effects and 3D effects. Destroying denim is as much an art as it is a technique. The direction in fashion is varied across all areas, particularly in denim dry finishing, from worn or torn to wrinkled or pressed; these trends can appeal to a wide range of consumers. Meanwhile, most of these processes are harmful to humans as well as the environment [60].

### 7.3.16 Denim hand sand/scraping

Hand sand is a step which is generally done to the rigid form of garments to achieve a distressed look. The location can be on the front thigh and back seat or it can be an overall or global application as per the standard. Emery paper is used to scrape the garments in a particular placement and design. The purpose of this process is to create a worn look to the garments. The most important factor is to select the right number of papers according to the fabric strength and intensity needed, feathering or merging the white sanded part of the dark blue area in such a way that it looks natural and not artificial. Scrubbing can be done on inflated rubber balloons for a better effect (horizontal or vertical: it is up to the operator's convenience); it can even be done on a plain wooden board of garment size, and hand pressure should be uniform to obtain better results (Fig. 7.5).

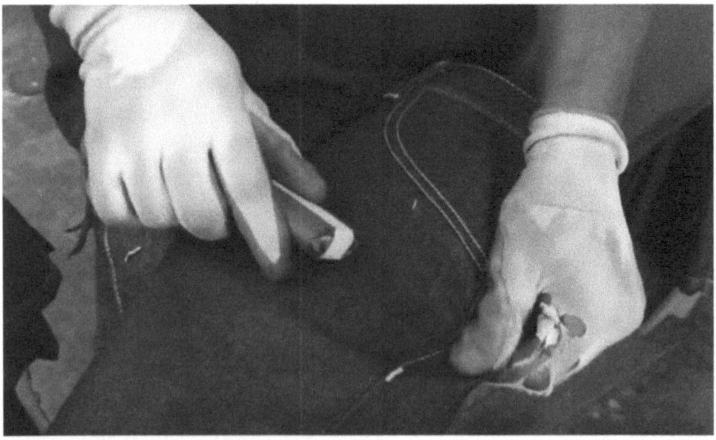

**Figure 7.5** Hand scraping process.
Image courtesy: Cold Stone Industries, Thailand.

### 7.3.17 Sandblasting

Sandblasting is the process of scrubbing the garment by blowing high-speed air mixed with very fine particles of sand. When the surface area of the garment is blasted, white cotton appears beneath the blasted area and the effect is similar to that of worn jeans [60].

During the sandblasting process, the denim is smoothed, shaped and cleaned by high-pressure microparticles generally called sand. This process is fast and cheap and produces an attractive effect (worn-out) compared with other methods. However this fashion comes at a price: the health and even the lives of sandblasting workers. Sandblasting is deadly.

In manual sandblasting, high-pressure sand passes through the denim surface, which leads to surface bleaching effects. Compressors are used to achieve optimal pressure; however, this process is done in the absence of sealed blasting cabinets and ventilation, and workers are exposed to the silica particles (tiny particles of blasted sand) (Fig. 7.6).

These silica particles are very tiny and if the workers inhaled them, the result may be severe respiratory problems. Long-term exposure may lead to fatal diseases such as silicosis and lung cancer. Most sandblasting units in Bangladesh and China use the manual type. The sandblasting process may control mechanical-based sandblasting units. Although it is controlled, many organizations do not follow proper safety practices, which results in exposure to silica dust because the work is carried out in an unsealed environment with little protection to workers, who are provided with inadequate safety equipment. Therefore, if workers are exposed for a long time, the procedure creates an extreme health hazard and can cause death within months or years of the employee starting work as a sandblaster [61]. The silica particles are so tiny they are invisible to the naked eye (Fig. 7.7). The body is unable to expel the silica particles, which cause disease such as silicosis; the particles penetrate the pulmonary alveoli and connective tissues. This begins to impair lung capacity and workers' ability to oxygenate blood. Symptoms may be shortness of breath and even coughing.

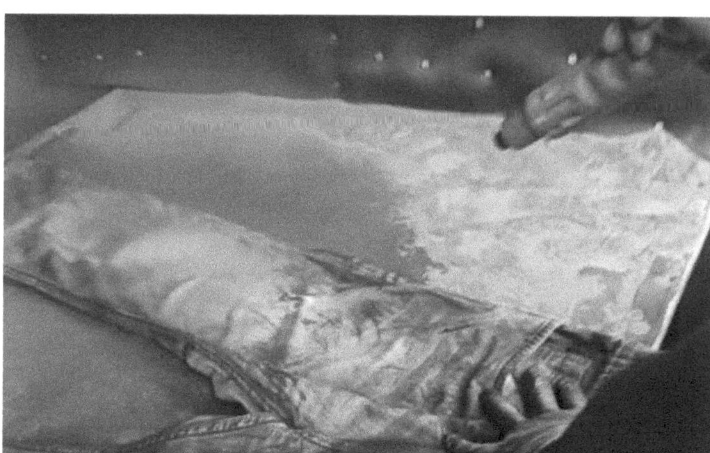

**Figure 7.6** Sandblasting process in a poor working environment.
Image courtesy: Cold Stone Industries, Thailand.

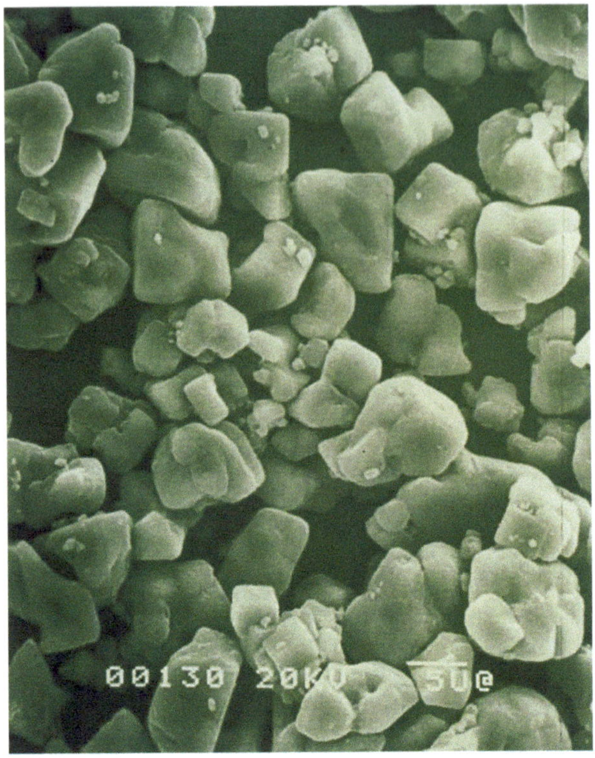


**Figure 7.7** Microstructure of silica particles used in sandblasting [63].
Cappuccio S, Why a global ban on sandblasting in the garment industry?, ETUI Conf. "Workers' Prot. Chem.; 2011.https://www.etui.org/content/download/1736/17375/file/Why+a+global+ban+on+sandblasting+in+the+garment+industry.pdf.

Sometimes the level of the disease increases even after workers have left the industry. This puts an additional strain on the heart, eventually leading to death. However, the progress of silicosis can be slowed if symptoms are diagnosed at an early stage [62].

In this situation, the European Union has strictly regulated sandblasting; the clothing industry has largely outsourced production to as yet unregulated regions such as Turkey, Bangladesh and China. In 2005, Turkish doctors diagnosed silicosis among denim workers. Akgun et al. studied this issue and published their results in 2005 [64]. In 2009, the Turkish government implemented a ban on sandblasting with the help of these results, and followed by more data. In autumn 2010, the Killer Jeans campaign was launched to create public awareness to abolish sandblasting process in the industry. However, few of brands provided clear enough information to implement a ban on the sandblasting process. None of them took responsibility for treating the workers under their supply chain. Figs 7.8 and 7.9 shows the X-ray results of normal and sandblasted workers.

### *7.3.17.1 Noise in sandblasting*

In the denim industry, the problem of noise in various departments such as sandblasting, PP spray and grinding is a cause for serious concern. The highest noise levels are to be

(a) (b)

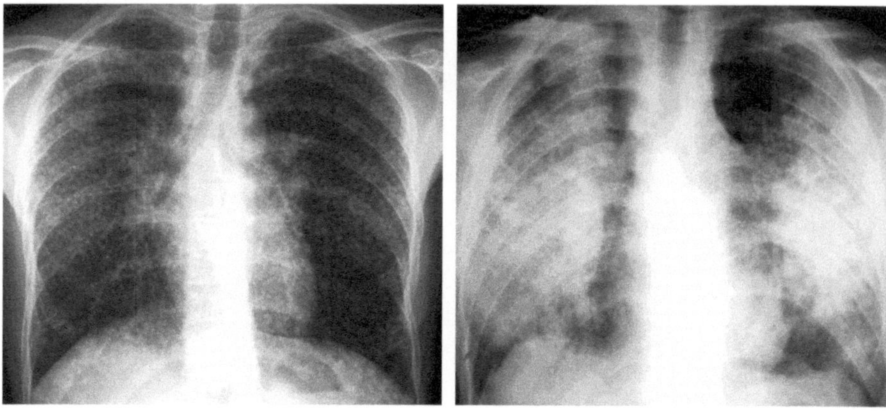

**Figure 7.8** Chest radiographs belonging to two different ex-denim sandblasters, showing (a) bilateral small, widespread nodules, and (b) large type C opacities.
Reproduced with permission of the European Respiratory Society ©. Eur Respir J November 2008;32(5):1295–303. http://dx.doi.org/10.1183/09031936.00093507.

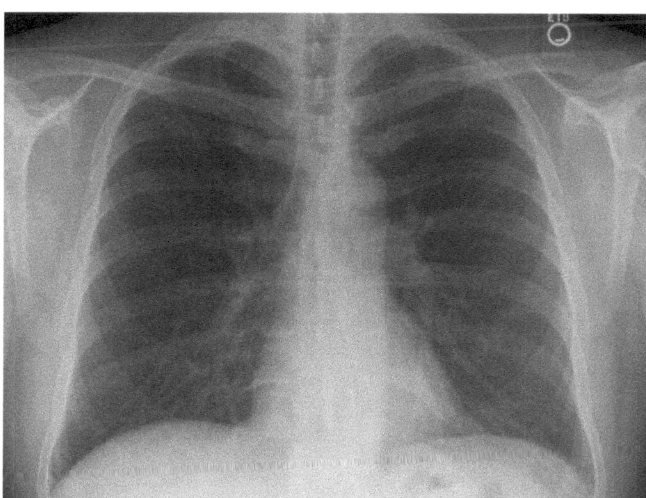

**Figure 7.9** Normal chest radiograph of a healthy person.
Image source: Wikipedia.

found in sandblasting, where operators are exposed to levels as high as 94–100 A-weighted decibels. However, it is possible to correct the acoustics of the working room.

### 7.3.18 Whiskers

The idea of whiskers is taken from the worn lines and impression patterns generated by natural wear on the hips and front thigh area. On old jeans, a number of patterns can be found consequential to fabric in the body shape of the user from a sitting posture.

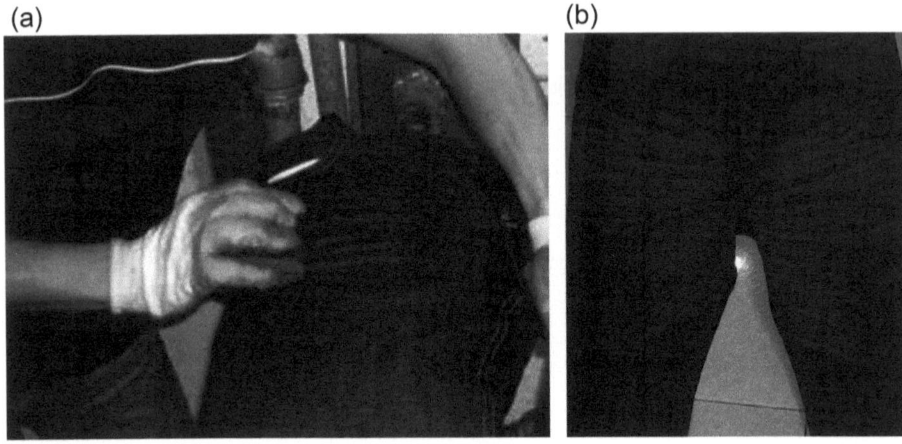


**Figure 7.10** (a) Typical methods for producing whiskering effects on jeans; (b) whiskering effects on denim.
Image courtesy: Cold Stone Industries, Thailand.

Garments are mounted on balloons and filled with air to obtain impressions of whiskers. The garment is scrubbed over the pattern carefully with sandpaper on engraved pattern lines. This operation requires highly skilled labourers who can handle uniformity and sequence in whisker line. However this method is more successful in industries where production is consistent with articles i.e., similar articles are produced in large quantities. This method is famous for high quality and cost-effectiveness [60] (Fig. 7.10).

### 7.3.19 Grinding

Edge distressing or grinding is the process of destruction on jeans particularly on bottom, pockets, and waistband. It is a kind of worn out style. This effect can be created on jeans by simple grinding machine. Mostly, the stone grinding machine is used as that of the wooden industry. Operators rub the edges of the rotating stone wheel to get the effect (Figs 7.11 and 7.12).

### 7.3.20 Ripping, pilling and cuts

These effects are created on denim by using sharp-edged tools such as line clippers, blades or even knives. To get this effect the warp yarn is cut and the undyed weft yarn is left in place, which creates a beautiful worn-out effect (Fig. 7.13) [60].

During the whiskering, ripping and grinding processes, the indigo dye and a small amount of fibres (short fibres) come out. Generally the denim warp is dyed with a rope-dyeing technology; dyes may be coated on the surface. The indigo has poor rubbing (abrasion) fastness properties. During these processes, microparticles of indigo dye and fibres will fly about, as discussed. Dyes contain heavy metals, which are highly hazardous to workers who inhale them. Thus companies must maintain proper ventilation facilities and provide protective wear, gloves, goggles, etc.

**Figure 7.11** Pen-type grinding process on denim.
Image courtesy: Cold Stone Industries, Thailand.

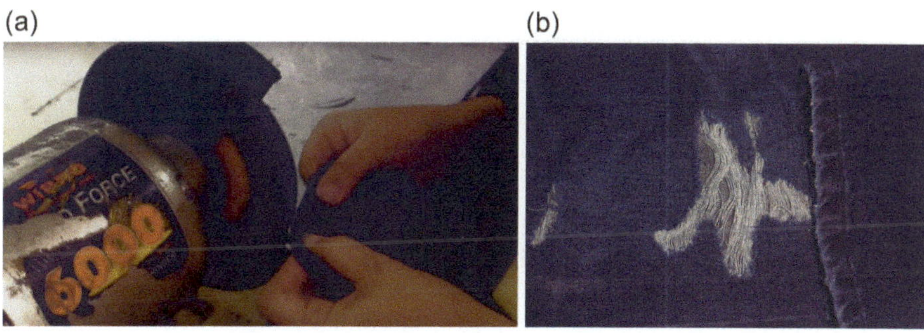


**Figure 7.12** (a) Grinding process (workers need to use two face masks and protective gloves for hands); (b) grinding effect on denim.
Image courtesy: Cold Stone Industries, Thailand.

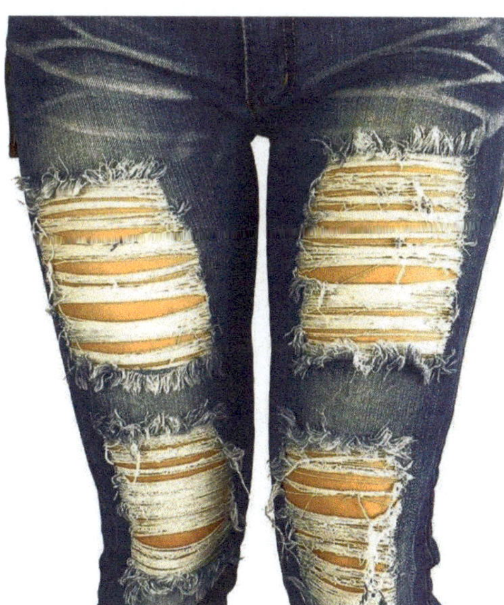

**Figure 7.13** Ripped effects on denim.
Image courtesy: Cold Stone Industries, Thailand.

### 7.3.21 Potassium permanganate spray

PP spray appears as pink on a garment when it is fresh and turns to muddy brown upon drying. The colour change results from the formation of magnesium dioxide. The garment is hung in the open to dry after PP spray, and when the PP turns colour completely it is considered to be ready for the next process. Magnesium dioxide has to be removed through a neutralization process. Sodium metabisulphate is the most commonly used neutralizer [60]. A number of products are available on the market for the neutralization process according to effect required on the blasted area, which depends on the neutralization. *Thio* groups cannot perform like *meta* groups on magnesium oxide (Fig. 7.14).

#### 7.3.21.1 Potassium permanganate brushing

In this method, the solution of PP is applied to the denim with the help of a paintbrush rather than a spray gun. The effect is different than that with the spray technique.

#### 7.3.21.2 Health hazards

PP is a powerful oxidizing agent; it can dye (colouration) most organic materials including skin. When the skin is in direct contacted with PP it may cause irritation or burning, redness, pain and itching. In the case of eyes it may cause damage or permanent loss of vision. Exposure to PP may irritate the respiratory tract and ingestion may cause nausea, diarrhoea, and other gastrointestinal issues. Doses of 10 g may cause severe health problems such as cardiovascular collapse, irritation in the nose and lungs, and kidney damage, or may cause death.

As industrial people have stated, 25-kg PP vessels (storage container) may be equal to one powerful bomb. Thus it is necessary to store it in properly. When storing PP, workers should be aware of safety procedures. Generally PP violently combustive when it is in contact with oxidizable substances including simple alcohols.

(a) (b)

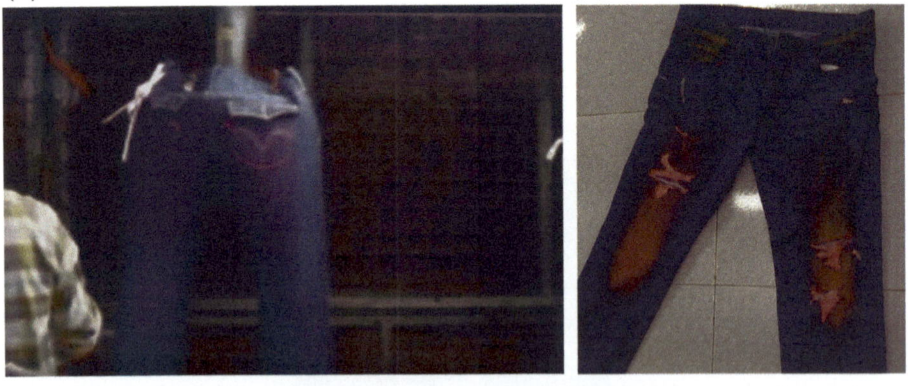

**Figure 7.14** (a) Typical potassium permanganate spray units without proper suction systems; (b) potassium permanganate spray denim.
Image courtesy: Cold Stone Industries, Thailand.

It has an unpleasant smell and is difficult to work with for a long time. Therefore this process requires proper ventilation systems, protective gear (such as goggles, masks and gloves) and adequate procedural training for factory workers; use of these chemicals is extremely risky to their short-term and long-term health. If the ventilation is not working properly, the spray fills the air and workers develop a rash. Yet there are no alternatives to the and research and development are progressing in the field.

### 7.3.21.3 *Sodium metabisulphite*

Sodium metabisulphite is another bleaching agent; it is used as a neutralizer to turn cloth white or close to white. The results are the appearance of sun bleaching or fading on denim [65]. Inhalation of sodium metabisulphite may irritate the respiratory system and initiate coughing and shortness of breath. In some case, it may cause an allergic reaction, asthma or related disease. Ingesting of sodium metabisulphite may cause nausea, vomiting, diarrhoea, abdominal pain, circulatory disturbance and central nervous system depression. In case of skin contact, it causes irritation and itching and the skin turn red with severe pain. Direct contact with the eyes may become worse, with irritation, pain, stinging, tearing, redness, swelling, corneal damage and blindness. These effects may be irreversible [66]. It becomes volatile when mixed with PP. Phosphoric acid has been used as an acid booster. It is a kind of inorganic acid and creates a bad odour when it is mixed with PP. If workers inhale these mixes, they feel burning in the respiratory system and immediately start coughing. The higher concentration may affect the throat and lead to lung cancer. There are alternative boosters but most of these use acids.

### 7.3.21.4 *Three-dimensional effects*

This process can be done by spraying or dipping garments into the resin and other additives. After applying resin solution in the right proportions, a manual design is made as needed on the thigh, hip and back knee area to get a 3D effect (Fig. 7.15). After it is made it is should be manually dried with a hot press or hair dryer, and then it must be cured in an oven at the correct temperature and time as mentioned in the resin product manual. Regarding health concerns, most resins are formaldehyde based; although the use of formaldehyde has been reduced, it has yet to be eliminated. Although the most popular resin used today is dimethylol dihydroxyethyleneurea, which has the least formaldehyde content, but it is made using formaldehyde. The exposure of the skin to formaldehyde may cause hypersensitivity, leading to the development of allergic contact dermatitis. This is caused by the accumulation of sweat which leaches out more formaldehyde from the compounds in contact with the skin, in addition to friction. Women are more often affected than men by this condition. Most European countries and Japan have national legislation restricting the presence of formaldehyde in textile and denim products [67,68].

## 7.4 Overall working conditions

Overall working conditions in denim industries are poor. Most people experience deplorable working conditions. (More than 1100 garment workers were killed in a factory collapse, Savar building collapse, occurred in Bangladesh, 2013, which was the worst

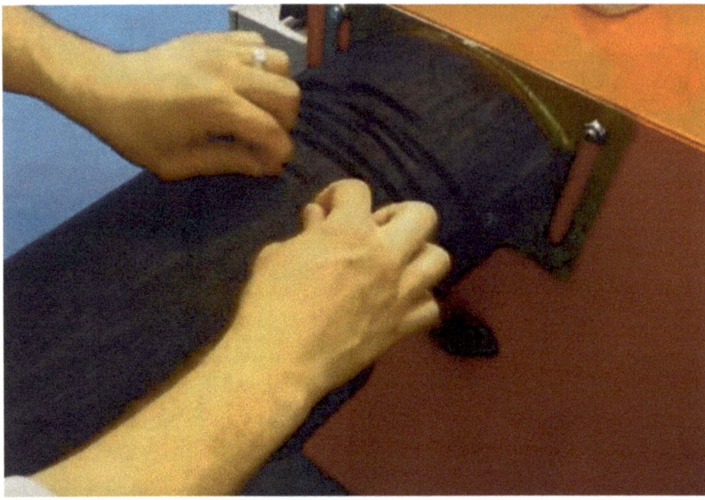

**Figure 7.15** Typical process for three-dimensional finishing process on denim (workers need to use protective gloves for hands).
Image courtesy: Cold Stone Industries, Thailand.

tragedy.) After 2013 it briefly focused global attention and many brands made some strict laws [69]. Although some denim factories provide a large proportion of orders to home workers, typically they receive salary based on piecework (i.e., pieces for money). In this case, mostly children and women are involved in this type of work in their home, which itself offers poor working conditions (Figs 7.16(a,b) and 7.17).

### 7.4.1 Ergonomics in denim industries

Ergonomics is a science that focuses on designing a job for workers. Too often employees perform job tasks that expose them to potential injuries and illnesses because of the poor design of a workstation or tool they are using. Generally, ergonomic changes are made to improve the fit between the demands of the tasks and the capabilities of the employees. An ergonomically designed job ensures that a tall worker has enough space to perform a job safely; similarly, a shorter worker can reach all of his tools and products without stretching beyond a comfortable and safe range. This may require employees to work in awkward postures, perform the same motion over and over or lift heavy loads, all of which may cause work-related musculoskeletal disorders. These injuries often start as minor aches and pains but can develop into disabling injuries that affect activities of daily, living such as doing laundry, enjoying hobbies (playing field games, trekking, etc.) and even the ability to pick up children. Ergonomics aims to prevent injuries by controlling risk factors such as force, repetition, posture and vibration that can cause injuries to develop. Ergonomic hazards have been identified as a major source of occupational injury and illness. Risk factors are present at varying levels for different jobs and tasks. Generally, the greater the exposure to a single risk factor or a combination of risk factors, the greater the probability is of an ergonomic injury or

(a)
(b)

**Figure 7.16** (a,b) Home workers working with denim. Child labourers are engaged in work in a poor work environment.
Image courtesy: Greenpeace.

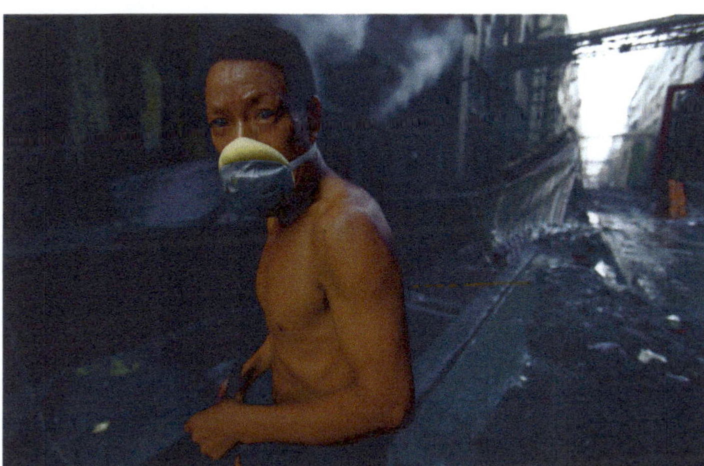


**Figure 7.17** Working conditions in Xintang.
Image courtesy: Greenpeace.

illness. Cumulative trauma disorders (CTDs) are a class of musculoskeletal disorders which arise from repeated biomechanical stress owing to ergonomic hazards. CTDs involve damage to the tendon sheaths and the related bones, muscles and nerves of the hands, wrists, elbows, shoulders, neck and back [70–72].

Injuries and illnesses occur among textile and apparel workers [73]:

- 70% of sewing machine operators using foot controls report back pain; 35% report persistent lower back pain.
- 25% have suffered a compensable CTD.
- 81% reported CTDs to the wrist.
- 14% reported CTDs to the elbow.
- 5% reported CTDs to the shoulder.
- 49% of workers experience neck pains.
- Absenteeism increases as working conditions worsen.
- Loss of the workforce as a result of injuries or high turnover is associated with working conditions.
- Hand sewing and trimming are stressful to upper limbs.
- Stitching tasks are associated with pain in the shoulders, wrists and hands.
- Ironing by hand is associated with elbow pain.
- Garment assembly tasks are associated with CTDs of the hands and wrists.
- Foot-operated sewing is associated with pain in the back.

## 7.5 Case studies

**Mr Nawaz** (32 years old, Chittagong, Bangladesh) has been working as a sandblaster for more than 6 years. He has experienced breathing problems for 3 years. Before entering this profession he was a farmer and his health was extremely good. He felt that the breathing problems increased daily and decided to consult with a doctor. In April 2016, he underwent an X-ray test of the chest and a spirometer test with at pulmonologist's office. The X-ray showed the slight formation of silicosis and the spirometry results (forced expiratory volume 1/forced vital capacity of around 53%) showed that he was affected by lung disease. The doctor said that he affected was by silicosis and very difficult to cure. He has two daughters and one wife; no one is working except himself, and now their life is unpredictable. There are so many workers like him. Although the sandblasting process is banned, such activities are continuing in small-scale industries.

**Mr Ravindra Gowda** (30 years old, Bangalore, India) was working in the cutting department of the garment unit; usually his factory exports denim garments. In his department, the workers do not commonly use masks. When an inspector inspects the units, they are forced to wear a mask. For 7 years he worked as a cutter man and no one advised him to use a face mask and protective gloves during his tenure. He was not aware of the need to use protective wear. The amount of dust particles in the cutting department is pretty high and is mixed with air. Therefore, it polluted the cutting room environment, which everyone who worked in the department was forced to inhale. As a result of the prolonged inhalation of dust particles, his lungs became affected.

He was eventually unable to breathe properly. Inhalation disorders may increase, if he continues in this profession, he is in a dilemma about changing the profession. Although he can change it, it is not possible to cure pulmonology problems.

**Mr Nguyen Van Ha** (43 years old, Biên Hòa, Viet Nam) worked for 12 years in the garment industry. He was transferred into the PP spray unit and became ill after 4 years of working as a PP sprayer. He experienced a lot of severe breathing problems. Whenever he coughed, he could feel a strong pain in the chest. He decided to consult with a doctor. He took medications for more than 9 months but there were no improvements in his health; rather, it made him worse than before. He was unable to afford to ask for a second opinion from another doctor. He has stopped working and stayed home unemployed, which has affected him economically. He continues to experience severe chest pain and weight loss.

**Mrs Shanti Ramanna** (28 years old, Bangalore, India) is from Hiriyur, Karnataka. She and her husband came to Bangalore 4 years previously. Their financial conditions were unfortunate and their neighbor helped them to get a job in a garment factory because her husband had an accident and was injured when they lived in the Hiriyur. After he regained his health, he joined the garment industry as a helper and she joined as a sewing machine operator. According to them, their family survived with the money both of earned from the garment industry. She was located on the fourth floor, where a minimum of 1500 workers work without proper ventilation and toilets. They were not given a mask to use. A tragedy occurred in 2014. The lunch break was 30 min, and within a stipulated time she had to go to the dining hall, which is located on the ground floor. The floor steps are narrow; she fell on the steps and needed to be admitted in the hospital. Her right side was paralyzed. Day by day she has become mentally ill, because these incidents made her physically, mentally, financially and so-called destruct situations.

**Mrs Hasana Salim** (31 years old, Dhaka, Bangladesh): Her husband was a farmer. They do not own agricultural land, but they were sharecroppers. They have two sons. One had accident, so they borrowed a huge amount of money from neighbors for his treatment. They were not able to pay the money because they earned less income from agriculture. This situation made them move to the garment industry and begin as garment workers. Both were working in the same profession but in different factories. One particular day, the electricity suddenly failed and they found out that the factory was on fire. Owing to improper ventilation, only a few of workers escaped and were rescued from the fire. Almost 10 people died. She was rescued with both external and internal injuries. When the doctor diagnosed her, they found that her uterus had become contaminated with sharp material as she tried to protect herself during the incident. The doctor advised surgery but she had no money to afford the treatment. This is just an example from so many people who were injured.

**Mr Somu Reddy** (26 years old, Bangalore, India) was working as a scraper for more than 5 years. He comes from a poor family from a neighboring state. He has to work hard to fulfil the financial needs of his family. Since his father passed away he has had to go to work to pay all the money borrowed for his father's treatment. Therefore, he moved to Bangalore and joined the denim industry. He started work 5 years previously, joining this industry as hand scraper; obviously it is a manual process which creates the particular effects on denim. While scrubbing takes place, the lint of fibres and microparticles of dyes are exposed and he has to inhale this contaminated air. The fibre dust creates breathing problems, and of course the dyes contain heavy metals, which are potentially life-threatening. He said that the company does not provide a mask consistently. One year earlier, he felt the symptoms of chest pain, burning red eyes and skin irritation. As a result of prolonged inhalation, he used to cough a lot. He was not in a condition to afford money for treatment for this issue, just like other workers. He experiences a lot of difficulty breathing. He is mentally, physically, emotionally and socially affected owing to this profession.

**Mr Azeem** (24 years old, Dhaka, Bangladesh) comes from a small village near Dhaka. He is working as an operator and rubs denim on a rotating stone to give it a grinding look. He has worked for a more than 3 years and also works overtime to pay for his wife's medical expenses. He never has time off and uses sharp-edged tools such as a line grinding wheel. He came to work as usual one day when an incident happened. While he was working at grinding, suddenly his leg moves toward the grinding machine and he met with an accident: his fingers were severely injured. He screamed, fell down and lost consciousness. When he opened his eyes he could not able to digest that he lost his two thumbs in this accident. He was depressed that no one was there to take care of his wife and family. He was not able to work anywhere to fulfil his basic needs; this affected him and his family very much.

In the denim-processing line, there are many processes which create environmental and health hazards. Researchers from major chemical manufacturers are trying to find alternatives. Although some alternatives are available, they are not economically feasible. Moreover, consumers should be ready to buy sustainable denims although they are costly. Per the product life-cycle philosophy, new sustainable products are always costly, owing to the huge investment in research and development for the invention of alternative processes. Therefore, consumers should encourage the purchase of such sustainable denims, which will start to reduce the process cost (i.e., when the market potential increases, the product price will decrease). Alternative processes for making sustainable denim are shown in Table 7.6, which describes practical limitations.

## 7.6 Challenges, future trends and conclusions

It is imperative to resolve all kinds of health problems, which require many challenges. In this case, the first responsibilities go to the designers, who should create dark denim. Of course, it does not require effects; therefore, there is no need to use a harmful process. The brands should work on this and motivate designers to develop dark denim. Also, they are responsible for broadcasting (marketing) among customers. Once customers know the facts, they will purchase this type of denim automatically.

Nevertheless if customers like to purchase such kinds of effects, it may be possible to use the robotics. Of course, it would require some time with more investment, which would increases the overall cost of the products. Therefore, customers must be ready to pay for it. Pizarro invented 'Eco-blast', which may replace the sandblasting process; they claimed that it is the first completely safe and certified replacement for the sandblasting process.

Dark denim has several advantages: It reduces the harmful process and saves the health of workers who are involved. Also, it reduces the pollution load, which saves the environment from unnecessary pollution which may cause global warming.

Polluted water needs more investment to be cleaned, and these treatments affect profit margins or the waste will go untreated and be discharged directly into rivers. The most serious problem in the denim is sulphur dyeing. The efficiency of sulphur dyes is around 50%, which remains in the trough and contaminates water sources. It is also difficult to remove from water. Researchers are working on the 'recovery of sulphur dyes and reuse' [77–79].

There are cold dyeing methods which can eliminate energy. However, sulphur dyes need reducing agents for reduction, so sodium sulphide-type reducers are commonly used, but the studies show sulphide can be replaced by hot dyeing with reducing sugars which include sugar, dextrins (corn sugar) and molasses [80].

## Abbreviations

| | |
|---|---|
| ARDS | Acute respiratory distress syndrome |
| AOX | Adsorbable organic halogens |
| BOD | Biological oxygen demand |
| COD | Chemical oxygen demand |
| DOT | Department of Transportation (United States) |
| EPA | Environmental Protection Agency |
| ETA | Enzyme Technical Association |
| IARC | International Agency for Research of Cancer |
| OSHA | Occupational Safety and Health Administration |
| ppm | parts per million |
| SOP | Standard operation procedure |
| VOC | Volatile organic compounds |

## Acknowledgement

The first author is thankful to the Greenpeace organization and European Respiratory Society for providing the images. We thank Mr Mohan Ram, Mr Thirumoorthy (Vietnam) and Mr K. Thavasiyappan (India) for their comments, which greatly improved this chapter.

## References

[1] Holkar CR, Jadhav AJ, V Pinjari D, Mahamuni NM, Pandit AB. A critical review on textile wastewater treatments: possible approaches. J Environ Manage 2016;182:351–66. http://dx.doi.org/10.1016/j.jenvman.2016.07.090.

[2] Angelis-Dimakis A, Alexandratou A, Balzarini A. Value chain upgrading in a textile dyeing industry. J Clean Prod 2016;138(Part 2):237–47. http://dx.doi.org/10.1016/j.jclepro.2016.02.137.

[3] Khan S, Malik A. Environmental and health effects of textile industry wastewater. In: Malik A, Grohmann E, Akhtar R, editors. Environmental deterioration and human health: natural and anthropogenic determinants. Dordrecht: Springer Netherlands; 2014. p. 55–71. http://dx.doi.org/10.1007/978-94-007-7890-0_4.

[4] Verma Y. Acute toxicity assessment of textile dyes and textile and dye industrial effluents using *Daphnia magna* bioassay. Toxicol Ind Health 2008;24:491–500. http://dx.doi.org/10.1177/0748233708095769.

[5] Muthu SS. 2-Ways of measuring the environmental impact of textile processing: an overview. In: Assessing the environmental impact of textiles and the clothing supply chain. Woodhead Publishing; 2014. p. 32–56. http://dx.doi.org/10.1533/9781782421122.32.

[6] Muthu SS. 10-Assessing the environmental impact of textiles: summary and conclusions. In: Assessing the environmental impact of textiles and the clothing supply chain. Woodhead Publishing; 2014. p. 180–6. http://dx.doi.org/10.1533/9781782421122.180.

[7] Hauser PJ. Reducing pollution and energy requirements in cotton dyeing. Text Chem Color Am Dyest Report 2000;32:44–8.

[8] Chequer FMD, de Oliveira GAR, Ferraz ERA, Cardoso JC, Zanoni MVB, de Oliveira DP. Textile dyes: dyeing process and environmental impact. In: Günay M, editor. Eco-friendly textile dyeing and finishing. 2013. p. 151–76. http://dx.doi.org/10.5772/3436.

[9] Roy Choudhury AK. Green chemistry and the textile industry. Text Prog 2013;45:3–143. http://dx.doi.org/10.1080/00405167.2013.807601.

[10] Shukla SR. 6-Pollution abatement and waste minimisation in textile dyeing. In: Christie RM, editor. Environmental aspects of textile dyeing. Woodhead Publishing; 2007. p. 116–48. http://dx.doi.org/10.1533/9781845693091.116.

[11] Claudio L. Waste couture: environmental impact of the clothing industry. Environ Health Perspect 2007;115:A449–54. http://www.ncbi.nlm.nih.gov/pmc/articles/PMC1964887/.

[12] Schrott W, Paul R. 20-Environmental impacts of denim manufacture. In: Denim. Woodhead publishing series in textiles. Woodhead Publishing; 2015. p. 563–80. http://dx.doi.org/10.1016/B978-0-85709-843-6.00020-2.

[13] Spengler JD, Sexton K. Indoor air pollution: a public health perspective. Science 1983; 221(4605):9–17. http://science.sciencemag.org/content/221/4605/9.abstract.

[14] Coppola L, Giunta R, Grassia A, Misso L, Verrazzo G, Violano PF, Grandillo F, Tirelli A. Air pollution by gasoline exhaust fumes: effect on platelet function and blood viscosity. Med Lav 1989;80:187–91.

[15] Prockop LD, Chichkova RI. Carbon monoxide intoxication: an updated review. J Neurol Sci 2007;262:122–30. http://dx.doi.org/10.1016/j.jns.2007.06.037.

[16] Cobb N, Etzel RA. Unintentional carbon monoxide-related deaths in the United States, 1979 through 1988. JAMA 1991;266:659–63.

[17] O EcoTextiles. Volatile organic compounds (VOCs). O EcoTextiles; 2010. https://oecotextiles.wordpress.com/2010/03/17/volatile-organic-compounds-vocs/.

[18] Igielska B, Wiglusz R, Sitko E, Nikel G. Release of volatile organic compounds from textile floor coverings in higher temperatures. Rocz Panstw Zakl Hig 2003;54: 329–35.

[19] Challa L. Impact of textiles and clothing industry on environment: approach towards eco-friendly textiles. Fiber2Fashion; 2015. http://www.fibre2fashion.com/industry-article/1709/impact-of-textiles-and-clothing-industry-on-environment?page=1.
[20] Fiber2Fashion. Various pollutants released into environment by textile industry. Fiber2Fashion; 2015. http://www.fibre2fashion.com/industry-article/6262/various-pollutants-released-into-environment-by-textile-industry?page=2.
[21] Manu B, Chaudhari S. Anaerobic decolorisation of simulated textile wastewater containing azo dyes. Bioresour Technol 2002;82:225–31.
[22] Karthik T, Gopalakrishnan D. Environmental analysis of textile value chain: an overview. In: Muthu SS, editor. Roadmap to sustainable textiles and clothing: environmental and social aspects of textiles and clothing supply chain. Singapore: Springer Singapore; 2014. p. 153–88. http://dx.doi.org/10.1007/978-981-287-110-7_6.
[23] Ramesh Babu B, Parande KA, Raghu A. Cotton textile processing: waste generation and effluent treatment. J Cotton Sci 2007;11:141–53.
[24] Slater K. 13-Protection of, or by, textiles from environmental damage. In: Environmental impact of textiles. Woodhead publishing series in textiles. Woodhead Publishing; 2003. p. 161–77. http://dx.doi.org/10.1533/9781855738645.161.
[25] Slater K. 10-Effects on textiles of natural exposure. In: Environmental impact of textiles. Woodhead publishing series in textiles. Woodhead Publishing; 2003. p. 115–38. http://dx.doi.org/10.1533/9781855738645.115.
[26] Slater K. 7-Fabric treatment processes. In: Environmental impact of textiles. Woodhead publishing series in textiles. Woodhead Publishing; 2003. p. 69–89. http://dx.doi.org/10.1533/9781855738645.69.
[27] Slater K. 6-Fabric production. In: Environmental impact of textiles. Woodhead publishing series in textiles. Woodhead Publishing; 2003. p. 61–8. http://dx.doi.org/10.1533/9781855738645.61.
[28] Slater K. 5-Yarn production. In: Environmental impact of textiles. Woodhead publishing series in textiles. Woodhead Publishing; 2003. p. 40–60. http://dx.doi.org/10.1533/9781855738645.40.
[29] Slater K. 4-Textile fibre production. In: Environmental impact of textiles. Woodhead publishing series in textiles. Woodhead Publishing; 2003. p. 23–39. http://dx.doi.org/10.1533/9781855738645.23.
[30] Chakraborty JN. 9-Dyeing with indigo. In: Fundamentals and practices in colouration of textiles. Woodhead Publishing India; 2014. p. 109–24. http://dx.doi.org/10.1016/B978-93-80308-46-3.50009-1.
[31] Greenpeace. The dirty secret behind jeans and bras. 2010. http://www.greenpeace.org/eastasia/news/stories/toxics/2010/textile-pollution-xintang-gurao/.
[32] World Bank. Environmental regulations and standards. World Bank; 2007. https://www.ifc.org/wps/wcm/connect/7dd5798048865849b52ef76a6515bb18/genenv_PPAH.pdf?MOD=AJPERES.
[33] Distressed denim trend costs Mexican farmers the earth. Tehuacán: Guardian; 2007.
[34] Kane C. Environmental and health hazards in spinning industry and their control. Indian J Fiber Text Res 2001;26:39–43. http://nopr.niscair.res.in/bitstream/123456789/24911/1/IJFTR 26(1-2) 39-43.pdf.
[35] Estevan C, Sogorb MA, Vilanova E. Chapter 29-air quality of textile and related industries. In: Scognamiglio V, Rea G, Arduini F, Palleschi G, editors. Comprehensive analytical chemistry. The quality of air. Elsevier; 2016. p. 785–800. http://dx.doi.org/10.1016/bs.coac.2016.02.016.
[36] Periyasamy AP, Dhurai B, Thangamani K. Salt-free dyeing – a new method of dyeing on lyocell/cotton blended fabrics with reactive dyes. Autex Res J 2011;11.

[37] Periyasamy AP, Dhurai B. Salt free dyeing: a new method of dyeing of lyocell fabrics with reactive dyes. Pak Text J 2011;60.
[38] Safety and health issues in the textile industry. Fiber2Fashion, 2010. www.fibre2fashion.com.
[39] Lai PS, Christiani DC. Long term respiratory health effects in textile workers. Curr Opin Pulm Med 2013;19:152–7. http://dx.doi.org/10.1097/MCP.0b013e32835cee9a.
[40] Occupational Safety and Health Administration. Occupational safety in textiles, 2016. https://www.osha.gov/pls/oshaweb/owadisp.show_document?p_table=STANDARDS&p_id=9858.
[41] Patil SL, Kasote PT, Bhala R. Occupational safety on spinning and weaving industry in Ichalkaranji. Kolhapur, India: Shivaji University; 2011.
[42] Fabric formation by weaving. In: Handbook of weaving. CRC Press; 2000. p. 1–7. http://dx.doi.org/10.1201/9781420031966.ch1.
[43] Manufacturing of special fabrics. In: Handbook of weaving. CRC Press; 2000. p. 311–60. http://dx.doi.org/10.1201/9781420031966.ch12.
[44] Mangat MM, Hes L. 15-Comfort aspects of denim garments. In: Paul R, editor. Denim. Woodhead publishing series in textiles. Woodhead Publishing; 2015. p. 461–79. http://dx.doi.org/10.1016/B978-0-85709-843-6.00015-9.
[45] Paul R, Genescà E. 8-The use of enzymatic techniques in the finishing of technical textiles. In: Gulrajani ML, editor. Advances in the dyeing and finishing of technical textiles. Woodhead Publishing; 2013. p. 177–98. http://dx.doi.org/10.1533/9780857097613.2.177.
[46] Colomera A, Kuilderd H. 12-Biotechnological washing of denim jeans. In: Paul R, editor. Denim. Woodhead Publishing; 2015. p. 357–403. http://dx.doi.org/10.1016/B978-0-85709-843-6.00012-3.
[47] Working safely with enzymes. Enzyme Technical Association, 2013. http://www.enzymeassociation.org/wp-content/uploads/2013/09/Working-Safely-With-Enzymes-English.pdf.
[48] Caddow AJ, Concoby B. Enzyme safety and regulatory considerations. In: Enzymes in industry. Wiley-VCH Verlag GmbH & Co. KGaA; 2003. p. 399–411. http://dx.doi.org/10.1002/3527602135.ch7.
[49] Odabasi M. Halogenated volatile organic compounds from the use of chlorine-bleach-containing household products. Environ Sci Technol 2008;42:1445–51.
[50] Hazardous substance fact sheet. 2008. http://nj.gov/health/eoh/rtkweb/documents/fs/1707.pdf.
[51] Iracheta-Cárdenas MM, Rocha-Peña MA, Galán-Wong LJ, Arévalo-Niño K, Tovar-Herrera OE. A *Pycnoporus sanguineus* laccase for denim bleaching and its comparison with an enzymatic commercial formulation. J Environ Manage 2016;177:93–100. http://dx.doi.org/10.1016/j.jenvman.2016.04.008.
[52] Piccoli HH, Ulson de Souza AA, Ulson de Souza SMAG. Bleaching of knitted cotton fabric applying ozone. Ozone Sci Eng 2015;37:170–7. http://dx.doi.org/10.1080/01919512.2014.939742.
[53] Kan CW. Colour fading effect of indigo-dyed cotton denim fabric by $CO_2$ laser. Fibers Polym 2014;15:426–9. http://dx.doi.org/10.1007/s12221-014-0426-2.
[54] Nayak R, Padhye R. The use of laser in garment manufacturing: an overview. Fash Text 2016;3:5. http://dx.doi.org/10.1186/s40691-016-0057-x.
[55] Štěpánková M, Wiener J, Rusinová K. Decolourization of vat dyes on cotton fabric with infrared laser light. Cellulose 2011;18:469–78. http://dx.doi.org/10.1007/s10570-011-9494-2.
[56] Kan CW, Yuen CWM, Cheng CW. Technical study of the effect of $CO_2$ laser surface engraving on the colour properties of denim fabric. Color Technol 2010;126:365–71. http://dx.doi.org/10.1111/j.1478-4408.2010.00270.x.

[57] Sliney DH. Laser safety. Lasers Surg Med 1995;16:215–25. http://dx.doi.org/10.1002/lsm.1900160303.
[58] Barkana Y, Belkin M. Laser eye injuries. Surv Ophthalmol 2000;44:459–78. http://dx.doi.org/10.1016/S0039-6257(00)00112-0.
[59] Stefan W, Baenninger PB, Schmid MK. Retinal injuries from a handheld laser pointer. N Engl J Med 2010;363:1089–91. http://dx.doi.org/10.1056/NEJMc1005818.
[60] Periyasamy AP, Chilukoti GR, Thavasiappan K. Dry finishing: enhancing value of denim. Indian Text J 2012;122. http://www.indiantextilejournal.com/articles/FAdetails.asp?id=4330.
[61] Akgun M, Araz O, Akkurt I, Eroglu A, Alper F, Saglam L, Mirici A, Gorguner M, Nemery B. An epidemic of silicosis among former denim sandblasters. Eur Respir J 2008;32:1295–303. http://dx.doi.org/10.1183/09031936.00093507.
[62] Hutyrova B, Smolkova P, Nakladakova M, Tichy T, Vítězslav K. Case of accelerated silicosis in a sandblaster. Ind Health 2015;53:178–83. http://dx.doi.org/10.2486/indhealth.2013-0032.
[63] Cappuccio S. Why a global ban on sandblasting in the garment industry?. In: ETUI conf. "workers" prot. chem.. 2011. https://www.etui.org/content/download/1736/17375/file/Why+a+global+ban+on+sandblasting+in+the+garment+industry.pdf.
[64] Akgun M, Gorguner M, Meral M, Turkyilmaz A, Erdogan F, Saglam L, Mirici A. Silicosis caused by sandblasting of jeans in Turkey: a report of two concomitant cases. J Occup Health 2005;47:346–9.
[65] Tellason. The chemical concerns in denim manufacturing. 2015. http://www.tellason.com/the-journal/the-chemical-concerns-in-denim-manufacturing/.
[66] Wasserman R. Dangers of sodium metabisulfite. 2013. http://www.livestrong.com/article/208572-dangers-of-sodium-metabisulfite/.
[67] Schindler WD, Hauser PJ. 5-Easy-care and durable press finishes of cellulosics. In: Schindler WD, Hauser PJ, editors. Chemical finishing of textiles. Woodhead Publishing; 2004. p. 51–73. http://dx.doi.org/10.1533/9781845690373.51.
[68] Latorre N, Silvestre JF, Monteagudo AF. Allergic contact dermatitis caused by formaldehyde and formaldehyde releasers. Actas Dermo-Sifiliogr Engl Ed 2011;102:86–97. http://dx.doi.org/10.1016/S1578-2190(11)70765-X.
[69] Burke J. Bangladesh garment workers suffer poor conditions two years after reform vows. Guardian; 2015. https://www.theguardian.com/world/2015/apr/22/garment-workers-in-bangladesh-still-suffering-two-years-after-factory-collapse.
[70] Colovic G. 4-Ergonomic principles. In: Ergonomics in the garment industry. Woodhead Publishing India; 2014. p. 105–73. http://dx.doi.org/10.1016/B978-0-85709-822-1.50004-7.
[71] Colovic G. 3-Ergonomic conditions of work. In: Ergonomics in the garment industry. Woodhead Publishing India; 2014. p. 61–103. http://dx.doi.org/10.1016/B978-0-85709-822-1.50003-5.
[72] Colovic G. 1-Ergonomics development. In: Ergonomics in the garment industry. Woodhead Publishing India; 2014. p. 1–16. http://dx.doi.org/10.1016/B978-0-85709-822-1.50001-1.
[73] Saravanan K. Importance and need of ergonomics in the apparel industry. Pak Text J 2011:57–8. http://www.ptj.com.pk/Web-2011/01-2011/PDF-January2011/Apparel-and-Knitwear-K-Saravan.pdf.
[74] Bechtold T, Berktold F, Turcanu A. The redox behaviour of CI sulphur black 1-a'basis for improved understanding of sulphur dyeing. Color Technol 2000;116:215–21. http://dx.doi.org/10.1111/j.1478-4408.2000.tb00041.x.
[75] Teli MD, Paul R, Sachin LM. Ecofriendly processing of sulphur and vat dyes – an overview. Indian J Fiber Text Res 2001;26:101–7.

[76] Roessler A, Dossenbach O, Rys P, Marte W. Direct electrochemical reduction of indigo: process optimization and scale-up in a flow cell. J Appl Electrochem 2002;32:647–51. http://dx.doi.org/10.1023/A:1020198116170.

[77] Nguyen TA, Juang R-S. Treatment of waters and wastewaters containing sulfur dyes: a review. Chem Eng J 2013;219:109–17. http://dx.doi.org/10.1016/j.cej.2012.12.102.

[78] Chakraborty JN, Dhiman G. Effectiveness of laccase in the oxidation and recovery of sulfur dyes. Text Res J 2013;83:441–9. http://trj.sagepub.com/content/83/5/441.abstract.

[79] Chakraborty JN. 1-Introduction to dyeing of textiles. In: Chakraborty JN, editor. Fundamentals and practices in colouration of textiles. Woodhead Publishing India; 2010. p. 1–10. http://dx.doi.org/10.1533/9780857092823.1.

[80] Blackburn RS, Harvey A. Green chemistry methods in sulfur dyeing: application of various reducing D-sugars and analysis of the importance of optimum redox potential. Environ Sci Technol 2004;38:4034–9.

# Chemicals and effluent treatment in denim processing

**8**

*R. Rathinamoorthy, T. Karthik*
PSG College of Technology, Coimbatore, India

## 8.1 Introduction

The textile industry is one of the largest industrial sectors in India and has an important role in the Indian economy. Denim production is one of the major subsectors in the textile industry. Denim products are some of the most highly used in textile clothing, with continuous choice of fashion and consumer preference, especially by young people. A survey by Cotton, Inc. showed that the global denim jeans market was projected to grow by 8% from $55 billion in 2015 to $59 billion by 2021, with Latin America and Asia expected to lead the increase [1]. The expected growth rate was 12% in Asia, 15% in Latin America, 10% in North America and 4% in Europe over 6 years [2]. Approximately 1.9 billion denim jeans were sold in 2015 alone [2]. It is also estimated that overall worldwide sales will cross 2 billion units by the end of 2021. In India, one of the fastest growing apparel sectors is denim. The production rate in 2010 was 700 million metres; in 2015, it was 1.2 billion metres with a growth rate of 500 million metres [2]. Denim fabrics alone make up 35% of total textile exports in India; this rate is expected to rise to approximately 45% of total exports by 2020. The production capacity is also expected to increase, to 1.5 billion metres by 2020 [2].

Denim manufacturing and other textile-based industries continue to be a source of specific environmental pollution because of considerable amounts of mineral substances that are used to process textile raw materials and semifinished products. Because of the enormous development of textile wet-processing industries such as dyeing, bleaching, printing, finishing and washing, the requirement of pure water for processing increased immensely. These industries use pure water for the process and they drain the water with wetting agents, soda ash, peroxides, sodium hypochloride, bleaching powders, common salt, dyestuffs, soap oils and fixing and finishing agents. Textile factories are among the largest industrial consumers of water: typically 0.2–0.5 $m^3$ of water is needed to produce 1 kg of finished product [3]. These wastewaters are harmful to the environment by directly discharging effluent and by disposing the sludge obtained from the treatment of effluent. The first step in a pollution prevention strategy for water is the thorough audit and characterization of wastewater from textile operations [4].

For denim manufacturing, mostly indigo dyes are used. Synthetic indigo dyes on the market today are usually sulphur based and have serious negative effects. Even if the colour is removed from wastewater, this does not necessarily mean that the water is clean: Sulphur dye residues tend to remain in wastewater even after treatment. Worldwide, the single biggest threat in textiles is wastewater effluent [5]. In addition to sulphur-based

**Sustainability in Denim. http://dx.doi.org/10.1016/B978-0-08-102043-2.00008-3**
Copyright © 2017 Elsevier Ltd. All rights reserved.

dyes, sodium hydroxide, hydrosulphate and formaldehyde are chemicals typically used in the denim manufacturing processes. In postmanufacturing finishing, at the end of the garment manufacturing process, jeans are chemically washed to achieve the faded look. Such chemicals in these finishing processes, such as potassium permanganate and sodium hydroxide, tend to appear in high concentrations in the wastewater effluent and have high basic pH levels [6]. Lots of developments are also taking place in the manufacture and finishing of jeans. It is well-known that denim jeans have an undisputed role in the fashion industry. Dyeing technologies for denim garments are also constantly adapting to rapid small-scale production catering to niche markets. Based on their high market potential and wastewater generation during the dyeing and finishing process, denim manufacturing possesses a significant threat to the environment.

Levi Strauss & Company, one of the world's leading denim manufacturers, started its self-analysis program in 2006. The company analyzed its internal and external environmental impacts throughout its product line [7]. In its life-cycle assessment, the life cycle of jeans from 'cradle to grave' was reported as a collection of seven independent systems. The analysis process started in the cotton production sector and covered denim material production steps such as fabric production, garment manufacturing, transportation and distribution, consumer use and recycling. Furthermore, the company listed the impact of the waste stream from the product manufacturing process in landfills. Its report, mentioned that making one pair of denim jean required almost 920 gallons of water and 400 MJ of energy and expelled 32 kg of carbon dioxide. It also pointed out that the major quantity of water consumed (68%; 2565 L) was in fiber production. The water consumption percentage is comparatively high in all phases of garment production, likely in fabric manufacturing (6%; 236 L), during sewing and finishing (1%; 34 L), in washing and packing (2%; 77 L) and at customer's end (23%; 860 L) [7].

The first part of this chapter deals with different steps involved in the denim manufacturing process and various effluent outputs at various process stages along with their impacts on the environment. The general chemicals and wastewater treatment methodologies adapted in the denim processing industry are also revealed. The second part of the chapter provides a case study conducted in the denim fabric processing industry, in which various effluent treatment methods followed in the industry, along with their specific pollution loads on the environment, are depicted. Common effluent analysis criteria such as the chemical oxygen demand (COD), biological oxygen demand (BOD), total dissolved solids (TDS), pH and conductivity levels are identified and analyzed. The last part of this chapter offers some remedial measures to reduce the impact of pollution using conventional pollution reduction strategies and some advancements in effluent treatment.

## 8.2 Denim processing stages

The general process flow of the denim manufacturing industry is provided in Fig. 8.1. The initial process of denim manufacturing consists of common activities from the mixing of cotton fiber to the spinning of fibers using either ring spinning or open end spinning. Generally, denim fabric is a 3/1 warp-faced twill fabric made from a yarn-dyed

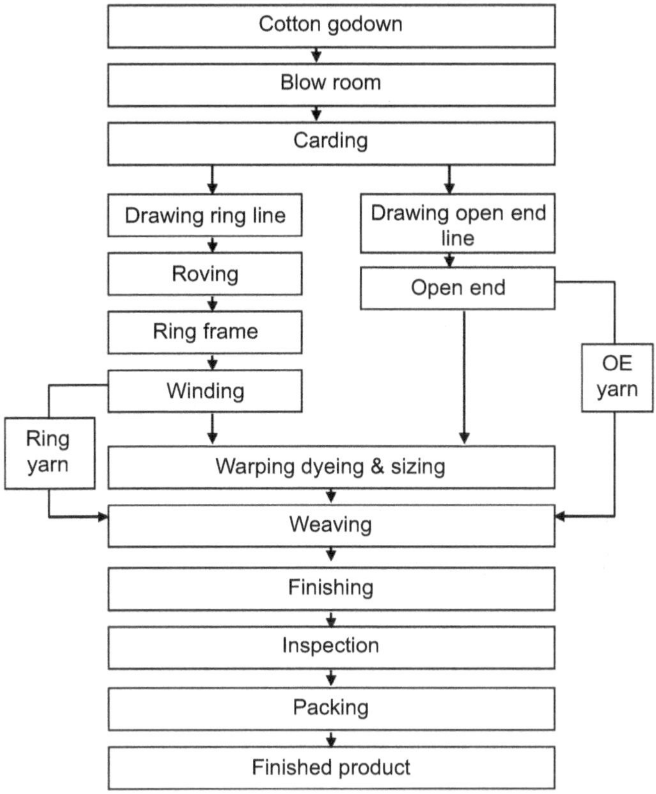


**Figure 8.1** General process flow of denim manufacturing from fiber to fabric.

warp and an undyed weft yarn. Normally dyed and grey ring or open-end yarns are used in the warp and weft, respectively. After the spinning process, the yarn is prepared for the warp as a beam in the warping process. The prepared warp yarns are dyed with indigo and sized with the help of two methods: (1) Threads from several back beams are combined to form a warp sheet and dyed and sized on the same machine; (2) about 350–400 threads are formed into ropes. Then 12 to 14 ropes are run adjacent to each other through the continuous dyeing unit. After dyeing, the ropes are dried on drying cylinders and then collected in a can. After that, a worker's beam is prepared.

Sizing is then done in the conventional manner. After sizing, weaving starts. After weaving, woven denim fabrics undergo various finishing processes such as brushing, singeing, washing, desizing, mercerization, finishing and drying before going through the garment manufacturing process.

## 8.3 Warp dyeing machines for denim manufacture

To start the warp dyeing process, the spun yarn needs to be prepared in the form of beams. This process can be done depending on the type of dyeing range the manufacturer uses for the warp dyeing process. Generally two types of dyeing machines are used in the warp dyeing process [8]:

1. Rope dyeing: For this machine, the ball-warping process is preferred, in which the yarn is arranged in small groups forming ropes. In this type of denim range, warp threads are extended along the range in the form of a rope. Yarn groups are composed of 300 and 400 ends which form a rope. The total width of rope ranges is taken up by a number of ropes ranging between 12 and 48, depending on the width of the machine. The total number of ends, depending on the number of ropes, can range between 3600 and 19,200.
2. Sheet or slasher dyeing: For this machine, beam warping is used, in which the beams are made up of yarn set parallel, flat and open. In this type of dyeing range, warp yarn is extended flat and separately and is arranged parallel to each other, with a millimetre gap between them in individual layers. The ranges' width corresponds to the sum of all warp yarn. The number of yarns per beam ranges between 300 and 750, whereas the total number of threads which make up the warp yarn vary between 3500 and 9500, according to the thickness and diameter of the yarn that is used, the type of warping range and the denim fabric design in the weaving process.

### 8.3.1 Rope dyeing machine

Rope dyeing accounts for a large percentage of yarn dyed for denim warps. The rope dyeing technique allows for the simultaneous production of more than one warp in one run. The process flow in the rope dyeing process is shown in Fig. 8.2. This maximizes the continuity of shades and minimizes shade variation. Yarn from the spinning process is fed directly to cross-winding and ball warping. The ball warping process forms a rope by merging 300 to 400 yarns together; this rope is wound on a beam or log. After ball warping, the desired number of beams is used in the dye range creel. From there, the rope runs through various stages of the indigo dyeing process at the same time. After this process, the rope of dyed yarn is opened into sheets of yarn and wound onto a warper or section beam. This process is known as long chain beaming. The beam developed by the long chain beaming process will be used in the sizing process.

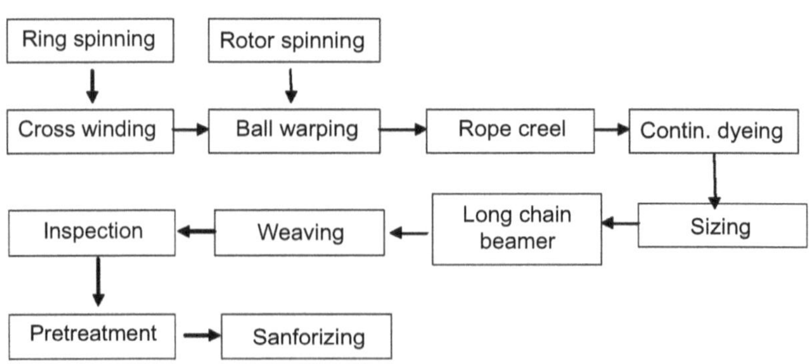


**Figure 8.2** Process flow of denim rope dyeing.

### 8.3.2 Slasher dyeing process

Open sheet or slasher dyeing machines were developed as a method to reduce the handling of yarn compared with dyeing in rope form. The necessity of first making the rope and then separating them after dyeing, beaming and rebeaming the warp could largely be reduced by passing the warp sheet of yarn directly into the dyeing range. The general process flow of the sheet dyeing operation is given in Fig. 8.3. This process eliminates the ball warping process, creel, rope dyeing range and associated slow-speed warping machines. In this method a sheet of single-end yarn sufficient for one piece of denim passes through the machine. During the operation, it is dyed, oxidized, dried, sized and dried to be assembled as warp.

### 8.3.3 Combined dyeing process

In addition to these techniques, sometimes denim fabrics may pass through different stages of a combined dyeing process:

- topping: This method is used to produce darker shades of denim fabric. Here the warp yarn passes through several indigo boxes, then two wash boxes and finally one or two sulphur black boxes. This process produces regular indigo layers on the yarn's surface and a black layer on top. This process is generally performed to improve the poor binding ability of indigo dyes.
- bottoming: This process is the reverse of topping. Here, the grey yarns pass directly through the black sulphur dyes and then are processed with the blue indigo dyes. After washing, even the indigo dye washes off. The bottom black colour can be seen, which produces a darker effect than the normal light colour.
- overdyeing: A similar effect is achieved when standard indigo-dyed denim (warp yarn dyed) is overdyed in a flat fabric form. Here, sulphur black or other colours can be also used for dyeing. The main issue with this process is the cost of the operation. The cost of the additional and separate dyeing process is higher than that for bottoming or topping.
- sandwich dyeing: The combination of bottoming and topping is known as sandwich dyeing, because the resulting dyeing structure shows the indigo layers between the bottom and top layers.

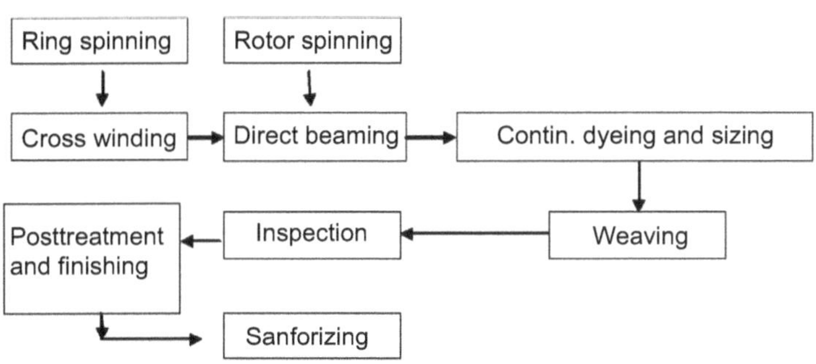


**Figure 8.3** Process flow of the sheet dyeing operation.

## 8.4 Chemicals used in denim processing

The chemicals used to manufacture denim products are initially used in the dyes: indigo, vat and sulphur. All of these need a reducing agent in the dyeing process to form the water-soluble 'leuco' form for the water-insoluble pigment that gives high fastness. The most commonly used reducing agent is sodium hydrosulphite. This reducing agent creates a potential fire risk and produces high wastewater residue. Along with the reducing agents, alkali is used in denim dyeing to enhance the reduction process. Common chemicals used in the denim processing are [9]:

- wetting agents: alkyl sulphates, alkyl sulphonates, phosphoric acid esters, fatty alcoholethoxylates, alkylphenolethoylates and fatty amine ethoxylates
- washing agents: detergents, tensides
- defoaming and deairing agents from many different chemical groups
- dispersing agents: lignosulphonate, condensation products of naphthalene sulphonic acid, fatty acid esters
- sequestering agents: phosphonates, polyphosphates, amino carboxylic acids, polyacrylates, gluconate
- rebeaming agents (for rope dyeing machines): cationic softeners, silicone emulsions, polyethylene (PE)emulsions, waxes, polyacrylates

In addition to these chemicals, a number of others such as size, softeners and finishing chemicals are used during production, to facilitate further processing (size) or provide an appropriate finish (softeners and finishing chemicals). In general, size is a natural or synthetic polymeric material that is put onto the dyed yarn before weaving to improve the strength of cotton yarn that will be used in a high-speed weaving process. Typically, 5%–15% size is applied to cotton yarn. When it is washed off, it ends up in the wastewater and can represent over 70% of the total COD value of wastewater from a denim mill [9]. Softeners are often used in the final fabric finishing process to give a smooth or soft hand to the fabric. In general, ester-based cationic and quaternary softeners are commonly used because they are readily biodegradable, but they can be toxic to aquatic life. Other than this, PE emulsions, silicone emulsions, wax and polyacrylates are also used as softeners. Most of these chemicals are nonbiodegradable and have potential environmental impacts.

Table 8.1 summarizes various chemicals used in different departments of denim manufacturing plants, with their requirements/purposes

## 8.5 Common effluent treatment methods

Wastewater from the textile processing industry has a huge quantity of pollutants with a high concentration of organic compounds and colour matter. Because of the high BOD/COD, colouration and salt load, wastewater drained from the textile processing industry is seriously polluted. Because aquatic organisms need light to develop and be sustained, coloured water delivered from the textile dyeing industry destroys properties and leads to an imbalance in the ecosystem [10].

**Table 8.1 Chemicals used in different departments of denim manufacturing plants**

| S.no | Section | Chemical used | Requirements/purpose |
|---|---|---|---|
| 1 | Dyeing | Buffering agent | For dyeing chemicals |
| 2 | | Dispersing agent | Maintains uniform dye dispersion |
| 3 | | Leveling agent | Develops even dyeing and dye fixation |
| 4 | | Dye-fixing agent | Improves dye fixation |
| 5 | | Washing-off agent | Removes unfixed dye particle from fabric |
| 6 | Desizing | Desizing agent | Size removal from fabric |
| 7 | | Wetting agent | Increases absorbency/reduces surface tension of water |
| 8 | | Sequestering agent | Reducing hardness of water |
| 9 | | Scouring agent | Removes wax, oil and other dirt |
| 10 | | Defoamers | Reduces foam generation |
| 11 | | Antiback-staining agent | Avoids redeposition of dye in fabric |
| 12 | Mercerization | Caustic solution | For mercerization |
| 13 | | Mercerizing and causticizing agent | Improves penetration of caustic soda and ensures removal of caustic soda during washing |
| 14 | | Neutralizing agent | Neutralize fabric pH before leaving operation |
| 15 | Finishing | Fixing agent | Improves fixation of finishing chemical |
| 16 | | Softener | Makes fabric feel soft and fluffy |
| 17 | | Binder | Improves fixation of finishing chemical |
| 18 | | Thickener | Water thickener for chemical preparation |
| 19 | | Wetting agent | Increases absorbency/reduces surface tension of water |
| 20 | | Fastness improver | Improves fastness of finishing process. |
| 21 | | Foaming agent | Develops foam during finishing and improves level of finishing |
| 22 | | Other finishing agent | Creates particular finishing effects such as antistatic, luster, etc. |

The preliminary treatment of wastewater decides whether it can be disposed of by surface or subsurface methods or dilution. However, additional treatment is necessary for that wastewater from a preliminary treatment plant to prevent further contamination based on its applications. Typically an effluent treatment plant (ETP) for a textile industry consists of primary, secondary and tertiary treatments

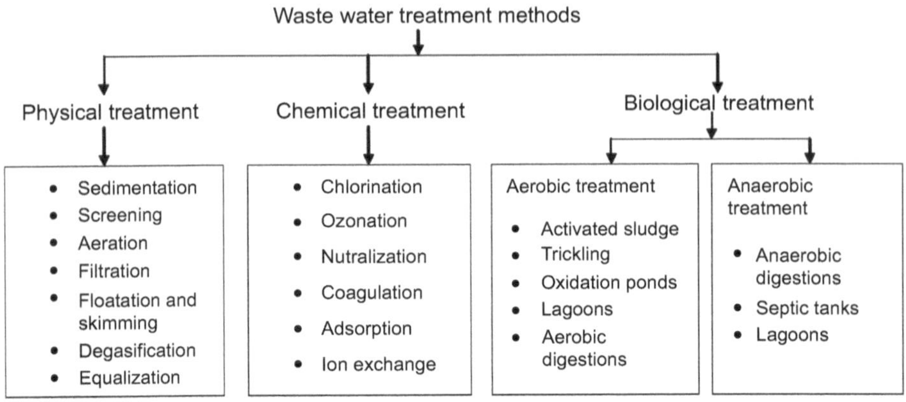


**Figure 8.4** Classification of effluent treatment methods.

to ensure the best performance [11]. Physical, chemical and biological methods are used to remove contaminants from wastewater. To achieve different levels of contaminant removal, individual wastewater treatment procedures are combined into a variety of systems, classified as primary, secondary, and tertiary wastewater treatment. More rigorous treatment of wastewater includes the removal of specific contaminants as well as the removal and control of nutrients. The general classification of different effluent treatment methods used in the textile industry is detailed in Fig. 8.4.

## 8.5.1 Physical treatment

Physical treatment, which is considered to be a primary effluent treatment of wastewater, represents the first stage in wastewater purification. Mainly suspended solids are removed during the operation that takes place at this stage. Among the first treatment methods used are physical unit operations, in which physical forces are applied to remove contaminants.

- sedimentation (clarification): This process is used to remove suspended solids from the wastewater with the help of gravity.
- screening: Floating matter such as paper and wood is removed from wastewater by this process with the help of mechanical bar screens.
- aeration: This process is generally applied to water bodies that experience pollution such as sewage mixing, agricultural runoff and overbaiting. It is commonly known as an anoxic condition of water bodies. Air infusion techniques are generally used for the aeration process from the bottom of the lake, pond or lagoon. Aeration can also be achieved by a surface agitation process with the help of a fountain or spray-like device, which allows wastewater to be exchanged with oxygen at the surface and noxious gases such as carbon dioxide methane and hydrogen sulphide to be released.
- filtration: Remaining suspended particles and unsettled flocks from wastewater are removed using a filtration process.
- flotation and skimming. Flotation is a unitary process in which solid particles in liquid suspension become attached to microscopic air bubbles, giving the air-solids agglomerate

buoyancy. Under the right conditions, the agglomerate will rise to the surface to join other particles and form a blanket that can be removed by mechanical means. In waste treatment, flotation is used primarily to remove light suspended solids (oils, fats, etc.) that have a tendency to float, as well as waste solids with a wide range of specific gravities

## 8.5.2 Chemical treatment

Chemical processes used in wastewater treatment are designed to bring about some form of change by means of chemical reactions. The chemical processes are combined and used with the physical unit operations and biological processes. The chemical treatment of wastewater is an additive process; it is the major disadvantage of a chemical process over a physical treatment. This process increases the net dissolved solid content of the treated water; hence it creates complexity when the wastewater is planned for reuse [12]:

- chlorination: This method of treatment is used in wastewater to disinfect or destroy pathogenic organisms and prevent water from decomposition.
- ozonation: Ozonation introduces an unstable molecule to wastewater and gives up one atom of oxygen to the water molecule, converting it into a powerful oxidizing agent against most waterborne organisms.
- neutralization: Neutralization can be performed for industrial wastewater containing acidic or base (alkaline) substances that need to be neutralized before discharge into water.
- coagulation: The main principle of coagulation treatment is to precipitate ions (heavy metals) and colloids (organic and inorganic) in the wastewater stream by electrical charge or by adding chemicals. In this process, the addition of ions with an opposite charge destabilizes colloids; the coagulation of wastewater can be achieved. Chemicals such as alum $[Al_2(SO_4)_3 \cdot 18H_2O]$ have been used in wastewater treatment for ages for this purpose.
- adsorption: The porous matrix of granular activated carbon is one of the most commonly used adsorbents. In this process, chemicals from wastewater are physically attached to the surface of the solid body by a suitable adhesion method. The effectiveness of the adsorbent is directly related to the amount of surface area available to attract the particles of the contaminant.
- ion exchange: This technique has been used extensively to remove hardness and iron and manganese salts from drinking water. It has also been used selectively to remove specific impurities and recover valuable trace metals such as chromium, nickel, copper, lead and cadmium from industrial waste discharge. The process takes advantage of the ability of certain natural and synthetic materials to exchange one of their ions.

## 8.5.3 Biological treatment

Biological unit processes are used to convert finely divided and dissolved organic matter in wastewater into flocculent settleable organic and inorganic solids. In these processes, microorganisms, particularly bacteria, convert colloidal and dissolved carbonaceous organic matter into various gases and cell tissue, which are then removed into sedimentation tanks. These processes are also usually used in conjunction with physical and chemical processes, with the main objective of reducing the organic content (measured as BOD, total organic demand or COD) and nutrient content (notably nitrogen and phosphorus) of wastewater [12].

**Aerobic:** Under aerobic ($O_2$) conditions, bacteria rapidly consume organic matter and convert it into $CO_2$.The following treatment methods are in this category:

- activated sludge
- trickling filtration
- oxidation ponds
- lagoons
- aerobic digestion

**Anaerobic:** This is a bacterial process for domestic sewage in septic tanks which normally retain sewage from 1 to 2 days, reducing BOD by about 35%–40%. The following treatment methods are in this category:

- anaerobic digestion
- septic tanks
- lagoons

## 8.6 General pollution assessment methods

COD and $BOD_5$ (BOD for 5 days) are two of the most common generic indices used to assess aquatic organic pollution in wastewater. BOD is often used to evaluate the biodegradable fraction, and COD, the total organic pollution load of water contaminated by reductive pollutants. The BOD/COD ratio is a measure of food value in wastewater. If the ratio is higher, it has higher food and less toxicity. The pollution level of textile wastewater is expressed in terms of pH, conductivity, BOD, COD, total suspended solids (TSS), TDS, heavy metals and the temperature of the discharging effluent. The testing methods for the pollution values are given subsequently. All analyses are performed according to standard methods for the examination of water and wastewater [13].

### *8.6.1 pH value*

pH is a term used to express the intensity of the acid or alkaline source. The measurement of pH represents the effective concentration (activity) of hydrogen ions ($H^+$) in water. This concentration can be expressed in the same kind of units as other dissolved species, but $H^+$ concentrations are much smaller than other species in most waters. The activity of $H^+$ can be expressed most conveniently in logarithmic units. pH is defined as the negative logarithm of the activity of $H^+$:

$$pH = -\log [H^+]$$

### *8.6.2 Electrical conductivity*

Electrical conductivity is a measure of water's capacity to convey electrical current. The electrical conductivity of water is directly proportional to its dissolved mineral matter content. The unit of conductivity is expressed in terms of micro mhos per centimetres. Electrical conductivity indicates the variation in dissolved solid content by periodical evaluation.

### 8.6.3 Chemical oxygen demand

In environmental chemistry, the COD test is commonly used to measure the amount of organic compounds in water indirectly. Most applications of COD determine the amount of organic pollutants found in surface water (e.g., lakes and rivers), which makes COD a useful measure of water quality. It is expressed in milligrams per litre (mg/L), which indicates the mass of oxygen consumed per litre of solution [14,15].

### 8.6.4 Biochemical oxygen demand

BOD is a chemical procedure for determining how fast biological organisms use up oxygen in a body of water. It is used in water quality management and assessment, ecology and environmental science. BOD is not an accurate quantitative test, although it could be considered an indication of the quality of water [16,17].

### 8.6.5 Total dissolved solids

TDS is an expression of the combined content of all inorganic and organic substances contained in a liquid which are present in a molecular, ionized or microgranular (colloidal sol) suspended form. Generally the operational definition is that the solids must be small enough to survive filtration through a sieve size of 2 μm. TDS are differentiated from TSS in that the latter cannot pass through a sieve of 2 μm and yet are indefinitely suspended in solution. The term 'settleable solids' refers to material of any size that will not remain suspended or dissolved in a holding tank not subject to motion, and exclude both TDS and TSS [18,19].

### 8.6.6 Total suspended solids

TSS is mainly organic in nature, is visible and can be removed from wastewater by physical/mechanical means, e.g., screening and sedimentation. TSS is measured by filtering a certain quantity of effluent and then drying the filtrate at certain temperature, e.g., 105°C, followed by weighing. TSS is expressed as parts per million or milligrams per litre. The pore size of the filter paper is important in estimating the TSS; the nominal pore size 1.58 μm [20].

## 8.7 Effluent analysis in a denim processing industry

Effluent analysis was performed in one of the leading denim manufacturing industries in India. The company either produces denim in from its own plant or sources the fabric from other manufacturers. The process flow of dyeing and finishing departments of the company is depicted in Fig. 8.5, in which the developed fabrics are mercerized, desized, top dyed with the help of a continuous dyeing range (CDR) and finished with

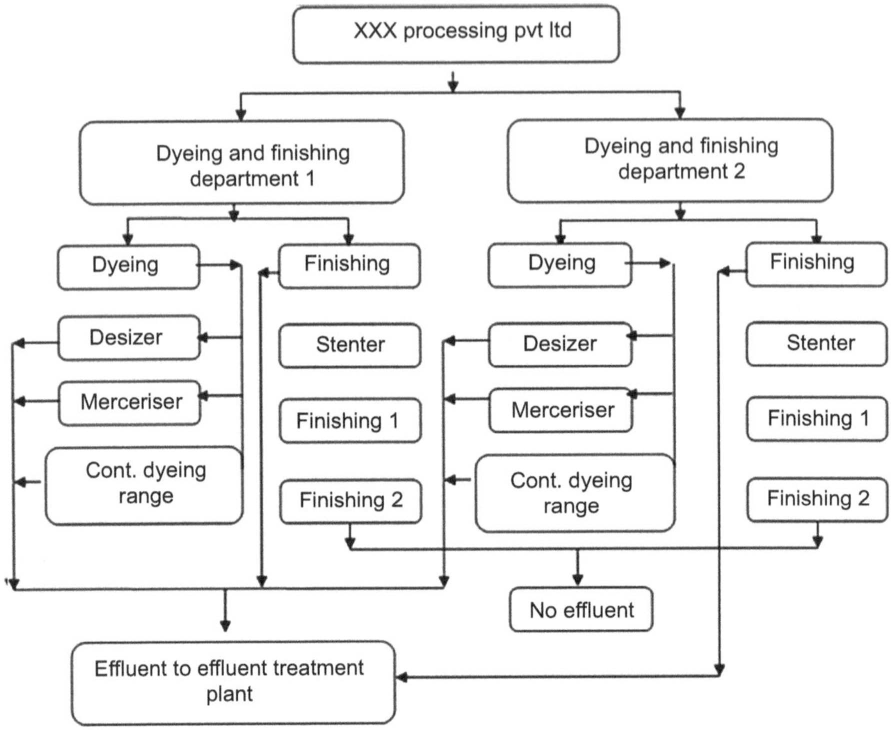


**Figure 8.5** Flow of effluent-generating areas in a denim manufacturing unit.

a softening chemical. In this work, effluents generated from the following processing machines are analyzed for their pollution content:

- desizer
- mercerizer
- CDR
- finishing machines

## 8.7.1 Desizing

After the weaving process, the sizes have to be removed from the fabric because they interfere with subsequent processing steps. In general, sizes have a high BOD and contribute significantly to the waste load of the mill's effluent. W. B. Achwal reported that the waste stream of a desizing operation can contribute up to 50% of the total pollution load of a mill's wastewater [21]. Three methods frequently used in textile processing are acid, enzyme and oxidative desizing.

The presence of sizing ingredients in fabric hinders processes such as dyeing, printing and finishing. For example, the presence of starch can hinder penetration of the dye into the fiber, which necessitates the removal of starch before dyeing or

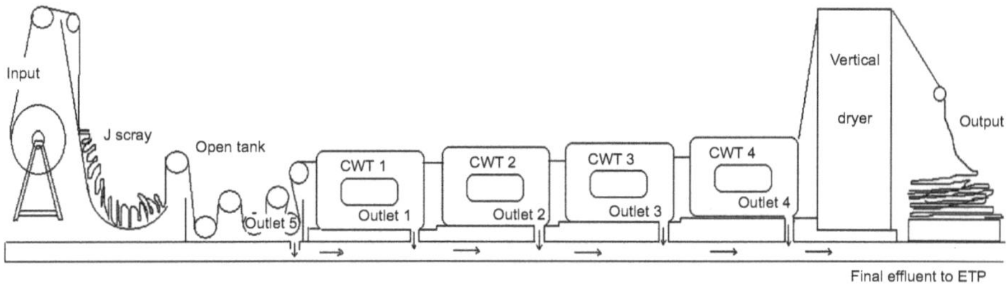


**Figure 8.6** Line diagram of desizing machine. *CWT*, closed washer tank; *ETP*, effluent treatment plant.

printing. Starch is removed or converted into simple water-soluble products by hydrolysis (enzymatic preparations or dilute mineral acids) or oxidation (sodium bromide, sodium chlorite, etc.) [22].

After it is put in, the fabric passes into an open tank and subsequently through a steamer tank where it is washed with continuous steam and water. After the steamer the fabric is passed to closed washer tank (CWT) 1 and then passes through CWT 2 and CWT 3. Hence the fabric becomes washed at all stages by CWT 4. the washing chemical is continually added at 1 L/min to increase the washability of the material. After that, the material passes through a vertical drying range for and output was removed. The effluent from washer tank 1–4 and open tank chemicals are drained into a common outlet, which is piped into an ETP, as shown in Fig. 8.6.

The pollution level of the effluent at different stages of the desizing process decreases step by step after every washer tank. The effluent analysis is performed in the final effluent of desizing (overall drain from the machine), which is pumped directly into the ETP. The pollution level analysis in the desizing effluent showed that the COD and TDS levels in the desizing effluent are high, with an average value of 8992 mg/L and average TDS of 7576 parts per million (ppm). The effluent has an alkaline pH at around 8.5 and the conductivity value is 12,770. The data were collected over a period of 30 days and are listed in Table 8.2.

### 8.7.1.1 Chemicals used in desizing process

General chemicals used in desizing and their applications are:

- enzymes for desizing
- emulsifying and wetting agents to improve the wettability of the fabric
- wetting and deaerating agents for the wettability of the fabric

### 8.7.1.2 Impact on environment

To analyze the actual properties of the chemicals in terms of their environmental impact, the TDS and COD levels of the chemicals were analyzed in the laboratory, in which 1 mL of chemical was mixed with 0.5 L of water and the test was performed. Test results are provided in Table 8.3.

**Table 8.2 Pollution load of effluent at the overall outlet of the desizer machine**

| S.no | pH | Conductivity | Total dissolved solids (parts per million) | Chemical oxygen demand (mg/L) |
|---|---|---|---|---|
| 1 | 8.87 | 2,050 | 5,600 | 7,000 |
| 2 | 8.94 | 2,110 | 7,350 | 8,400 |
| 3 | 8.20 | 2,660 | 7,400 | 9,480 |
| 4 | 9.64 | 23,900 | 8,150 | 10,960 |
| 5 | 8.24 | 14,400 | 7,300 | 10,800 |
| 6 | 8.44 | 13,600 | 7,800 | 8,320 |
| 7 | 8.1 | 17,400 | 7,200 | 8,000 |
| 8 | 7.56 | 16,800 | 8,700 | 13,040 |
| 9 | 7.98 | 9,000 | 6,700 | 8,320 |
| 10 | 7.14 | 14,900 | 8,900 | 9,280 |
| 11 | 7.89 | 17,700 | 8,900 | 8,960 |
| 12 | 9.82 | 17,100 | 4,950 | 5,080 |
| 13 | 8.24 | 10,240 | 8,400 | 8,200 |
| 14 | 9.12 | 16,500 | 8,250 | 10,600 |
| 15 | 8.54 | 13,200 | 8,050 | 8,450 |
| Average | 8.44 | 12,770.67 | 7,576.667 | 8,992.667 |

**Table 8.3 Desizing chemicals and their pollution loads**

| S.no | Chemical used | Total dissolved solids (parts per million) | Chemical oxygen demand (mg/L) | Use (kg) (total consumption per month) |
|---|---|---|---|---|
| 1 | Desizing enzyme | 417,300 | 468,000 | 1,536 |
| 2 | Emulsifying agent | 200,000 | 1,464,000 | 970 |
| 3 | Wetting agent | 900,000 | 1,720,000 | 951 |

The results in Table 8.3 show that for a high-level dilution of 1 mL in 0.5 L of water, the chemical load in the water is high for all chemicals. However, of chemicals that are used, the wetting agent in the desizing process is one most loaded with pollution. The TDS value of the wetting agent is around 900,000 ppm and the COD value is 1,720,000 mg/L. These values show the influence of these chemicals on the pollution level of the final effluent. Next to the wetting agent, the emulsifying agent possesses the highest load; the lowest values are for the desizing enzymes.

### 8.7.2 *Mercerization*

Mercerization is the treatment of pure cotton fabric or yarn with a strong caustic soda solution to improve the strength of the dye substantively and the strength and smoothness of the yarn. Sufficient washing is required after this step to remove any traces

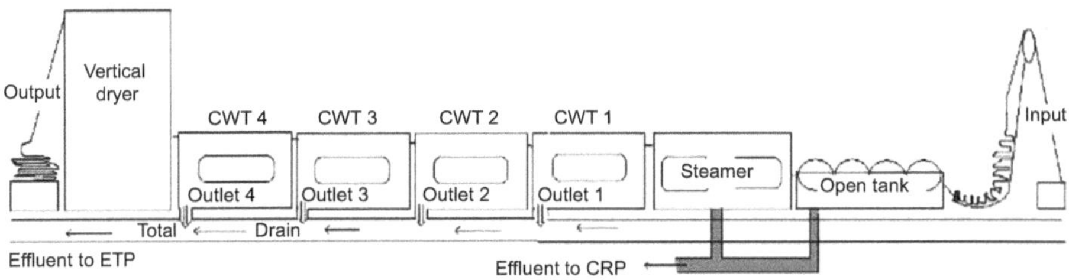


**Figure 8.7** Line diagram of a mercerizing machine. *CRP*, caustic recovery plant; *CWT*, closed washer tank; *ETP*, effluent treatment plant.

of caustic soda. To impart luster and improve strength and dye uptake, cotton fiber and fabric are mercerized in the grey state after bleaching. Essentially, mercerization is carried out by treating cotton material with a strong solution of sodium hydroxide (about 18%–24%) and washing the caustic out after 1–3 min while holding the material under tension. Cotton is known to undergo longitudinal shrinkage upon impregnation with this solution. This can be prevented by stretching it or holding it under tension. The material acquires the desired properties of luster, increased strength, dye uptake, and increased absorbency. Large concentrations of NaOH in the wash water can be recovered by membrane techniques. The general process flow of mercerizing equipment is provided in Fig. 8.7.

Effluent from the stabilizer and from the open caustic tank passes separately and is treated in a caustic recovery plant (CRP) to recover 99.99% of the sodium hydroxide. The CRP is used to recycle the used caustic chemical from the mercerizer. The general process flow is provided in Fig. 8.7.

After the stabilizer tank the fabric is passed to CWT 1. Here the fabric is washed at 90°C and subsequently is passed through a set of squeezing rollers with the Nip pressure of 2–3 bar. From this the fabric subsequently passes through CWT 2 and CWT 3. Because of the caustic impregnation the fabric will be alkaline even after three washes. So, to neutralize the fabric at CWT 4, neutralizing acid is added to the tank and the temperature is maintained at room temperature. After the neutralization process the material passes over the vertical drying cylinder and finally to plaiter before the output. Here the effluent from Washer Tanks 1–5 are drained into the common outlet, which are piped into the ETP, as shown in Fig. 8.8.

The properties of the effluent at different stages of the mercerizing process undergo reasonable changes. With respect to the pH of the effluent, it is high in the case of CWT 1 and is reduced step by step until CWT 3, but the introduction of a neutralizing agent in CWT 4 tremendously reduces the pH from 10.55 (average at CWT 3) to 3.5 (average at CWT 4). In the case of TDS, there is a decreasing trend from the mercerizing tank to CWT 3, but the addition of a neutralizing agent increases the TDS level slightly at CWT 4. The COD values in the mercerizing process are much less from CWT 1 to CWT 3 (around 2800 at CWT 3); however, the addition of the neutralizing agent in CWT 4 increases the COD level in the CWT 4 effluent to a maximum of 10,800 (average at CWT 4).

Table 8.4 lists the data for effluent collected from the drain of the mercerizer. The values indicate that the effluent from the mercerizer is completely alkaline and the values indicating the conductivity of the effluent are very high, around 57,928 (average).

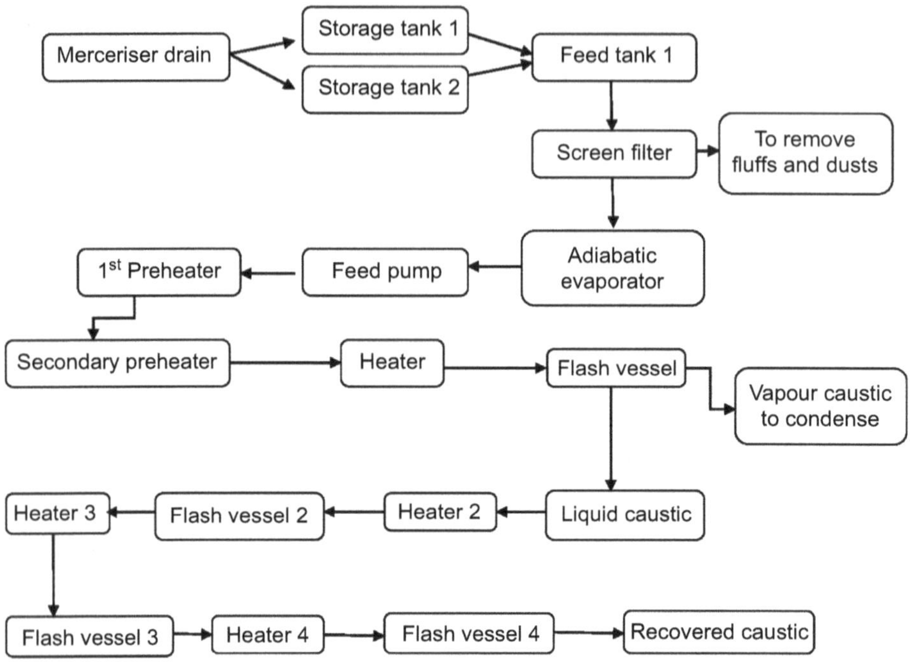


**Figure 8.8** Process flow of caustic recovery plant.

## Table 8.4 Pollution load of effluent in overall outlet of mercerizing machine

| S.no | pH | Conductivity | Total dissolved solid (parts per million) | Chemical oxygen demand (mg/L) |
|---|---|---|---|---|
| 1 | 11.60 | 82,800 | 35,950 | 6,400 |
| 2 | 12.64 | 15,100 | 41,200 | 3,000 |
| 3 | 12.24 | 10,300 | 35,700 | 9,280 |
| 4 | 14.08 | 77,200 | 34,100 | 7,760 |
| 5 | 12.79 | 53,900 | 37,350 | 4,690 |
| 6 | 12.99 | 17,220 | 16,050 | 3,800 |
| 7 | 11.52 | 27,500 | 17,750 | 5,440 |
| 8 | 13.22 | 81,900 | 23,250 | 1,800 |
| 9 | 12.25 | 14,140 | 51,000 | 2,739 |
| 10 | 13.87 | 121,400 | 37,900 | 5,320 |
| 11 | 13.77 | 170,000 | 73,500 | 3,040 |
| 12 | 14.6 | 9,800 | 25,790 | 6,400 |
| 13 | 13.98 | 144,000 | 35,950 | 3,000 |
| 14 | 12.6 | 15,100 | 27,800 | 4,800 |
| 15 | 12.10 | 28,560 | 17,800 | 5,420 |
| Average | 12.95 | 57,928 | 34,072.67 | 4,859.26 |

The COD is 4859 mg/L and the TDS is 34,072 ppm. Of all values, the mercerizer effluent produces the highest amount of TDS in the effluent.

General chemicals used in the mercerization section and their applications are:

- caustic soda for mercerizing activity
- mercerizing and caustic agent
- dispersing, emulsifying and leveling agent to improve wetting and leveling
- neutralizing agent to reduce the fabric's pH

### 8.7.2.1 Impact on environment

To analyze the actual property of the chemicals in terms of their environmental impact, the TDS and COD levels of the chemicals used in the mercerizer were analyzed in the laboratory, in which 1 mL of chemical was mixed with 0.5 L of water and a test was performed. Test results are provided in Table 8.5.

The results in Table 8.5 show that for a high level dilution of 1 mL in 0.5 L of water, the chemical load on the water is higher than for all other chemicals. However, of chemicals that were used, the caustic agent used in the mercerization process is more loaded with pollution. The TDS value of the caustic agent is around 1,000,000 ppm and the COD value is 844,000 mg/L. These values show the influence of these chemicals on the pollution level of the final effluent. Next to the wetting agent, caustic soda possesses the highest load; the least values are for the leveling agent, and the neutralizing agent possesses the second highest value of COD, which is 260,000 mg/L.

## 8.7.3 Continuous dyeing range

In general there are three processes in CDR: They are caustisizing, overdyeing, and desizing. In desizing, the fabric passes through the open tank, where the wetting agent and enzymes are used to desize the fabric. Mostly α-amylase enzyme is used as an enzyme for desizing and nonionic wetting agents are used with it. In an open tank, in addition to nonionic wetting agents, a nonionic emulsifying agent is used. As mentioned, the effluent sample was collected from the final outlet of the machine.

Caustisizing is a process similar to mercerization. Here the fabric is treated with a caustic solution. The CDR machine consists of a chemical tank and steam stabilizer

**Table 8.5 Mercerizing chemicals and their pollution loads**

| S.no | Chemicals | Total dissolved solids (parts per million) | COD in mg/L | Usage (kg) (total consumption per month) |
|---|---|---|---|---|
| 1 | Caustic soda | 755,000 | 26,800 | 118,200 |
| 2 | Caustic agent | 1,000,000 | 844,000 | 2,300 |
| 3 | Leveling agent | 600 | 15,200 | 1,165 |
| 4 | Neutralizing agent | 100,000 | 520,000 | 25,000 |

tank. After this it has a continuous washing range from Tanks 1–7. Finally it has three vertical drying ranges. The effluent samples were collected from the common outlet, which was piped to the ETP. During this process the effluent from the caustisizing tank were bypassed to a CRP.

#### *8.7.3.1 Overdyeing*

Overdyeing is a technique which takes advantage of the fact that two primary colours mixed together produce a secondary colour. It uses the layering of colours to produce unique hues. As shown in Fig. 8.3, material is fed into the machine from the input trolley, but instead of a caustic chemical, dye is used. Mostly sulphur dyes used for overdyeing purposes. In the dye bath, along with antioxidants, a wetting agent is also used.

From the dye bath the fabric passes through a stabilizer tank where the fabric is washed with continuous steam and water. After the stabilizer tank the fabric passes to CWT 1–7. In the process, acetic acid and hydrogen peroxide are added to Washer Tank 2 at a constant flow of 500 mL/min, and at Washer Tank 4 sequestering agents are added at a constant rate (450 mL/min). After the washing process, the processed material passes over the vertical drying cylinder and the finally to plaiter before the output. Data were collected from the final drain of the CDR for all three of these operations and are listed in Table 8.6.

The results in Table 8.6 shows that, out of all activities, the COD values were highest in the case of the desizing effluent and the TDS values were highest in the case of the caustisizing effluent. The overdyeing effluent had TDS and COD values both comparatively less than the other two operations did. The effluent pH was slightly acidic in the overdyeing process, and for the caustisizing and desizing processes it was highly alkaline.

#### *8.7.3.2 Impact on environment*

To analyze the actual properties of the chemicals in terms of their environmental impact, the TDS and COD levels of the chemicals used in the CDR were analyzed in the laboratory, where 1 mL of chemical was mixed with 0.5 L of water and a test was performed. Test results are provided in Table 8.7.

To know the effect of different chemicals and their TDS and COD, both tests were conducted for all chemicals used in the CDR. For the desizing process, three chemicals were used; their pollution assessment results are shown in Table 8.7. The results confirmed that the nonionic wetting agent in an open tank causes high COD and TDS in the final effluent and the nonionic emulsifying agent has higher COD values than does the enzyme. Regarding the amount of use, the nonionic wetting agent caused a high TDS and COD.

In the case of the caustisizing process, results confirmed that the neutralizing agent (acid) had the highest amount of COD compared with the antiback-staining agent. Furthermore, the study showed that at the washer tank where the antiback-staining agent was added, no increase in COD value was noted. However, there is a great increase in COD in the washer tank in which the neutralizing agent (acid) was added. In the case of TDS, the caustic solution was the main reason for a high TDS in the output effluent. Although the antiback-staining agent had a high TDS near the caustic agent, its effect on the final effluent was marginal because of its lower use, as shown in Table 8.7.

**Table 8.6** **Pollution load of effluent at the overall outlet of the continuous dyeing range machine**

| S.no | pH | Conductivity | Total dissolved solids (parts per million) | Chemical oxygen demand (mg/L) |
|---|---|---|---|---|
| **For desizing** | | | | |
| 1 | 7.89 | 17,700 | 8,900 | 8,960 |
| 2 | 10.09 | 18,700 | 5,700 | 12,320 |
| 3 | 9.44 | 11,330 | 3,850 | 7,920 |
| 4 | 9.2 | 12,700 | 9,650 | 9,720 |
| 5 | 8.94 | 11,080 | 5,600 | 7,000 |
| Average | **9.112** | **14,302** | **6,740** | **9,184** |
| **For caustisizing** | | | | |
| 1 | 11.95 | 77,900 | 31,000 | 2,720 |
| 2 | 11.80 | 82,500 | 39,500 | 3,120 |
| 3 | 12.07 | 88,300 | 37,600 | 3,440 |
| 4 | 12.05 | 72,000 | 36,000 | 4,200 |
| 5 | 12.18 | 22,200 | 30,600 | 5,680 |
| Average | **12.01** | **68,580** | **34,940** | **3,832** |
| **For overdyeing** | | | | |
| 1 | 5.45 | 6,100 | 6,500 | 3,920 |
| 2 | 4.67 | 5,000 | 4,000 | 5,440 |
| 3 | 4.30 | 6,500 | 5,800 | 5,520 |
| 4 | 4.20 | 4,600 | 1,500 | 4,240 |
| 5 | 5.12 | 5,520 | 4,850 | 5,420 |
| Average | **4.748** | **5,544** | **4,530** | **4,908** |

In the case of the overdyeing process, test results confirmed that the antioxidant in the dye pad and the addition of acetic acid caused a high COD in final effluent. Table 8.7 shows that sequestering agents, wetting agents and antioxidants had higher TDS values than dyes. Regarding the amount of use, the antioxidant caused a high TDS and COD; acetic acid had second place in COD, and sequestering agents in TDS. Although, wetting agents had high COD and TDS values, because of their lower consumption, their effect on the final effluent is lesser.

### *8.7.4 Finishing*

Finishing is the last manufacturing step in the production of textile fabrics. Finishing is not restricted to wet processing alone; any operation to improve the appearance or usefulness of a fabric after it leaves the loom can be considered a finishing step. The finish can be achieved by either chemicals that change the fabric's aesthetics and/or physical properties or changes in the texture or surface characteristics brought

Table 8.7 **Continuous dyeing range chemicals and their pollution loads**

| Chemicals | Chemical oxygen demand (mg/L) | Total suspended solids (parts per million) | Use (kg) (total consumption per month) |
|---|---|---|---|
| **Desizing** | | | |
| α-Amylase enzyme | 468,000 | 417,300 | 30 |
| Nonionic wetting agent | 1,720,000 | 900,000 | 14 |
| Nonionic emulsifying agent | 1,464,000 | 200,000 | 14 |
| **Caustisizing** | | | |
| Caustic solution | 26,800 | 895,000 | 22,530 |
| Wetting agent | 424,000 | 260,000 | 17 |
| Antiback-staining agent | 1,888,000 | 870,000 | 354.45 |
| Neutralizing agent | 280,000 | 40,000 | 4,400 |
| **Overdyeing** | | | |
| Acetic acid | 312,000 | 10,000 | 742 (rate 500 mL/min) |
| Hydrogen peroxide | 8,000 | 5,000 | 416 (rate 500 mL/min) |
| Sequestering agents | 352,000 | 495,000 | 155.4 (rate 450 mL/min) |
| Antioxidant | 256,000 | 425,000 | 851.5 |
| Wetting agent | 832,000 | 585,000 | 109 |

about by physically manipulating the fabric with mechanical devices. It can also be a combination of the two. From the finishing department's final outlet, the effluent was collected at the main time of the effluent outlet for a month. The collected samples were analyzed for COD, TDS, pH and conductivity values. Results are shown in Table 8.8.

#### 8.7.4.1 *Impact on environment*

To analyze the actual properties of the chemicals in terms of their environmental impact, the TDS and COD levels of the chemicals used in the finishing process were analyzed in the laboratory, where 1 mL of chemical was mixed with 0.5 L of water and the test was performed. The test results are provided in Table 8.9.

Test results of the chemical analysis are provided in Table 8.9. Results showed that the highest COD value of 1,328,000 mg/L was noted for alkali stable wetting, rewetting and deaerating agent. In the case of TDS, the same chemical was found to have a greater impact on the finishing effluent. Compared with the quantity used, even the COD and TDS values were less; the impact of cationic softening agent was also nonnegligible.

**Table 8.8** **Pollution load of effluent at overall outlet of finishing machine**

| S.no | pH | Conductivity | Total dissolved solid (parts per million) | Chemical oxygen demand (mg/L) |
|---|---|---|---|---|
| 1 | 3.18 | 3,200 | 2,800 | 11,200 |
| 2 | 4.1 | 9,100 | 4,900 | 13,440 |
| 3 | 4.69 | 12,200 | 8,100 | 14,720 |
| 4 | 4.55 | 11,800 | 7,000 | 12,160 |
| 5 | 4.68 | 12,400 | 9,800 | 5,440 |
| 6 | 4.25 | 14,200 | 7,500 | 19,040 |
| 7 | 4.35 | 7,300 | 8,900 | 20,000 |
| 8 | 4.8 | 9,200 | 2,200 | 3,040 |
| 9 | 4.18 | 18,700 | 9,200 | 17,200 |
| 10 | 4.53 | 9,600 | 4,000 | 10,640 |
| 11 | 6.69 | 11,300 | 9,800 | 15,040 |
| 12 | 5.28 | 15,400 | 2,200 | 10,840 |
| 13 | 5.92 | 11,800 | 4,000 | 14,000 |
| Average | **4.71** | **11,246.15** | **6,184.62** | **12,827.69** |

**Table 8.9** **Finishing machine chemicals and their pollution loads**

| S.no | Chemicals | Total dissolved solids (parts per million) | Chemical oxygen demand (mg/L) | Use (kg) (total consumption per month) |
|---|---|---|---|---|
| 1 | Cationic softening agent | 210,000 | 352,000 | 3,441 |
| 2 | Alkali stable wetting, rewetting and deaerating agent | 900,000 | 1,328,000 | 1,286.85 |
| 3 | Neutralizing agent | 75,000 | 300,000 | 783.75 |
| 4 | Dry and wet crock fastness improving agent for sulphur over dyed fabric | 460,000 | 688,000 | 2,400 |

## 8.7.5 Discussion

The study was performed over 6 months in the denim processing industry. Results are mentioned in terms of the pollution levels in the effluent from each section of the factory. The results indicated that most of the time the pollution in the effluent resulted from the higher load of the individual chemicals used. Table 8.10 lists the overall pollution levels of the individual departments. The table shows that the higher COD

**Table 8.10 Overall results of chemicals and their pollution loads**

| Machine | pH | Conductivity | Total dissolved solids | Chemical oxygen demand |
|---|---|---|---|---|
| Desizer | 8.44 | 12,770.67 | 7,576.667 | **8,992.667** |
| Mercerizer | **12.95** | **57,928** | **34,072.67** | 4,859.26 |
| CDR–desizer | 9.112 | 14,302 | 6,740 | **9,184** |
| CDR–mercerizer | **12.01** | **68,580** | **34,940** | 3,832 |
| CDR–overdyeing | 4.748 | 5,544 | 4,530 | 4,908 |
| Finishing machines | 4.71 | 11,246.15 | 6,184.62 | **12,827.69** |

*CDR*, continuous dyeing range.

levels in the effluent were found in the finishing and desizing processes. The highest level of TDS noted in the mercerization process. This was also noted in both the CDR and desizing machine. The highest levels of conductivity and pH were noted in the mercerizing effluent.

Data from the study (Table 8.10) showed the overall contributing factors for COD, TDS, conductivity and pH in the denim effluent. The effluent from the finishing and desizing processes possessed a higher amount of COD than for the other processing departments. From the chemical analysis process it was found that the wetting agent used in the finishing and desizing processes might be the reason. If we consider the quantity of the use, the desizing enzyme was highest, and so the desizing enzyme was a secondary factor for the high COD generation in the effluent. The effluent from the mercerization process possessed a higher TDS content than any other process. This is because of the use of caustic and neutralizing agents, as confirmed in the chemical analysis results. The caustic agents are used in this process to improve the wetting power of the lye solution and accelerate their even penetration into the fiber and at the final stage of the washing process during mercerizing; the neutralizing agent is added to neutralize the pH of the fabric before it used for the next step of the finishing process. The values of pH, conductivity and TDS are interrelated; hence the mercerizer effluent had the highest pH and conductivity values.

## 8.8 Effluent treatment process

The effluent treatment process in the factory is performed by two different means. First, the effluent from the dyeing process was treated separately, and second, the effluent from finishing (overdyeing and finishing) and pretreatment (desizing, mercerization, desizing in CDR and caustisizing in CDR) was treated separately. This is because of the higher colour content in the dye effluent than the finishing effluent. The effluent process flows in both treatment plants (finishing and dyeing) are given in Figs. 8.9 and 8.10.

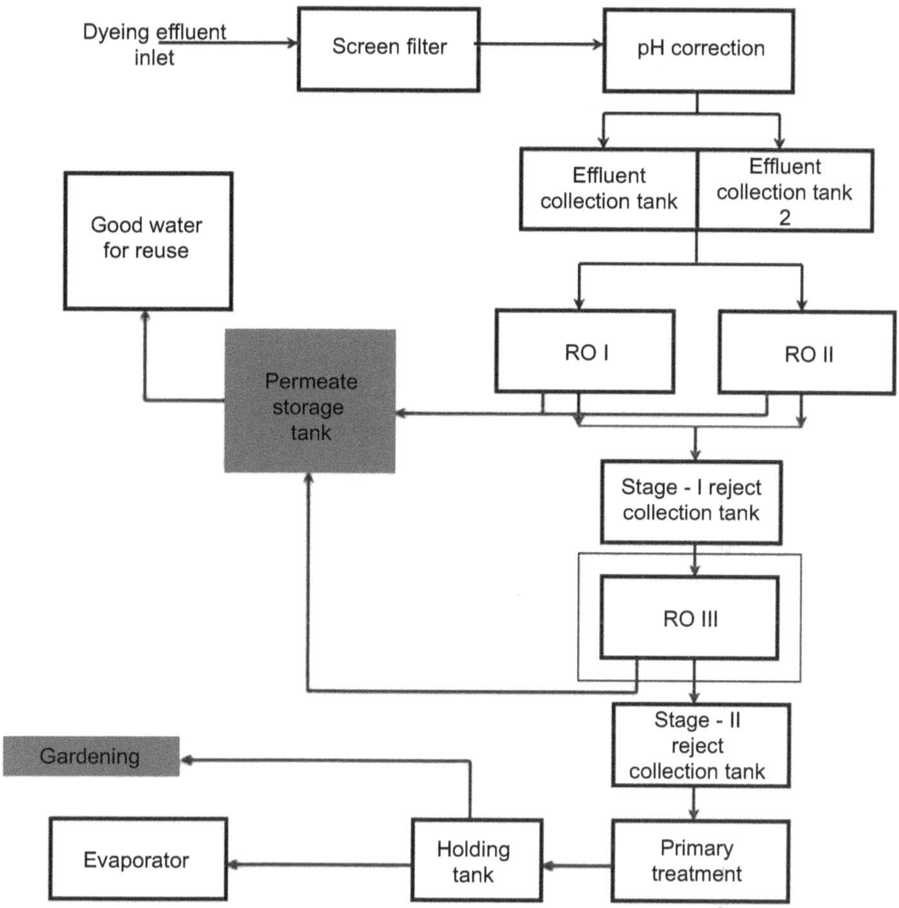


**Figure 8.9** Effluent treatment plant flow process for dyeing effluent. *RO*, reverse osmosis.

In this process, the final rejected effluents from both plants were connected to a centralized holding tank. All of the rejected effluent was stored in a holding tank for a particular time and then passed to the evaporator. After the evaporation process, the remaining solid sedimentation or sludge was stored or disposed of separately. In this analysis, the study was performed to evaluate the efficiency of the finishing ETP. To identify this, inlet effluent pollution levels (overall from the factory) were calculated for 3 months and the output effluent (final accepted/treated) were also analyzed for that period. The calculated data regarding the pollution analysis of the industry before and after treatment are given in Appendix Tables 1 and 2.

The collected data were averaged and the percent pollution removal efficiency of the ETP was calculated. The result indicated that the treatment plant removed or reduced the pH of the effluent at maximums of 84.31% and 65.8%, respectively, of the wastewater conductivity. Next to the pH and conductivity, the maximum percentage of cleaning was noted in the TDS, which was 75.11%. However, the removal of COD,

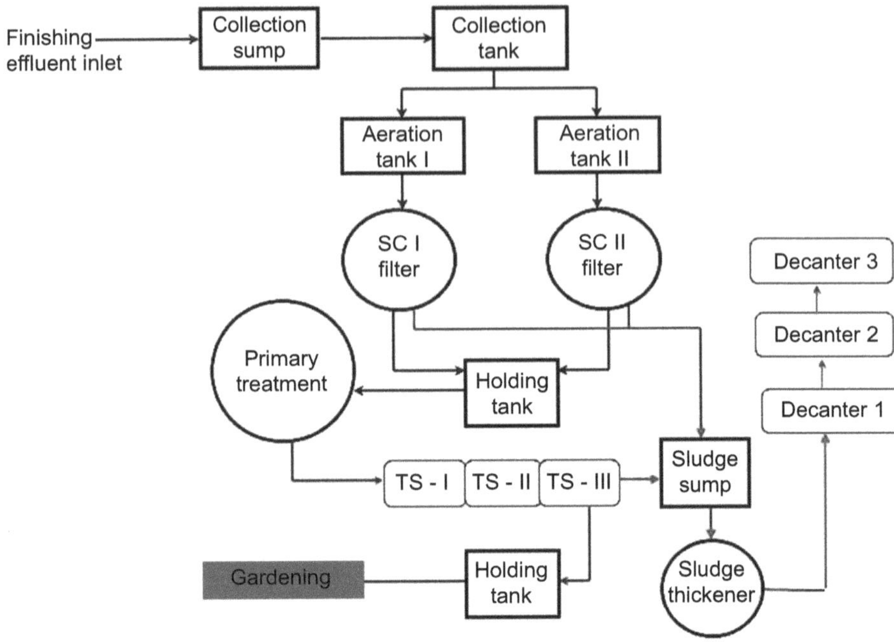


**Figure 8.10** Effluent treatment plant flow process for finishing effluent. *SC*, secondary clarifier; *TS*, tube settler

## Table 8.11 Calculation of pollution removal efficiency of effluent treatment plant (%)

| Pollution removal efficiency of effluent treatment plant (%) | |
|---|---|
| pH removal efficiency | 84.318 |
| Conductivity removal efficiency | 65.800 |
| Total dissolved solids removal efficiency | 75.113 |
| Chemical oxygen demand removal efficiency | 16.600 |
| Total suspended solids removal efficiency | 7.482 |
| Colour removal efficiency | 12.433 |
| **Average** | **43.624** |

TSS and colour reduction percentages was much less at 16.1%, 7.42% and 12.433%, respectively. Table 8.11 describes the overall performance of ETP in the industry.

This analysis from the denim processing industry provides a basic idea of the approximate pollution level in the effluent from the industry. The study also highlights the possible chemicals with high levels of pollution load used in the process. The ETP analysis results showed that even after the ETP, the quality of the wastewater did not fall under the norms of the pollution control board. As per the pollution control board norms and regulations for industrial wastewater that is discharged into the environment [23–25], standard allowable limits are described in Table 8.12.

Table 8.12 **Standards and allowable limits for various types of pollution**

| Parameter | Value range | Examination method |
|---|---|---|
| pH value | 5.5–9.0 | pH meter |
| Total dissolved solids (TDS) | • Not more than **3000 mg/L**, depending on receiving water or type of industry under consideration of pollution control committee (PCC) but not exceeding 5000 mg/L<br>• Not more than **5000 mg/L** exceed TDS of receiving water with salinity of more than 2000 mg/L or TDS of sea if discharged to sea | Dry evaporation 103–105°C, 1 h |
| Suspended solids | • Not more than **50 mg/L**, depending on receiving water or type of industry or waste-water treatment system under consideration of PCC but not exceeding 150 mg/L | Glass fiber filter disc |
| Temperature | Not more than 40°C | Thermometer during sampling |
| Colour and odour | Not objectionable | Not specified |
| Fat, oil and grease | Not more than **5.0 mg/L**, depending on receiving water or type of industry under consideration of PCC but not exceeding 15.0 mg/L | Solvent extraction by weight |
| Formaldehyde | Not more than **1.0 mg/L** | Spectrophotometry |
| Chemical oxygen demand | Not more than **250 mg/L**, depending on receiving water or type of industry | Potassium dichromate digestion |
| Biochemical oxygen demand | Not more than **20 mg/L**, depending on receiving water or type of industry under consideration of PCC but not exceeding 60 mg/L | Azide modification at 20°C, 5 days |

The results show that the average TDS, TSS, COD and colour values of the treated effluent are still higher than permissible limits. The conductivity value is also considerably high; this is an easy measurement to make and relates closely to the TDS content of water. In general, the TDS is about 70% of the conductivity. Fig. 8.11 compares the inlet and outlet parameters of the effluent along with the standard values.

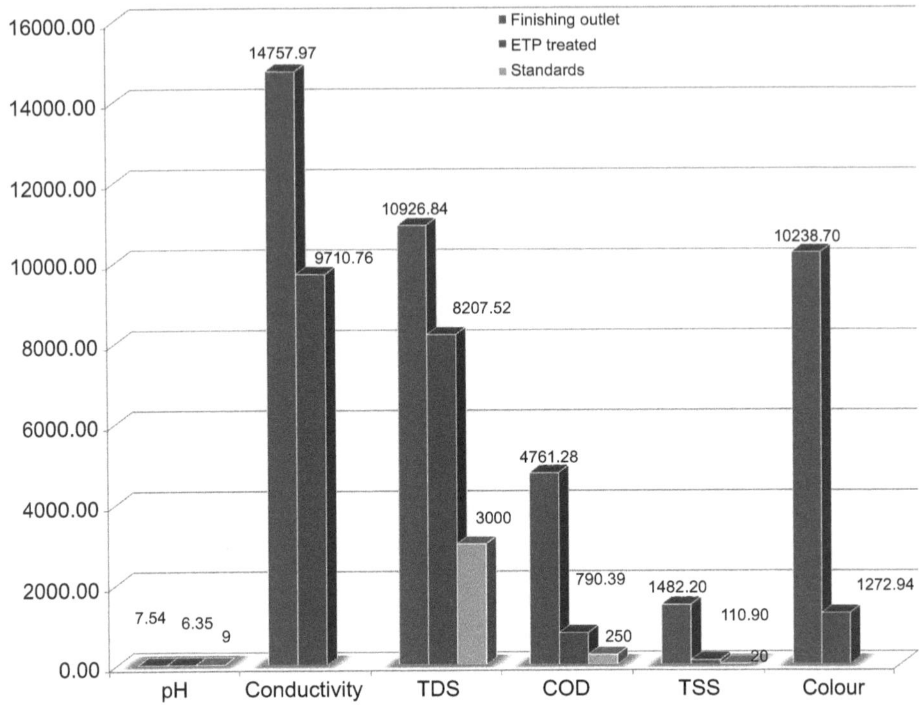


**Figure 8.11** Comparison of average inlet and outlet pollution level in the effluent treatment plant (ETP). *COD*, chemical oxygen demand; *TDS*, total dissolved solids; *TSS*, total suspended solids.

## 8.9 Suggestions to reduce effluent pollution

Based on observations from the different pretreatment and finishing departments, the results indicated that the main reason for the high-level pollution in the final outlet effluent is that different chemicals are used in the processing stages. It is impossible to avoid using those chemicals for the process from the point of view of a good-quality end product. Hence, the pollution assessment committee suggested some basic pollution reduction strategies to the company to adopt, to reduce overall effluent contamination. The following simple methods are suggested and should be followed to reduce the pollution load in the final effluent.

### *8.9.1 Waste segregation*

Effluent from different parts of the industry is segregated based on its pollution load and concentration.

- For example, effluent from the desizing machine consists of a higher COD value than other effluent from the industry. The study results identified that enzymes used for the desizing process and the wetting agent used in the desizing (bath) process are the main reason. Hence, effluent from the first washer tank (CWT 1) usually consists of a higher amount of desizing and wetting agent content compared with the rest of the washer tanks can be treated separately.

- In the case of the mercerizer, this particular machine effluent was identified as a main source of TDS. This is because of the use of sodium hydroxide and a caustic agent in this process. Hence, effluent from the next washer tank after the open tank (CWT 1) can be taken separately for treatment without mixing it with the other effluent. This will reduce the TDS of the overall factory effluent.
- In the same manner, with the mercerizer, at CWT 4 the neutralizing agent was added to neutralize the fabric's pH from alkali to acidic. By redirecting or treating this effluent separately, it may reduce the COD level of the final effluent of the machine

## 8.9.2 Recovery and reuse

- Recovery of some valuable material from the effluent stream substantially reduces pollution, especially the COD and TDS load. Recovered materials can be reused for another process, which will leads to cost savings.
- For example, in the mercerizer, the used caustic chemical from the open tank is taken for recovery and reused; in the same manner, the effluent from the CWT can also be passed to a recovery plant (considering technical requirements).

## 8.9.3 Substitution of alternative chemicals

- An overall wetting agent was identified as the main cause for both TDS and COD in finishing machine and desizer. In the case of the desizer, an enzyme has a second major role in effluent contamination.
- In the mercerizer, a caustic chemical causes high TDS, and acid, which is used for neutralizing in CWT 4, cause high COD in the final effluent.
- In the case of the finishing machine, the major pollutant is the wetting and softening agents. Because of the amount at which they are used, their COD and TDS levels are heavily affected.
- Recommendations were suggested for the industry based on the following aspects:
  - The industry generally using a particular brand chemical (manufacturer) for all machines irrespective of their specialization, owing to the price range.
  - Hence, the suggestion was made to opt for alternative chemicals from different manufacturers with a low pollution level (by considering the best price and quality).
  - The suggestions was also made to conduct regular pollution level assessments at different machines and ETP sites, to monitor fluctuations in the effluent pollution load.
  - The suggestion was made to conduct optimization studies to minimize the use of chemicals in every operation to obtain the same end results in terms of cost and quality (on a long-term basis).
  - The study also suggested that the recycling plant in the mercerizer for caustic recovery is effective. In the same manner, the possibility of recovering different chemicals may be identified at the particular washer tanks; those effluents can treated separately, and after recovery, can be passed to the ETP (e.g., in the mercerizer, CWT, neutralizer can be used). It is an acid for neutralizing the fabric's pH, the major COD in mercer effluent caused by this acid. Hence, the possibility of recovering the acid from that particular washer tank effluent can be analyzed. This helps in two ways: The recovered material can be reused and the pollution load in the final effluent to ETP can be reduced.

## 8.10 Developments in effluent treatment methods

Indigo dye and indigo dye mixed denim production effluent such as from industrial output is toxic and causes injury to human health. These kinds of toxic dyes resist biodegradation and decrease the efficiency of biological wastewater treatment plants. In addition, physical and chemical methods traditionally used to remove colour and chemicals from textile effluents have operational problems such as sludge generation, membrane fouling, and phase changes in the pollutants. This section of the chapter details various advanced oxidation processes that can be adopted in the denim effluent treatment process.

### 8.10.1 Ozonation

Ozonation is a possible method to treat denim effluent. This process involves the development of hydroxyl radicals and their subsequent reaction with denim wastewater. This reaction is the most powerful oxidizing reaction after fluorine [26]. Chemical oxidation with ozone is one of the most suitable chemical oxidation processes for effective colour removal from textile effluents, with simultaneous interaction and breakdown of refractory organic matter resistant to biodegradation, without leading to sludge production [27]. Research conducted on ozonation efficiency in effluent treatment by Doğruel et al. detailed the effect of the ozone flow rate and contact time, and found that a flow of ozone around 60 mg/min was determined to be optimal, providing almost complete colour removal [27].

### 8.10.2 Electrolysis

Electro-oxidation is the most promising alternative way to treat aqueous streams containing heavy organic compounds via the simultaneous evolution of oxygen at an anode surface. It is probably the most adequate tool in aqueous effluent treatment and is ideally suited for the current era in which environmental considerations are always of concern [25]. In electro-oxidation, the main reagent used to treat the water stream is electrons, which react with organic waste in the textile wastewater and remove organic matter without generating secondary pollutants [28]. The unique features of the electrochemical oxidation process are its simplicity and robustness of structure and operation. It is possible that this process could be developed as a cost-effective technology for textile wastewater treatment [29].

### 8.10.3 Photocatalysis

Photosensitization occurs as a result of the initial absorption of radiation by coloured molecules in wastewater, which act as photosensitizers. Photosensitization is based on the use of wavelengths longer than those in the UV region, in the existence of a suitable conductor to degrade the organic compounds to $CO_2$, $H_2O$ and mineral acid [30]. Titanium dioxide and zinc oxide have good photocatalytic properties; both catalysts are promising substrates for photodegradation of water pollutants because they demonstrate appropriate activity in the range of solar radiation. In general, the photo-degradation of denim industry effluent/indigo dye depends on the initial concentration of the dye, the amount of $TiO_2$ used for the photocatalytic process, the pH of the solution, the presence of inorganic anions, the temperature, and the addition of different concentrations of hydrogen peroxide. The working efficacy of the photocatalytic process strongly depends on the geometry of the photoreactor, which should enable all photocatalyst particles to be fully illuminated [31].

### 8.10.4 Sonolysis

Sonication has been widely known to induce radical reactions. This process applies ultrasound with a frequency ranging between 20 kHz and 500 MHz. When ultrasound propagates in a liquid, it promotes the formation of cavitational bubbles whose collapse is associated with both physical and chemical effects. Wastewater from the textile industry contains a high organic load and toxicity. Owing to the characteristics of these effluents, biological processes cannot be used directly because no COD removal is achieved biologically if the effluent remains same. The application of ultrasound waves causes sonolytic degradation of aqueous organic pollutants and the sonochemical synthesis of chemicals. These phenomena lead to sonolytic splitting of water as well as pyrolysis of a vaporized molecule [32].

### 8.10.5 Radiolysis

Ionizing radiation is a promising method for the treatment of textile wastewater because the effect of radiation can be intensified in an aqueous solution, in which dye molecules are degraded effectively by the primary products formed through the radiolysis of water [33]. The radiation used in the radiolysis process has sufficient energy to dislodge electrons from atoms and molecules and convert them to electrically charged particles called ions. Further reactions of these species lead to the formation of free radicals which are usually reactive and eventually will lead to chemical reactions [34].

### 8.10.6 Other processes

Harrelkaset al. analyzed the effects of photocatalysis and combined anaerobic–photocatalytic treatment of textile dye mixed wastewater [35]. For this study they used a photocatalytic process based on immobilized titanium dioxide. The treatment was applied on both the autooxidized chemically reduced azo dyes and normal dyes. The results were promising in that photocatalysis process able to remove more than 90% of colour from crude as well as autooxidized chemically reduced dye solutions. UV absorbance and COD were also removed up to 50% on average.

Several scientific researchers reported that a biological process consisting of an anaerobic and aerobic process in series is promising for treating textile wastewater [36–38]. In this method, in the first stage, anaerobic microorganisms reduced the colour and increased the biodegradability of the wastewater; in the second stage, the reduction in bacterial count rendered the effluent more amenable to subsequent aerobic treatment [39].

Other than these methods, several hybrid treatment systems are available to treat textile effluent [40]:

- a combination of advanced oxidation processes
  - different photochemical processes
  - photochemical/electrochemical
  - sonolysis and other advanced oxidation processes
- combination: advanced oxidation processes and other physicochemical processes
  - coagulation-based combinations
  - adsorption-based combinations
  - membrane-based combinations
- biological treatment-based combinations
  - combination of biological processes
  - hybrid technologies based on biological processes

## 8.11 Summary

This chapter summarizes various process methods, materials and chemicals used in the denim manufacturing process and their hazardous impact on the environment. The case study presented in this chapter details the common manufacturing technology used in the denim processing industry. The study clearly depicts the various chemicals used in each stage of denim fabric production and pollution loads in terms of COD, BOD, pH, TDS, conductivity and colour. The study also analyzes an existing ETP and its flow process, along with the pollution level of effluent at different manufacturing processes. Finally, it analyzes the efficiency of pollution removal at the treatment plant in terms of these factors and suggests potential strategies and new methods that can be adopted in the denim industry to reduce the pollution load of the effluent.

## References

[1] Driving demand for denim jeans. http://www.cottoninc.com/corporate/Market-Data/SupplyChainInsights/Driving-Demand-For-Denim-Jeans/.

[2] M. Saini. India's denim market seen primed for growth. http://wwd.com/markets-news/denim/india-denim-market-growth-10388888/.

[3] Marcucci M, Nosenzo G, Capanelli G, Ciabatti I, Corrieri D, Ciardelli G. Treatment and reuse of textile effluents based on new ultrafiltration and other membrane technologies. Desalination 2001;138:75–82.

[4] Wood KN, Bishop AL. Effluent guidelines compliance through waste minimization. Water Sci Technol 1992;26(1–2):301.

[5] Snyder RL. Fugitive denim: a moving story of people and pants in a borderless world. 1st ed. New York (NY): W.W. Norton & Company, Inc.; 2008.

[6] Wambuguh D, Chianelli RR. Indigo dye waste recovery from blue denim textile effluent: a by-product synergy approach. New J Chem 2008;32(12):2189–94.

[7] Levi Strauss & Co. The life cycle of a jean – understanding the environmental impact of a pair of Levi's 501 jeans. http://levistrauss.com/wp-content/uploads/2015/03/Full-LCA-Results-Deck- FINAL.pdf.
[8] Parmar MS, Satsangi SS, Prakash J. Denim – a fabric for all. Delhi: Northern India Textile Research Association; 1996.
[9] Schrott W, Paul R. Environmental impacts of denim manufacture. In: Paul R, editor. Denim manufacture, finishing and applications. Cambridge (UK): Woodhead Publisher; 2015.
[10] Allegre C, Moulin P, Maisseu M, Charbit F. Treatment and reuse of reactive dyeing effluents. J Membr Sci 2006;269:15–34.
[11] Lash D, Kominek E. Primary-waste-treatment methods. In: Industrial wastewater and solid waste engineering. New York: McGraw-Hill; 1980. p. 81–123.
[12] A report on waste-water treatment technologies: a general review, Economic and Social Commission for Western Asia. New York: United Nations; 2003.
[13] Standard methods for the examination of water and wastewater. 18th ed. Washington, DC: PHA, AWWA, WEF; 1992.
[14] Clair NS, McCarty PL, Parkin GF. Chemistry for environmental engineering and science. 5th ed. New York: McGraw-Hill; 2003.
[15] Clescerl LS, Greenberg AE, Eaton AD. Standard methods for examination of water & wastewater. 20th ed. Washington, DC: American Public Health Association; 1999.
[16] Method 507. Standard methods for the examination of water and wastewater. 16th ed. 1985. p. 525–32.
[17] Method 218B. Standard methods for the examination of water and wastewater. 13th ed. 1971. p. 477–81.
[18] Total Dissolved Solids. www.wikipedia.org/wiki/Total_dissolved_solids.
[19] Howard CS. Determination of total dissolved solids in water analysis. Ind Eng Chem Anal Ed 1993;5:4.
[20] Joseph MM, Michael DM, James HW. Introduction to environmental science. 2nd ed. New York: W.H. Freeman and Company; 1980.
[21] Achwal WB. Environmental aspects of textile chemical processing (part I). Colourage 1990;37(9):40.
[22] Batra SH. Other long vegetable fibers: abaca, banana, sisal, henequen, flax, ramie, hemp, sunn, and coir. In: Lewin M, Pearce EM, editors. Handbook of fiber science and technology. Fiber chemistry, vol. IV. New York: Marcel Dekker; 1985. p. 15–22.
[23] Notification the Ministry of Science, Technology and Environment, No. 3, B.E.2539 issued under the Enhancement and Conservation of the Bibliography, National Environmental Quality Act B.E.25 35(1992), published in the Royal Government Gazette, vol. 113, Part 13 D, dated February 13, 1996.
[24] Environment Protection Act, S.O. 844 (E), dated the 1986. Central Pollution Control Board, India. http://mpcb.gov.in/relatedtopics/pdf/CETP%20Standards.pdf.
[25] www.cpcb.nic.in. (GSR 801 (E), EPA, 1986, dated December 31, 1993).
[26] Neamtu M, Yediler A, Siminiceanu I, Macoveanu M, Kettrup A. Decolourization of disperse red 354 azo dye in water by several oxidation processes – a comparative study. Dyes Pigment 2004;60:61–8.
[27] Doğruel S, Germirli Babuna F, Kabdaşl I, Güçlü I, Orhon D. Effect of stream segregation on ozonation for the removal of significant COD fractions from textile wastewater. J Chem Techol Biotechnol 2002;78:6–14.
[28] Rajeshwar K, Ibanez JG. Fundamentals and applications of pollution sensors and pollutant treatment. In: Environmental electrochemistry. USA: Academic Press; 1997.

[29] Mohan N, Balasubramanian N, Basha CA. Electrochemical oxidation of textile wastewater and its reuse. J Hazard Mater 2007;147:644.
[30] Li M, Feng C, Hu W, Zhang Z, Sugiura N. Electrochemical degradation of phenol using electrodes of Ti/RuO2-Pt and Ti/IrO2-Pt. J Hazard Mater 2008;162:455.
[31] Fujishima A, Honda K. Electrochemical photolysis of water at a semiconductor electrode. Nature 1972;238:37–8. http://dx.doi.org/10.1038/238037a0.
[32] Mao K, Li Y, Zhang H, Zhang W, Yan W. Photocatalytical degradation of 17-ethinylestradiol and inactivation of *Escherichia coli* using Ag-modified TiO2 nanotube ar- rays. Clean – Soil Air Water 2013;41:455–62. http://dx.doi.org/10.1002/clen.201100698.
[33] Dahlan KZM. Radaition sciences. Advances in tissue banking, vol. 5. Delwhich;World Scientific Publishing Co. Pte. Ltd.; 2001.
[34] Solphan D. Decolouration and degradation of some textile dyes by gamma irradiation. Ankara (Turkey): Hacettepe University, Department of Chemistry; 2001.
[35] Harrelkas F, Paulo A, Alves MM, El Khadir L, Zahraa O, Pons MN, Van der Zee FP. Photocatalytic and combined anaerobic–photocatalytic treatment of textile dyes. Chemosphere 2008;72:1816–22.
[36] Loyd CK, Bordman GD, Michelson DL. Anaerobic/aerobic treatment of a textile dye wastewater, pre-printed for mid-Atlantic industrial waste conference, July 15–17, Morgantown, WV. 1991.
[37] Seshadri S, Bishop PL. Anaerobic/aerobic treatment of selected azo dyes in wastewater. Waste Manag 1994;14(2):127–317.
[38] Zaoyan Y, Ke S, Guangliang S, Fan Y, Jinshan D, Haunian M. Anaerobic-aerobic treatment of a dye wastewater by combination of RBC with activated sludge. Water Sci Technol 1992;26(9–11):2093–6.
[39] Minke R, Rott U. Anaerobic treatment of split flow wastewater and concentrates from textile processing industry. Water Sci Technol 1999;40(1):169–76.
[40] Ibney HF, Yamamoto K, Fukushi K. Hybrid treatment systems for dye wastewater. Crit Rev Environ Sci Technol 1999;37(4):315–77.

# Appendix

**Table 1 Pollution analysis of factory's overall effluent before treatment from finishing process**

| Date (2015–2016) | pH | Conductivity (micro mhos/cm) | Total dissolved solids (mg/L) | Chemical oxygen demand (mg/L) | Total suspended solids (mg/L) | Colour (Pt/Co units) |
|---|---|---|---|---|---|---|
| 8 November | 7.1 | 15,410 | 12,200 | 5,440 | 1,710 | 14,620 |
| 9 November | 7.21 | 12,610 | – | 5,500 | – | – |
| 10 November | 7.41 | 11,200 | 8,900 | 5,970 | 1,230 | 16,410 |
| 11 November | 7.6 | 10,630 | 7,400 | 4,440 | 1,360 | 12,610 |
| 12 November | 6.9 | 13,850 | – | 5,040 | 1,010 | 8,960 |

Table 1 **Continued**

| Date (2015–2016) | pH | Conductivity (micro mhos/cm) | Total dissolved solids (mg/L) | Chemical oxygen demand (mg/L) | Total suspended solids (mg/L) | Colour (Pt/Co units) |
|---|---|---|---|---|---|---|
| 13 November | 6.91 | 10,680 | – | 4,480 | 1,090 | 13,620 |
| 14 November | 7.4 | 11,380 | 8,100 | 5,680 | 1,310 | 12,630 |
| 15 November | 7.1 | 5,900 | 5,500 | 5,360 | 1,040 | 8,960 |
| 16 November | 8.1 | 10,560 | – | 5,400 | – | – |
| 17 November | 7.6 | 11,100 | 10,000 | 5,600 | 1,310 | 14,630 |
| 18 November | 7.3 | 13,450 | 9,620 | 5,310 | 1,240 | 11,620 |
| 19 November | 7.11 | 17,400 | 13,460 | 4,960 | 1,310 | 16,470 |
| 20 November | 7.6 | 15,580 | 13,500 | 5,400 | 1,410 | 15,610 |
| 21 November | 7.6 | 16,460 | 12,550 | 6,400 | 1,610 | 17,620 |
| 22 November | 7.66 | 14,860 | 12,600 | 5,410 | 1,060 | 12,760 |
| 23 November | 7.3 | 14,500 | – | 5,860 | – | – |
| 24 November | 7.9 | 16,700 | 13,950 | 5,600 | 1,670 | 14,670 |
| 25 November | 7.4 | 19,600 | 14,550 | 4,670 | 1,810 | 16,420 |
| 26 November | 7.2 | 18,840 | 13,100 | 4,960 | 1,040 | 12,610 |
| 27 November | 7.31 | 29,600 | 18,000 | 5,320 | 1,000 | 8,920 |
| 28 November | 7.5 | 15,600 | 13,900 | 5,480 | 1,470 | 12,620 |
| 29 November | 7.1 | 17,510 | 12,550 | 4,650 | 1,670 | 18,410 |
| 30 November | 7.2 | 10,200 | 8,550 | 4,600 | – | – |
| 1 December | 7.4 | 13,200 | 11,700 | 5,016 | 1,360 | 12,610 |
| 2 December | 7.6 | 12,710 | 11,750 | 5,380 | 2,000 | 10,400 |
| 3 December | 7.84 | 19,750 | 15,350 | 5,643 | 1,910 | 16,940 |
| 4 December | 7.9 | 26,410 | 18,100 | 5,448 | 1,860 | 17,690 |
| 5 December | 7 | 19,220 | 15,700 | 5,300 | 2,650 | 12,700 |
| 6 December | 7.8 | 14,350 | 16,350 | 5,400 | 880 | 10,460 |
| 7 December | 7.65 | 12,450 | – | 5,851 | – | – |
| 8 December | 7.7 | 12,120 | 10,800 | 5,470 | 2,230 | 10,920 |
| 9 December | 7.4 | 16,360 | 13,300 | 5,200 | 2,100 | 12,680 |
| 10 December | 7.19 | 10,460 | 7,950 | 4,240 | 3,040 | 28,610 |
| 11 December | 7.64 | 19,100 | 15,750 | 5,160 | 1,860 | 12,680 |
| 12 December | 7.6 | 15,300 | 14,250 | 5,480 | – | – |
| 13 December | 7.1 | 15,960 | 12,450 | 5,221 | 1,230 | 6,800 |
| 14 December | 7.4 | 17,230 | – | 5,840 | – | – |
| 15 December | 7.6 | 12,210 | 8,900 | 5,400 | 1,070 | 7,800 |
| 16 December | 7.1 | 11,580 | 9,750 | 4,480 | 2,240 | 12,160 |
| 17 December | 7.9 | 18,500 | 15,900 | 660 | 2,870 | 16,410 |
| 18 December | 7.6 | 11,210 | 8,400 | 4,880 | 2,210 | 13,610 |
| 19 December | 7.9 | 11,350 | 7,700 | 3,920 | 2,940 | 19,310 |
| 20 December | 7.59 | 12,610 | 8,550 | 4,400 | 790 | 11,350 |
| 21 December | 7.9 | 13,420 | – | 5,240 | – | – |
| 22 December | 7.8 | 15,670 | 13,650 | 5,040 | 1,960 | 10,740 |

*Continued*

**Table 1** Continued

| Date (2015–2016) | pH | Conductivity (micro mhos/cm) | Total dissolved solids (mg/L) | Chemical oxygen demand (mg/L) | Total suspended solids (mg/L) | Colour (Pt/Co units) |
|---|---|---|---|---|---|---|
| 23 December | 7.1 | 22,610 | 14,450 | 5,260 | 1,660 | 7,940 |
| 24 December | 7.8 | 13,590 | 12,850 | 5,080 | 1,040 | 9,670 |
| 25 December | 7.64 | 17,240 | 12,400 | 3,920 | 1,440 | 7,560 |
| 26 December | 7.9 | 14,300 | 11,150 | 4,640 | 1,060 | 7,640 |
| 27 December | 7.56 | 11,140 | 10,550 | 5,040 | 1,700 | 8,380 |
| 28 December | 7.86 | 12,520 | – | 3,120 | – | – |
| 29 December | 7.8 | 10,590 | 9,100 | 3,760 | 760 | 6,400 |
| 30 December | 7.04 | 14,110 | 12,700 | 4,520 | 1,380 | 8,080 |
| 31 December | 7.45 | 13,970 | 10,700 | 4,600 | 1,380 | 7,200 |
| 1 January | 7.1 | 17,660 | 13,750 | 4,280 | 1,840 | 8,920 |
| 2 January | 7.08 | 10,420 | 9,000 | 4,520 | 2,120 | 8,820 |
| 3 January | 8.98 | 6,950 | – | – | – | – |
| 4 January | | | | | | |
| 5 January | 7.88 | 13,040 | 10,000 | 4,320 | 1,760 | 7,840 |
| 6 January | 7.4 | 11,040 | 7,300 | 6,000 | 1,620 | 7,760 |
| 7 January | 7.04 | 9,630 | 6,750 | 3,320 | 2,210 | 12,160 |
| 8 January | 7.1 | 10,670 | 7,450 | 3,600 | 1,870 | 10,960 |
| 9 January | 7.95 | 17,670 | 13,000 | 4,320 | 600 | 5,670 |
| 10 January | 7.06 | 10,670 | 7,600 | 4,600 | 1,040 | 8,970 |
| 11 January | 7.4 | 11,480 | – | 4,350 | – | – |
| 12 January | 7.3 | 20,610 | 13,450 | 3,440 | 1,040 | 7,970 |
| 13 January | 7.8 | 17,140 | 11,700 | 4,400 | 820 | 6,480 |
| 14 January | 7.11 | 11,630 | 7,400 | 5,040 | 4,340 | 17,480 |
| 15 January | 7.02 | 11,460 | 7,100 | 4,120 | 1,680 | 8,420 |
| 16 January | 7.87 | 18,960 | 11,600 | 7,460 | 1,040 | 7,860 |
| 17 January | 7.11 | 10,610 | 6,750 | 5,520 | 910 | 5,170 |
| 18 January | 9.33 | 6,220 | | 5,960 | | |
| 19 January | | | | | | |
| 20 January | 7.8 | 13,690 | 9,050 | 4,040 | 700 | 5,610 |
| 21 January | 7.98 | 13,070 | 9,800 | 4,160 | 4,300 | 18,500 |
| 22 January | 7.28 | 10,440 | 6,700 | 3,160 | 1,690 | 12,640 |
| 23 January | 7.66 | 16,710 | 10,350 | 3,600 | 1,040 | 7,970 |
| 24 January | 7.23 | 10,470 | 8,500 | 3,440 | 1,360 | 8,970 |
| 25 January | 7.4 | 18,630 | 12,450 | 4,040 | 1,000 | 10,280 |
| 26 January | 7.94 | 16,410 | 9,650 | 4,320 | 1,360 | 5,960 |
| 27 January | 7.23 | 14,210 | 10,240 | 2,350 | 1,440 | 5,960 |
| 28 January | 7.4 | 6,400 | 4,300 | 3,320 | 890 | 7,960 |
| 29 January | 7.9 | 14,610 | 9,500 | 3,000 | 660 | 5,680 |
| 30 January | 7.84 | 14,930 | 9,450 | 4,880 | 1,040 | 9,500 |
| 31 January | 7.64 | | 8,550 | 2,480 | 640 | 12,560 |

**Table 1 Continued**

| Date (2015–2016) | pH | Conductivity (micro mhos/cm) | Total dissolved solids (mg/L) | Chemical oxygen demand (mg/L) | Total suspended solids (mg/L) | Colour (Pt/Co units) |
|---|---|---|---|---|---|---|
| 1 February | 7.14 | 12,000 | 11,300 | 3,920 | 2,000 | 9,500 |
| 2 February | 7.9 | 6,960 | 6,900 | 2,200 | 2,410 | 12,560 |
| 3 February | 7.4 | 12,610 | 7,650 | 3,160 | 1,810 | 7,970 |
| 4 February | 7.62 | – | 7,610 | 2,240 | 430 | 5,670 |
| 5 February | 7.2 | 9,670 | 6,000 | 2,400 | 630 | 4,070 |
| 6 February | 7.7 | 15,670 | 10,500 | 4,210 | 2,600 | 13,000 |
| 7 February | 7.58 | | 11,600 | 4,000 | 1,260 | 6,460 |
| 8 February | | | | | | |
| 9 February | 7.9 | 8,960 | 6,600 | 3,600 | 850 | 4,440 |
| 10 February | 7.76 | | 11,210 | 5,400 | 740 | 4,940 |
| 11 February | 7.64 | 13,610 | 9,000 | 4,320 | 1,940 | 8,520 |
| 12 February | 7.69 | 14,640 | 9,900 | 4,320 | 910 | 4,850 |
| 13 February | 7.24 | | 11,600 | 4,440 | 1,210 | 7,960 |
| 14 February | 7.2 | 10,610 | | 4,610 | 1,310 | 7,960 |
| 15 February | 7.5 | 5,600 | | 3,280 | | |
| 16 February | 7.45 | 16,170 | 10,900 | 4,200 | 1,310 | 8,960 |
| 17 February | 7.9 | 14,610 | 10,875 | 4,760 | 890 | 7,360 |
| 18 February | 7.38 | 15,360 | 10,300 | 4,360 | 1,020 | 6,410 |
| 19 February | 7.12 | 12,640 | 7,150 | 4,680 | 1,100 | 6,040 |
| 20 February | 7.69 | 18,610 | 12,250 | 6,260 | 1,540 | 8,120 |
| 21 February | 7.76 | | 12,500 | 5,600 | 1,130 | 7,960 |
| 22 February | 8.05 | 13,600 | 8,740 | 6,000 | | |
| 23 February | 7.62 | 18,610 | 13,950 | 3,760 | 680 | 5,160 |
| 24 February | 7.22 | 18,610 | 14,700 | 4,680 | 2,740 | 14,610 |
| 25 February | 7.92 | 30,170 | 17,250 | 4,680 | 820 | 6,400 |
| 26 February | 7.9 | 18,470 | 12,350 | 3,720 | 480 | 5,170 |
| 27 February | 7.7 | 13,610 | 9,400 | 3,960 | 690 | 3,960 |
| 28 February | 7.34 | 19,460 | 13,950 | | 880 | 5,290 |
| 1 March | 7.52 | 16,610 | 9,800 | 4,280 | | |
| 2 March | 7.94 | 13,400 | 8,650 | 7,320 | 1,170 | 7,960 |
| 3 March | 7.9 | 18,690 | 12,750 | 6,000 | 920 | 5,200 |
| 4 March | 7.1 | 17,600 | 11,550 | 5,240 | 970 | 6,490 |
| 5 March | 7.94 | | 19,900 | 5,360 | 3,800 | 17,900 |
| 6 March | 8.57 | 14,500 | | 6,440 | | |
| Average | 7.54 | 14,757.97 | 10,926.84 | 4,761.28 | 1,482.20 | 10,238.70 |

Table 2 **Pollution analysis of overall factory effluent after treatment**

| Date (2015–2016) | pH | Conductivity (micro mhos/cm) | Total dissolved solids (mg/L) | Chemical oxygen demand (mg/L) | Total suspended solids (mg/L) | Colour (Pt/Co units) |
|---|---|---|---|---|---|---|
| 8 November | 6.69 | 9,610 | 6,900 | 890 | 160 | 1,830 |
| 9 November | – | – | – | – | – | – |
| 10 November | 6.96 | 9,710 | 7,450 | 879 | 120 | 1,710 |
| 11 November | 6.92 | 9,150 | 8,160 | 864 | 140 | 1,470 |
| 12 November | 7.04 | 8,610 | – | 916 | 130 | 1,640 |
| 13 November | 6.73 | 9,003 | – | 932 | 122 | 1,630 |
| 14 November | 6.97 | 9,450 | 9,300 | 913 | 137 | 1,710 |
| 15 November | 6.87 | 9,300 | 7,550 | 920 | 130 | 1,630 |
| 16 November | – | – | – | – | – | – |
| 17 November | 7.04 | 9,780 | 9,000 | 908 | 140 | 1,740 |
| 18 November | 7.04 | 9,780 | 8,560 | 912 | 135 | 1,790 |
| 19 November | 6.66 | 10,460 | 8,580 | 896 | 160 | 1,860 |
| 20 November | 6.6 | 10,360 | 10,300 | 914 | 130 | 1,660 |
| 21 November | 6.1 | 10,960 | 9,600 | 945 | 160 | 1,860 |
| 22 November | 6.2 | 10,430 | 9,850 | 930 | 140 | 1,730 |
| 23 November | – | – | – | – | – | – |
| 24 November | 5.9 | 12,230 | 10,650 | 930 | 139 | 1,690 |
| 25 November | 6.99 | 10,980 | 10,650 | 980 | 180 | 2,270 |
| 26 November | 6.49 | 12,710 | 11,150 | 994 | 190 | 1,670 |
| 27 November | 6.57 | 12,560 | 10,800 | 949 | 192 | 2,040 |
| 28 November | 6.6 | 12,690 | 12,000 | 936 | 140 | 1,830 |
| 29 November | 6.1 | 12,610 | 10,050 | 913 | 160 | 1,910 |
| 30 November | 6.2 | 12,240 | 10,950 | 872 | – | – |
| 1 December | 6.15 | 11,140 | 10,100 | 914 | 140 | 1,790 |
| 2 December | 6.2 | 11,000 | 9,400 | 886 | 110 | 1,540 |
| 3 December | 6.1 | 10,490 | 10,600 | 867 | 130 | 1,710 |
| 4 December | 6.1 | 11,380 | 11,950 | 980 | 130 | 1,890 |
| 5 December | 6.3 | 12,670 | 11,600 | 963 | 190 | 2,200 |
| 6 December | 6.3 | 11,310 | 11,400 | 954 | 130 | 1,870 |
| 7 December | 6.1 | 10,880 | – | 969 | – | – |
| 8 December | 6.08 | 10,010 | 9,950 | 996 | 110 | 1,560 |
| 9 December | 6.1 | 9,670 | 8,250 | 824 | 97 | 1,470 |
| 10 December | 6.2 | 10,670 | 8,150 | 874 | 110 | 1,630 |
| 11 December | 5.78 | 9,340 | 7,300 | 840 | 110 | 1,530 |
| 12 December | 5.6 | 9,460 | 10,150 | 808 | – | – |
| 13 December | 5.83 | 1,270 | 9,350 | 842 | 190 | 1,640 |
| 14 December | 5.9 | 10,370 | – | 776 | – | – |
| 15 December | 6.15 | 10,280 | 7,550 | 818 | 90 | 1,340 |
| 16 December | 6.1 | 10,640 | 8,100 | 888 | 81 | 1,360 |
| 17 December | 6 | 10,810 | 8,400 | 920 | 94 | 1,210 |
| 18 December | 6.02 | 10,870 | 9,850 | 896 | 102 | 1,460 |

Table 2 **Continued**

| Date (2015–2016) | pH | Conductivity (micro mhos/cm) | Total dissolved solids (mg/L) | Chemical oxygen demand (mg/L) | Total suspended solids (mg/L) | Colour (Pt/Co units) |
|---|---|---|---|---|---|---|
| 19 December | 6.13 | 9,670 | 10,550 | 840 | 94 | 1,230 |
| 20 December | 6.4 | 10,040 | 7,250 | 672 | 79 | 1,190 |
| 21 December | 6.03 | 8,970 | – | 740 | – | – |
| 22 December | 6.3 | 10,240 | 8,150 | 672 | 92 | 1,130 |
| 23 December | 6.34 | 10,200 | 8,950 | 722 | 103 | 1,270 |
| 24 December | 6.16 | 12,230 | 10,500 | 790 | 97 | 1,310 |
| 25 December | 5.41 | 10,670 | 9,100 | 814 | 74 | 1,060 |
| 26 December | 5.84 | 10,100 | 10,400 | 824 | 55 | 970 |
| 27 December | 5.92 | 11,370 | 9,850 | 890 | 62 | 899 |
| 28 December | 5.91 | 10,350 | – | 879 | – | – |
| 29 December | 5.92 | 9,010 | 8,200 | 790 | 64 | 840 |
| 30 December | 6.17 | 9,810 | 7,300 | 815 | 84 | 1,040 |
| 31 December | 5.74 | 10,210 | 7,950 | 832 | 44 | 860 |
| 1 January | 6.17 | 11,230 | 7,450 | 784 | 74 | 1,040 |
| 2 January | 6.24 | 11,760 | 9,500 | 804 | 88 | 1,090 |
| 3 January | – | – | – | – | – | – |
| 4 January | – | – | – | – | – | – |
| 5 January | 5.96 | 9,040 | 7,450 | 824 | 84 | 1,070 |
| 6 January | 6.2 | 8,940 | 6,900 | 794 | 94 | 1,170 |
| 7 January | 6.02 | 8,330 | 6,400 | 740 | 84 | 970 |
| 8 January | 6.1 | 8,100 | 5,950 | 680 | 89 | 990 |
| 9 January | 6.06 | 8,910 | 7,100 | 710 | 45 | 870 |
| 10 January | 6.3 | 9,940 | 8,000 | 760 | 65 | 890 |
| 11 January | 6.41 | 9,820 | – | 790 | – | – |
| 12 January | 6.4 | 9,670 | 8,400 | 768 | 84 | 820 |
| 13 January | 6.3 | 12,460 | 7,650 | 752 | 52 | 764 |
| 14 January | 6.36 | 10,670 | 8,400 | 794 | 51 | 794 |
| 15 January | 6.35 | 9,660 | 7,500 | 752 | 42 | 810 |
| 16 January | 6.14 | 9,940 | 6,700 | 776 | 44 | 690 |
| 17 January | 6.02 | 8,440 | 6,200 | 730 | 52 | 710 |
| 18 January | 6.49 | 8,640 | – | 720 | – | – |
| 19 January | – | – | – | – | – | – |
| 20 January | 5.64 | 8,200 | 5,050 | 704 | 90 | 1,070 |
| 21 January | 5.75 | 8,440 | 6,050 | 694 | 84 | 940 |
| 22 January | 5.92 | 8,210 | 5,800 | 710 | 84 | 980 |
| 23 January | 6.8 | 7,240 | 6,050 | 664 | 105 | 1,250 |
| 24 January | 6.4 | 8,750 | 6,350 | 670 | 94 | 1,160 |
| 25 January | 6.15 | 8,640 | 6,840 | 696 | 110 | 1,350 |
| 26 January | 6.34 | 10,740 | 7,500 | 720 | 96 | 1,210 |
| 27 January | 6.2 | 10,140 | 7,630 | 584 | 91 | 980 |

*Continued*

Table 2 **Continued**

| Date (2015–2016) | pH | Conductivity (micro mhos/cm) | Total dissolved solids (mg/L) | Chemical oxygen demand (mg/L) | Total suspended solids (mg/L) | Colour (Pt/Co units) |
|---|---|---|---|---|---|---|
| 28 January | 6.62 | 6,470 | 5,670 | 610 | 89 | 1,040 |
| 29 January | 6.72 | 7,970 | 5,500 | 622 | 129 | 1,230 |
| 30 January | 6.33 | 8,890 | 6,700 | 590 | 54 | 760 |
| 31 January | 6.35 | 7,960 | 6,300 | 608 | 113 | 1,090 |
| 1 February | 6.31 | – | 5,750 | 752 | 188 | 1,300 |
| 2 February | 5.8 | 7,010 | 6,900 | 699 | 91 | 1,040 |
| 3 February | 6.1 | 6,640 | 7,150 | 64 | 94 | 1,110 |
| 4 February | 6.38 | 7,410 | 7,250 | 672 | 150 | 1,074 |
| 5 February | 6.2 | 6,810 | 7,410 | 660 | 120 | 927 |
| 6 February | 6.6 | 7,200 | 6,400 | 682 | 151 | 1,060 |
| 7 February | 6.56 | 7,810 | 6,800 | 672 | 254 | 1,650 |
| 8 February | – | – | – | – | – | – |
| 9 February | 6.449 | 7,960 | 5,700 | 648 | 195 | 1,260 |
| 10 February | 6.33 | 7,410 | 5,500 | 620 | 130 | 890 |
| 11 February | 6.1 | 7,690 | 7,700 | 640 | 158 | 1,010 |
| 12 February | 6.8 | 8,260 | 8,100 | 648 | 58 | 870 |
| 13 February | 6.94 | 9,740 | 7,700 | 680 | 64 | 890 |
| 14 February | 5.77 | 7,960 | – | 662 | 110 | 1,340 |
| 15 February | 6.5 | 7,210 | – | 654 | – | – |
| 16 February | 6.08 | 7,870 | 7,640 | 680 | 54 | 870 |
| 17 February | 6.69 | 7,960 | 6,300 | 654 | 59 | 910 |
| 18 February | 6.54 | 8,790 | 5,900 | 640 | 122 | 906 |
| 19 February | 5.38 | 7,960 | 6,200 | 604 | 45 | 670 |
| 20 February | 6.44 | 7,810 | 6,900 | 598 | 87 | 599 |
| 21 February | 6.44 | 8,690 | 7,100 | 696 | 59 | 790 |
| 22 February | 7.43 | 9,100 | 6,200 | 688 | – | – |
| 23 February | 7.04 | 10,860 | 8,900 | 710 | 109 | 780 |
| 24 February | 6.72 | 11,940 | 8,300 | 732 | 126 | 940 |
| 25 February | 6.52 | 11,920 | 7,700 | 716 | 137 | 890 |
| 26 February | 6.66 | 10,060 | 8,200 | 740 | 121 | 780 |
| 27 February | 6.71 | 10,470 | 9,200 | 719 | 102 | 810 |
| 28 February | 6.99 | 10,690 | 8,150 | – | 179 | 1,260 |
| 1 March | 6.63 | 9,510 | 9,600 | 784 | – | – |
| 2 March | 6.59 | 11,690 | 10,100 | 736 | 76 | 809 |
| 3 March | 6.81 | 11,700 | 11,050 | 762 | 82 | 840 |
| 4 March | 6.56 | 11,910 | 9,750 | 794 | 86 | 1,060 |
| 5 March | 6.8 | 10,980 | 11,100 | 840 | 63 | 760 |
| 6 March | 6.8 | 12,500 | | 832 | – | – |
| Average | 6.35 | 9,710.76 | 8,207.52 | 790.39 | 110.90 | 1,272.94 |

# Chemicals and effluent treatment in indigo denim processes

**9**

*V. Buscio, C. Gutiérrez-Bouzán*
Universitat Politècnica de Catalunya-BarcelonaTech (UPC), Terrassa, Spain

## 9.1 Introduction

The story of the denim industry began simultaneously in two cities: Nîmes in France and Genoa in Italy. Denim was first used as work clothes because of its toughness, especially for sailors and dockworkers, replacing classical brown and beige cloth [1]. Later, Levi Strauss, a young German who immigrated to the United States, decided to reinforce the pockets and exposed areas of the fabric with copper rivets. Since then, jeans have become increasingly popular and have solidified their reputation over the years. Nowadays, about 1240 million jeans are sold worldwide annually [2].

Because denim is generally made of 100% cotton, different types of dyes can be applied to provide colour: direct, reactive, vat or sulphur dyes. In the last years, denim is dyed with indigo and sulphur dyes, being the indigo the most used (67%) In general, 100% indigo is not applied; it is commonly combined with sulphur dyes to reduce the cost [3]. Despite that, indigo remains synonymous with the denim industry. No other dye provides the characteristic bright blue and properties of indigo dye in the final product.

Indigo dye is a dark blue crystalline powder. Its molecular formula is $C_{16}H_{10}N_2O_2$ with a molecular weight of 262.27 g/mol. It is insoluble in water, alcohol or ethyl ether but soluble in chloroform, nitrobenzene or concentrated sulphuric acid [4,5]. The reduced form of indigo, called leuco, is yellow and soluble in water. It is mainly applied in the dyeing of cellulosic fibres, but it can also be used for polyamide and wool. Around 80,000 tons of indigo dye were produced in 2010 [6].

Before synthetic indigo dye was developed in the 19th century, it was extracted from plants. The oldest process for obtaining natural indigo was to extract the dye from *Indigofera tinctoria*. It is a primitive process in which the plants, with their stems and roots, were placed in a tank with water. The fermentation process was carried out using the enzyme known as indimulsin. Once the fermentation process was completed, the solution was stirred to facilitate oxidation of the dye. The generated precipitate was filtered and dried [5].

In 1870, Adolph von Baeyer determined the structure of indigo dye and performed the first synthesis of indigo from isatin. However, isatin was extremely expensive and this process could not be applied on an industrial scale. In 1878, Baeyer developed a new synthesis from phenylacetic acid. Five years later, two new routes were created by BASF and implemented industrially, but none was economically viable.

Sustainability in Denim. http://dx.doi.org/10.1016/B978-0-08-102043-2.00009-5
Copyright © 2017 Elsevier Ltd. All rights reserved.

Subsequently, in 1882, von Baeyer and Drewson developed a way to synthesize indigo from *o*-nitrobenzaldehyde. It was discarded owing to its high cost, although nowadays it could be implemented.

In 1890, Karl Heumann discovered a new route from aniline. The reaction had a low yield so it could not be used on an industrial scale. Later, he developed a second method for synthesizing indigo from anthranilic acid that had great success.

The production of indigo has been improved over the years and its synthesis has been adapted to obtain a competitive marketable product. The current route for synthesizing indigo dye was developed in 1925 by BASF. It consists of generating *N*-phenylglycine from the *N*-methylation of aniline with formaldehyde and hydrogen cyanide. Then, saponification of the resulting nitrile intermediate is carried out to produce indigo [3].

# 9.2 Chemicals in indigo dyeing process

## 9.2.1 Indigo dye structure and redox properties

The oxidized form of indigo dye (blue) is insoluble in water and does not have affinity for cellulose [1]. As for the rest of the vat dyes, to carry out the dyeing process indigo must be reduced to its leuco form to be water soluble [3,7]. As shown in Fig. 9.1, the reduction involves two electron transfers, the reduced form of which is easily reoxidized [8,9].

Depending on the pH of the dye bath, reduced indigo can exist in three forms (Fig. 9.2): (1) acid leuco indigo, (2) monophenolate ion, and (3) biphenolate ion [10–14].

The dyeing process is carried out at pH 11, where the higher proportion of monophenolate ions occurs. The highest sorption and colour strength is observed at that pH [12]. At lower pH values, shades on the fabric can be observed as a result of the

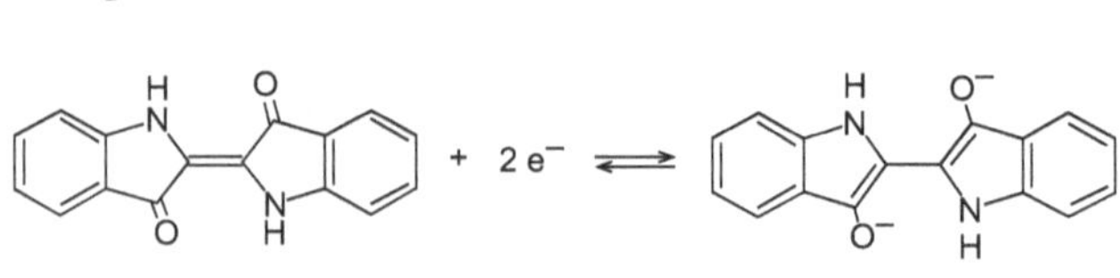


**Figure 9.1** Equilibrium of oxidized form and reduced form of indigo.

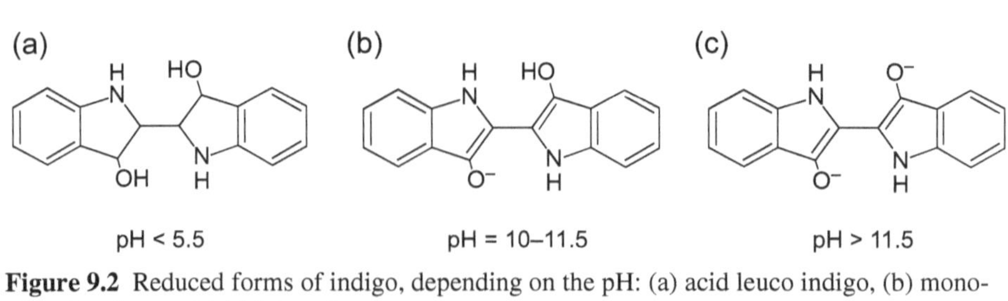


**Figure 9.2** Reduced forms of indigo, depending on the pH: (a) acid leuco indigo, (b) monophenolate ion, and (c) biphenolate ion.

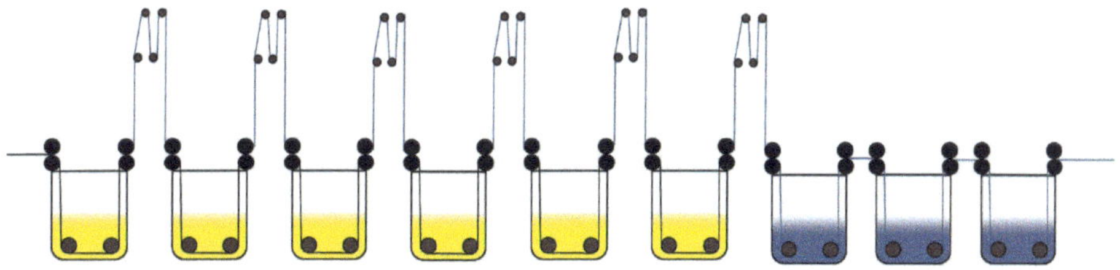


**Figure 9.3** Indigo dyeing process.

chemical form of indigo dye called acid leuco, which is insoluble and without affinity. At higher pH, greater diffusion and lower intensities are obtained [1].

The process of dyeing consists of passing cotton yarns through a bath containing the yellow reduced dye. After impregnation, the yarns are exposed to air and the dye is oxidized to the blue form. It becomes insoluble and fixed on the yarns by means of weak bonds. The dipping/exposing steps are repeated many times to obtain the different shades [15]. It is important to highlight that the deposition of indigo on yarns is superficial; therefore it has limitations in fastness [1]. The dye that has not been fixed on the fibre is removed in subsequent washing steps. Fig. 9.3 shows a schema of the dyeing process with indigo dye.

## 9.2.2 *Indigo reduction*

Indigo reduction is one of the most important steps in the dyeing process. There are different types of reduction processes:

- biological reduction
- chemical reduction
- catalytic hydrogenation
- electrochemical reduction

### 9.2.2.1 *Biological reduction*

The use of biological processes to reduce indigo dye is the most primitive mechanism known. There is evidence that indigo dye has been used to dye cellulose and wool for centuries. The first complete description of the dyeing process with indigo dye dates to 700 BC [12].

Although biological reduction was first used for dyeing, it was not until recently that the bacteria involved in indigo reduction were investigated. It was found that the main bacterium responsible for reducing indigo dye was *Clostridium isatidis*, which was isolated from a woad vat following the procedure carried out in medieval Europe [16]. The mechanism of bacterial indigo reduction is still unknown [11].

Enzymatic reduction with dehydrogenases has also been studied. However, currently there is no enzymatic process that can replace chemical reduction with sodium dithionite [11].

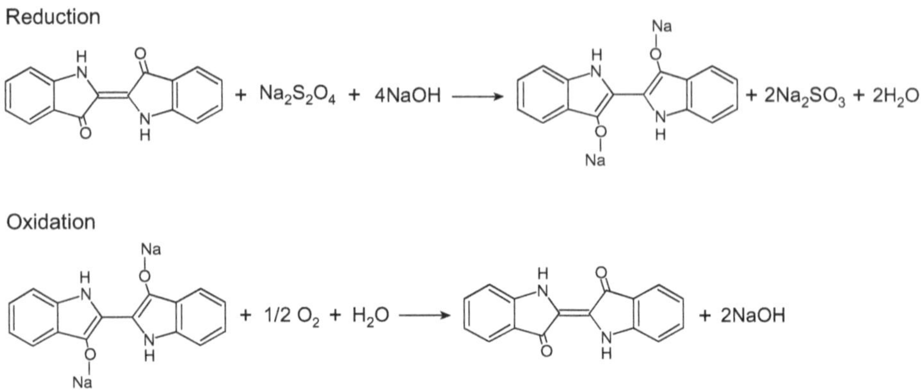


**Figure 9.4** Chemical reduction and oxidation reactions of indigo dye.

### 9.2.2.2 *Chemical reduction*

Sodium dithionite ($Na_2S_2O_4$) is the most widely used reducing agent in the dyeing process with indigo [3,11]. Fig. 9.4 shows the reduction–oxidation reactions of indigo dye that occur during the dyeing process.

In addition to its economic properties, sodium dithionite produces a rapid reduction of dye at temperatures from 30°C to 60°C. However, it also reacts rapidly with atmospheric oxygen which competes with the reduction of the dye [3,11]. As a result, large amounts, over stoichiometric, of sodium dithionite are required and continual addition to the dye bath is needed. Oxidation of sodium dithionite generates sulphate ($SO_4^{2-}$), sulphite ($SO_3^{2-}$) and thiosulphate ions ($S_2O_3^{2-}$). Sulphite and thiosulphate ions are not only toxic, they have corrosive effects. In addition, an excess of sodium dithionite discharged in wastewater can affect subsequent treatment and hydrogen sulphide [$H_2S$], which is also toxic, may be anaerobically formed [3,11,17,18].

To minimize the environmental impact of the dyeing process with indigo dye, other reagents such as iron complexes [19,20], glucose [21], sodium borohydride [15] and α-hydroxy ketones [8] have been tested as reducing agents. Considering its price and the kinetics of the indigo reduction, there is still no other reducing agent that can replace sodium dithionite.

### 9.2.2.3 *Catalytic hydrogenation*

Catalytic hydrogenation is the direct reaction of the indigo oxidized form with hydrogen gas, using an appropriate catalyst. This option has the main drawback of using $H_2$, which increases the risk of explosion and fire. For this reason, it is only advisable in a very controlled atmosphere and generally is not used onsite in the industry [17]. Currently, catalytic hydrogenation is used to obtain a prereduced indigo solution (40%) that is commercialized by DyStar.

#### 9.2.2.4 Electrochemical reduction

A promising alternative to chemical reduction with sodium dithionite is the use of electrochemical techniques, which are able to generate the required reagent in situ. There are different types of electrochemical processes: direct, indirect and electrocatalytic hydrogenation.

**Direct electrochemical** reduction is the most attractive option from an environmental point of view. Because indigo dye is insoluble in water, this process is limited by contact between the cathode surface and the particles of indigo, which hinders the transfer of electrons [17]. Therefore, direct electrochemical reduction requires previous generation of the indigo radical by a conventional chemical reduction process with sodium dithionite. Then the electrochemical reduction of this radical is carried out [22]. Consequently, the amount of sodium dithionite used to reduce the indigo dye is diminished because it is only used to initiate electroreduction by generating the indigo-soluble radical. An alternative to direct electrochemical reduction after generation of the indigo radical is to use fixed or fluidized beds of graphite granules as a cathode [23]. The company RedElec Technologie SA commercializes electrochemical cells for reducing indigo dye. In this case, the cathode consists of a bed of carbon particles.

As previously discussed, the main limitation of direct electrochemical reduction is contact between the cathode surface and the particles of indigo. To resolve this inconvenience, **indirect electrochemical** reduction can be used. In this process there is no direct contact between the electrode and dye surfaces [17,24]. The reduction in dye takes place via a mediator. Several chemicals have been studied as mediators in indirect electrochemical reduction; the iron–triethanolamine complex is the most promising [24]. The iron(II)-amine complex has demonstrated a higher reduction rate and faster kinetics than sodium dithionite. However, the reduction takes place with low current density, which requires a high cathode surface. A new design based on a multicathode cell connected to one or two anodes allows operation with the maximum cathode and minimum anode areas [17,25].

Finally, **electrocatalytic hydrogenation** produces hydrogen in situ by electrolysing water. Then the hydrogen generated reacts with the indigo dye on the electrode surface [3,26].

## 9.3 Treatments of indigo effluents

### 9.3.1 Textile wastewater

In general, characterization of the pollution caused by the textile industry is complex owing to the variety of fibres and production methods used [27]. The different processes carried out in the textile industry generate gaseous, liquid and solid waste. Therefore, the textile industry has two major problems: its high water consumption (up to 100L of water is required to produce a kilogram of textile product) and the complexity and variability of its wastewater [28–30].

The main characteristics of textile wastewater depend mainly on the process and machinery used by the company [31]. Effluents often contain subproducts, residual dyes, salts, acids/alkalis, surfactants and solvents [32,33]. Wastewater from the dyeing process is characterized by high values of colour, biological oxygen demand, chemical oxygen demand (COD), pH and salinity [34–42].

### 9.3.2 Treatments for colour removal from textile effluents

Different techniques have been investigated to treat textile effluents [43]. Most published studies focused on removing dyes because it involves great difficulty (Fig. 9.5).

**Biological processes** are widely applied to treat wastewater from the dyeing process because they are simple and have low operational cost. Biological methods show high turbidity and COD removal but low efficiency in colour removal owing to the chemical stability and resistance to microbiological attack of the dyes [44]. In general, in aerobic processes, colour removal observed mainly results from the adsorption of dyes on the biomass [45,46]. To improve the efficiency of the biological treatments, aerobic and anaerobic processes can be combined into one or more stages [47–56].

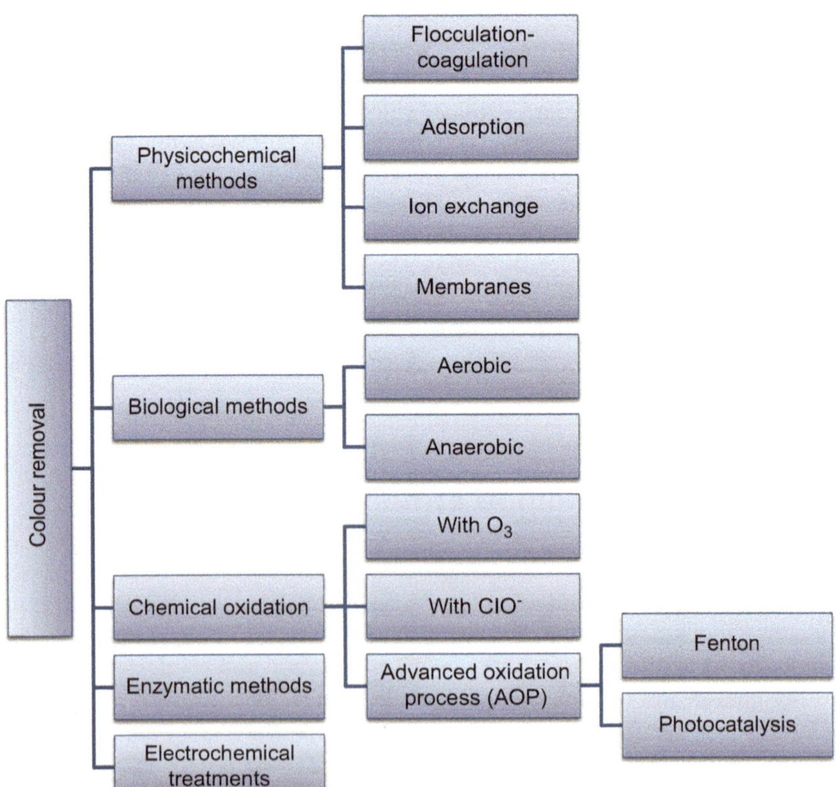


**Figure 9.5** Treatments for colour removal from textile effluents.

The treatment of textile effluents by **electrochemical techniques** has been studied by various authors with satisfactory results in terms of colour removal [35,57–64]. However, this technique generally has a high operational cost [43].

One the emerging treatment for the decolourization of textile wastewater is **enzymatic methods**. Its main disadvantage is that the process variables must be carefully controlled (especially temperature and pH). In addition, separation and purification of the enzyme are delicate processes [26].

Regarding **physicochemical treatments**, coagulation–flocculation is the most used. Although it shows high efficiency in colour removal, this method produces a sludge that requires further treatment [26,31]. Adsorption (usually with activated carbon) is influenced by various parameters such as interactions between the dye and the adsorbent, the adsorbent surface area, particle size, etc. [65]. It exhibits high efficiency for all kind of dyes, but with a high operational cost because the adsorbent material must be regenerated after several treatments. Ion exchange treatments also have the problem of resin regeneration and are not effective for all dyes. Different authors have studied the possibility of applying membranes to treat textile wastewater [66–75]. Membrane treatment enables permeate free of dyes to be obtained and a concentrate that contains them. Both can be reused in new dyeing processes [76–78]. One main disadvantage of membranes is fouling, which causes an increase in maintenance cost [79]. To reduce this phenomenon and increase treatment efficiency, membrane processes are often combined with other processes such as biological treatments [38,80–83] or coagulation–flocculation methods [37,84–88].

**Chemical oxidation treatments** involve the addition of oxidizing compounds such as ozone ($O_3$), hydrogen peroxide ($H_2O_2$), hypochlorite ($ClO^-$) or permanganate ($MnO_4^-$). Ozone is the most used owing to its high efficiency in removing dyes. However, it is not effective when the effluent contains insoluble dyes such as vat or dispersal dyes [31].

## 9.3.3 *Indigo effluents*

As explained previously, effluents from the textile industry are characterized by their high colouration because of the residual dyes. In the case of effluents from the denim industry, an alkaline pH can also be observed resulting from the reducing step previous to the dyeing process.

The main characteristics of wastewater from the denim dyeing process are shown in Table 9.1 [3].

As previously discussed, the most common treatment for textile wastewater is the aerobic biological process with activated sludge. These treatments are able to remove biodegradable organic matter, which is present in greater proportion in textile wastewater. However, the full removal of dyes by biological methods is not possible because most have low biodegradability.

Residual salinity is also an important concern. High water salinity has a notorious environmental impact and can even be toxic to plants and animals. Taking into consideration its effects on aquatic life, current policies are focused on reducing the use of salts. In addition, discharge limits are becoming lower and the cost of disposal is

Table 9.1 **Characteristics of wastewater from denim dyeing process**

| Parameter | Indigo dyeing | Sulphur dyeing | Denim plant |
|---|---|---|---|
| Chemical oxygen demand (mg/L) | 750–950 | 2500–3500 | 1500–3100 |
| Colour (Pt-Co) | 5500–7000 | 24,000–34,000 | 1147–3547 |
| Total suspended solids (mg/L) | 50–300 | 100–340 | 150–300 |
| Total dissolved solids (mg/L) | – | – | 4000–8000 |
| pH | 10.5–11.5 | 10–11 | 9–13 |
| Conductivity (mS/cm) | – | – | 6–12 |
| Total phosphorus (mg/L) | – | – | 0–3 |
| Total nitrogen (mg/L) | – | – | 17–23 |

Roshan P, editor. Denim: manufacture, finishing and applications. Cambridge; 2015.

increasing. In general, salt removal can be achieved by reverse osmosis but its high operational cost limits its use.

Regarding colour, the colour reduction observed when biological treatments are applied is usually the result of the adsorption of dyes on the sludge. Residual colour must be removed by means of tertiary treatments, such as the ones described in Section 9.3.2.

In particular, the aerobic biodegradation of textile effluents containing indigo dye has been investigated [89]. Because of the alkaline pH of the effluents, they were found not to be suitable for biological treatment. If the effluent pH is adjusted to 7.5 before biological treatment, up to 97% of colour removal can be reached. However, the content of indigo adsorbed on the sludge and the amount of dye subsequently desorbed was not investigated.

On the other hand, among physicochemical methods, coagulation–flocculation is the most conventional. Different coagulants such as aluminium sulphate [$Al_2(SO_4)$] and ferric chloride ($FeCl_3$) have been tested [90]. However, it was observed that a high coagulant dose is required, which increases the cost of the treatment. To decrease operational costs, the use of wastes such as residual bittern containing $Mg^{2+}$ has been studied [91].

Membrane techniques are particularly interesting for the treatment of wastewater containing water-insoluble compounds because they can easily be separated from the effluent. This is the case of indigo (or other vat dyes) and sulphur dyes. In addition, membrane techniques allow concentrated dye to be recovered.

Treatment of textile effluents containing indigo dye by means of inorganic ceramic membranes was first studied in the 1980s. The company named TIA in Belgium carried out a study to recover indigo after membrane filtration processes. Up to 98% of indigo dye was recovered. However, that method was discarded because of fouling of the membranes and the decrease in permeate flux during the concentration phase [92]. The treatment of textile wastewater containing indigo using different membranes techniques based on organic membranes was studied by several authors [90,93]. Porter et al. [94] studied the recovery of indigo dye with a vinyl-sulphone membrane using a multistage system in which the indigo concentration became progressively higher. New types of membranes and materials have appeared allowing studies to be undertaken based on the recovery of indigo dye [78]. Results of these studies demonstrated that membrane technology is a promising technique for treating this type of wastewater.

## 9.4 Case study: treatment of indigo effluents by ultrafiltration membranes and reuse of dye

An exhaustive study of the treatment of real indigo dyeing effluents by means of polyvinylidene fluoride (PVDF) ultrafiltration membranes was carried out to recover and reuse the dye in new dyeing processes.

Four different membranes were selected for this work. Effluents were treated and indigo was concentrated up to 20 g/L. Then 100% of the dye recovered was reused in new dyeing procedures. Finally, fabrics dyed with the recovered indigo were evaluated with respect to fabrics dyed with commercial dye in terms of colour differences, dry rubbing and washing fastness tests. Details of this study are presented subsequently.

### *9.4.1 Experimental procedure*

Real industrial effluents were used for this study, collected in a Spanish mill from the first washing step after the dyeing process. Three effluents from dyeing of different types of fibres and production periods were selected to determine the behaviour of the membranes under different conditions.

As can be seen, effluents generated by the dyeing process with indigo are characterized by an alkaline pH and high conductivity (Table 9.2). The concentration of indigo dye and content of organic matter in the effluents depend on the type of fibre and dyeing conditions.

The effluents were prefiltered (500 μm) before the ultrafiltration membrane process.

The concentration of indigo was determined after the reduction of dye with a solution containing sodium dithionite, sodium hydroxide and 1-methyl-2-pyrrolidine, and determined by UV-visible spectrophotometry according to the method described by Buscio et al. [7] COD was measured according to Standard Methods 21st ed. 5220 [95] Fabrics dyed with recovered indigo were evaluated with respect to references in terms of colour differences [$DE_{CMC(2:1)}$] and fastness in rubbing and washing tests.

Table 9.2 **Characterization of indigo effluents**

| | Dyed fibre | Indigo concentration (mg/L) | pH | Chemical oxygen demand (mg/L) | Conductivity (μS/cm) |
|---|---|---|---|---|---|
| Effluent 1 | Cotton | 58 | 8.6 | 320 | 6751 |
| Effluent 2 | Lyocell | 118 | 9.5 | 800 | 7266 |
| Effluent 3 | Cotton | 82 | 10.4 | 572 | 4230 |

## 9.4.2 Ultrafiltration membrane treatments

Four PVDF hollow-fibre membranes were used (Zenon ZeeWeed-1, Motimo UOF-1b, UOF-4 and FP-T0008). Pore size range was between 0.03 and 0.1 µm. According to the geometry and specific requirements of each membrane, three pilot plants were built. The pilot plants worked in cycles of 15 min filtration and 30 s backwashing with permeate. After each filtration process, membranes were cleaned with a sodium dithionite solution at pH 11, followed by rinsing with a sodium hypochlorite solution (5 mg/L).

The efficiency of the different membranes in the treatment of effluents containing indigo dye was determined by means of permeate flux measurements and an analysis of permeate parameters such as pH, COD, and indigo dye removal (%$R_{indigo}$).

### 9.4.2.1 External cross-flow membrane configuration

First, the effluent one was treated with an external membrane configuration. The UOF-1b membrane supplied by Motimo Membrane Technology (China) had a pore size of 0.04 µm and a membrane surface of 0.5 $m^2$.

Fig. 9.6 shows the scheme of the pilot plant used in this study. The pilot plant was fed by a 100-L tank and was equipped with a peristaltic pump for feed, permeate and concentrated effluents.

Membrane treatment resulted in 96% dye retention and about 40% COD removal. Both indigo retention and COD removal increased at the end of the experiment owing to the formation of a porous layer of dye on the membrane surface. This layer acted as a barrier increasing the efficiency of the process, which was reported by other authors [74,96–98].

The pH and conductivity remained constant during the membrane treatment. Ultrafiltration membranes cannot retain ions present in the indigo effluents such as sodium or sulphates. For this purpose, nanofiltration or reverse osmosis membranes should be used.

Regarding membrane behaviour, a remarkable decrease in membrane flux was observed after 6 days of operation (Fig. 9.7). Backwashing was not able to restore the initial conditions of membrane flux.

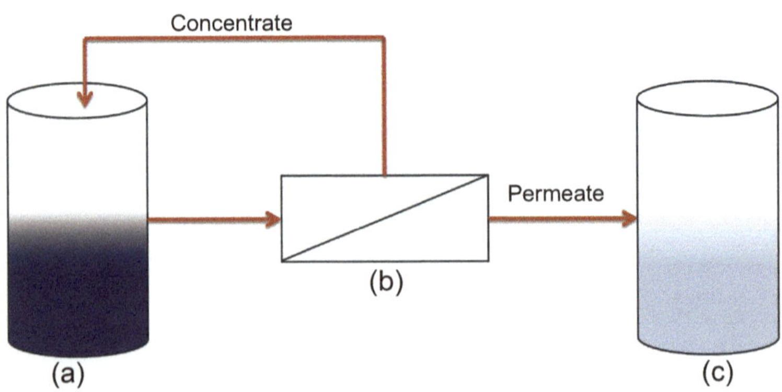


**Figure 9.6** Pilot plant with UOF1-b external membrane: (a) feed tank, (b) membrane and (c) permeate tank.

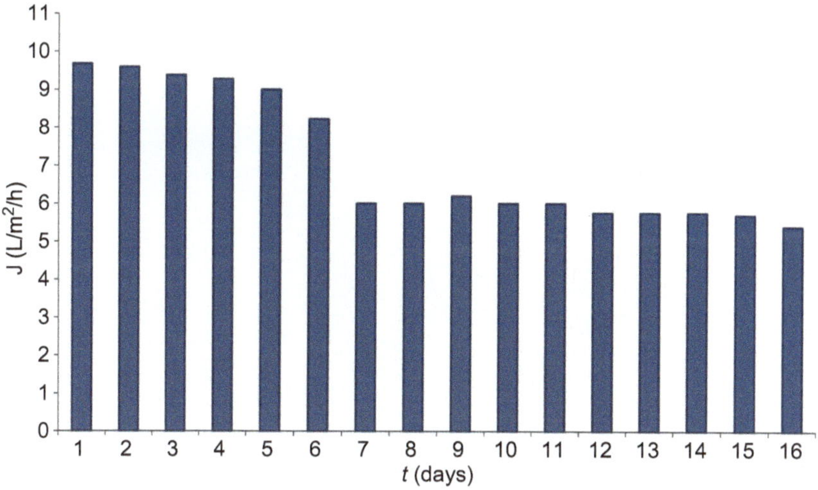


**Figure 9.7** Permeate flux values of the UOF-1b membrane. *t*, time.

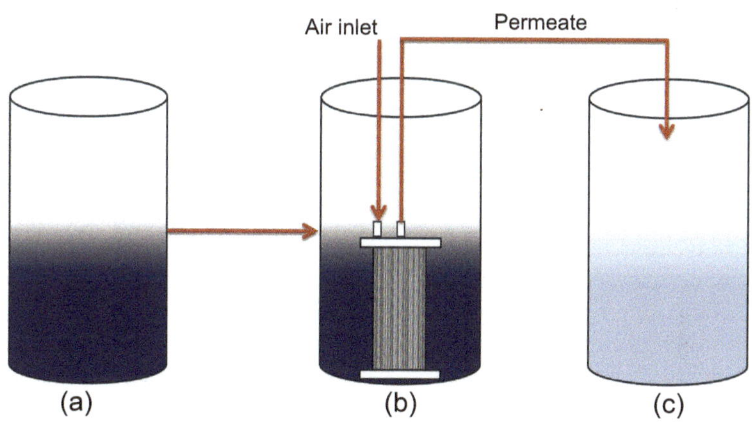


**Figure 9.8** Pilot plant with ZeeWeed-1 submerged membrane: (a) feed tank, (b) membrane reactor and (c) permeate tank.

This type of configuration was able to obtain a concentrate that contained up to 3 g/L indigo dye. Although this concentration is enough for reuse in a new dyeing process, most mills have automated systems which require at least 20 g/L indigo dye. In addition, an external configuration has high operating costs [99]. Taking this into account, this configuration was not considered usable for indigo dye recovery.

### 9.4.2.2 Submerged membrane configuration

Effluent 2 was treated with a submerged membrane configuration. The ZeeWeed-1 membrane provided by GE Power and Water (Canada) had a pore size of 0.04 μm and a 0.05-$m^2$ membrane surface.

The scheme of the pilot plant used with this membrane is presented in Fig. 9.8.

**Figure 9.9** Result of indigo effluent treatment with ZeeWeed-1 membrane: (a) permeate and (b) concentrate.

The pilot plant was equipped with a 20-L membrane reactor fed from a 20-L tank by a centrifugal pump. A peristaltic pump was used for permeate effluent. The membrane module had an air inlet to prevent fouling.

Membrane treatment showed 99% indigo dye retention. As can be seen in Fig. 9.9, no colouration was observed in permeate. In addition, 80% COD removal was achieved. Both pH and conductivity remained constant along the experiment, because inorganic ions were not retained. To remove salts, additional treatment such as a reverse osmosis membrane is required.

A main limitation of using ultrafiltration membranes to treat wastewater is fouling [100], which occurs when particles present in the effluent are deposited either on or within the pores of the membrane [101]. It affects membrane permeability, increasing the maintenance cost of the membrane and reducing its lifetime [79]. In this case, backwashing was able to maintain the initial membrane flux value. In addition, the membrane enabled a concentrated effluent containing 20 g/L indigo dye to be obtained, the minimum dye concentration required for an automated dyeing process. This result evidenced the feasibility of applying a submerged membrane configuration to recover indigo dye.

#### *9.4.2.3 Semiindustrial application*

This study was carried out with Effluent 3, consecutively using the two membranes configurations previously described, to observe their behaviour on a semiindustrial scale.

The semiindustrial system was designed on the basis of the two laboratory pilot results; it was able to place two membrane modules supplied by Motimo Membrane Technology (China). The effluent was concentrated to 3 g/L with an external membrane configuration. The volume of the feeding tank was 1000 L. The concentrate

**Table 9.3 Semiindustrial pilot plant characteristics**

| Membrane | Configuration | Pore size (μm) | Membrane surface ($m^2$) |
|---|---|---|---|
| UOF-4 | External | 0.03 | 40 |
| FP-T0008 | Submerged | 0.1 | 1 |

obtained was then applied as a feed to a submerged membrane module with a volume of 100 L, which concentrated the indigo dye to 20 g/L.

Table 9.3 shows the main characteristics of the membranes used on a semiindustrial scale.

The study showed that the combination of external and submerged membrane configurations was able to treat and recover indigo dye. Up to 98% of dye retention and a 67% decrease in COD was observed. The behaviour of the membranes in terms of dye removal, COD reduction and permeate flux was similar to those in tests conducted in the laboratory.

## 9.4.3 Reuse studies

### 9.4.3.1 Indigo dye reuse

Indigo dye reuse tests were performed with 100% recovered indigo dye. The study was carried out with the indigo solution recovered from the submerged and the semiindustrial (external plus submerged) configuration. The configuration concentrated from the external membrane was discarded because it did not meet the minimum concentration required for automated mills (20 g/L).

Fabrics dyed with recovered indigo were compared with references with commercial indigo dye in terms of colour differences [$DE_{CMC(2:1)}$]. Colour difference is composed of three chromatic coordinates: differences in lightness (DL), differences in chroma and differences in hue. Table 9.4 shows the colour differences ($DE_{CMC(2:1)}$) and chromatic coordinates obtained.

In general, after dyeing fabrics are submitted to bleaching or sandwashing processes. For this reason, in the denim industry the colour difference acceptance limit is 1.2. As shown in Table 9.4, the indigo dye reuse tests exhibited a $DE_{CMC(2:1)}$ within the established acceptance criteria. In addition, the main differences between reference fabrics and the ones dyed with reused indigo mainly resulted from the influence of DL. The negative value of DL means that results of dyeing with reused

**Table 9.4 Colour differences and chromatic coordinate values**

| Membrane | $DE_{CMC(2:1)}$ | DL | DC | DH |
|---|---|---|---|---|
| Submerged | 0.84 | −0.76 | −0.26 | 0.22 |
| Semiindustrial | 1.19 | −1.19 | −0.31 | −0.35 |

*DL*, differences in lightness; *DC*, differences in chroma; *DH*, differences in hue.

Table 9.5 **Washing and rubbing fastness values**

| Membrane | Washing | Rubbing |
|---|---|---|
| Reference | 4 | 2 |
| Submerged | 5 | 2 |
| Semiindustrial | 5 | 2 |

indigo were darker than the reference, which may be due to the concentration of residual auxiliary products during membrane filtration. Therefore a lesser amount of recovered indigo is required to obtain the same lightness as references with commercial dye, which represents an important advantage from an economic point of view.

Regarding fastness properties, Table 9.5 lists values for colour fastness after washing and rubbing. Fastness was evaluated on a scale of 1 to 5, in which 1 indicates low fastness and 5 corresponds to high fastness.

Results of dyeing with reused indigo showed similar fastness as with fabric dyed with commercial dye. Washing fastness values were even higher when recovered indigo was used.

#### *9.4.3.2 Permeate reuse*

Although this study focused on the recovery and reuse of indigo dye, the quality of permeate obtained after membrane treatments is also important. The low dye concentration (<2 mg/L) and low organic matter (<250 mg/L) content in permeate allowed us to entertain the possibility of reusing permeate in new processes.

It is estimated that the annual consumption of freshwater in the textile industry at the European level is 600 million $m^3$ [102], in which 3000 $m^3$ of water is consumed per pair of jeans produced (from fibre to fabric production) [103]. Despite the clear benefits of water reuse, its implementation is still not a common practice in the textile sector.

### *9.4.4 Achievements and challenges*

Indigo dye from wastewater was removed by means of different ultrafiltration membranes using different configuration systems. External configuration can be used to concentrate the dye only up to 3 g/L. A higher concentration produces fouling on the membrane surface. The submerged configuration enabled us to obtain a concentrate with 20 g/L of indigo dye, which is the minimal concentration required for automated dyeing processes. The tests performed on a laboratory scale were scalable to a semi-industrial system.

In addition, the indigo dye reuse tests showed the feasibility of the membrane technology for recovering indigo dye. Dyeing with recovered dye resulted in similar characteristics as reference dyeing with commercial dye with respect to colour differences, washing and rubbing fastness. The reuse of permeate is also a promising alternative for denim mills.

The reuse of water and indigo dye in a new dyeing process has a positive impact from both an environmental and economic point of view, reducing not only the water and dye costs but also the cost of wastewater discharge. Taking into account results obtained from the case in this study, the next step is to optimize the membrane process (membrane material and configuration) to manufacture a prototype. In this case, the main challenge is to obtain a process that provides a relation between the reuse of water or dye and the cost of treatment that is favourable with respect to the methods currently applied.

## 9.5 Future trends

A major problem in the textile industry, and particularly in denim processes, is high water consumption. Its scarcity in certain regions has caused an increase the cost of water. For this reason, environmental policies are focusing on water recycling and reuse. The reuse of water in new processes reduces water consumption but also the cost of wastewater treatment.

Another important environmental concern in the textile industry is residual dyes. Approximately 800,000 tons of dyes are produced annually worldwide. Depending on the dyeing processes, between 10% and 50% of residual dyes are discarded into the environment. Some are supposed to be highly toxic and mutagenic. In addition, the presence of a low concentration of dyes in wastewater is highly visible and has a notoriously visual impact. On the bases of these considerations, the recovery and reuse of dyes, especially indigo dye, are the focus of many research studies. In the case of indigo, reuse of the dye is particularly advantageous. Owing to the size of the particles in an oxidized form, which are insoluble in water, they can easily be retained through ultrafiltration and nanofiltration membranes and subsequently reduced into a water-soluble form for reuse.

The use of sodium dithionite to achieve the chemical reduction of indigo dye is a problematic step in the denim industry. The oxidation of sodium dithionite generates sulphate, sulphite and thiosulphate ions, which increases the salinity of the discharged wastewater. Sulphite and thiosulphate are toxic and have corrosive effects; they affect subsequent effluent treatment. Although reduction with sodium dithionite is widely used on an industrial level, studies have focused on the possibility of replacing this compound by electrochemical reduction. In fact, its application on an industrial scale has become a reality.

Finally, reverse osmosis is a good option for dealing with the high salinity of denim effluents. To avoid fouling problems, it should be applied only once organic matter has been almost fully removed from the effluents. The cost and operational difficulties of reverse osmosis keep it from being attractive. However it is a good alternative to explore in the future to increase the rate of water reuse.

In general, the trend in treating effluents is to achieve zero discharge. In particular, the denim industry aims to explore any initiative focused on reducing the amount of pollutants discharged into the environment. Evidence of that is increasing efforts made to reuse water and dyes. In our opinion, the treatment of effluents to reuse water and chemicals is the best way to approach a cyclic process. This approach will involve both environmental and economic benefits.

## 9.6 Conclusions

The denim industry is one of the most important worldwide and its growth is expected to continue in coming years.

Indigo dye is the most used dye for denim garments. It is insoluble in water and must be reduced to its leuco form before the dyeing process. Chemical reduction with sodium dithionite is the most widely used procedure. However, its environmental problems (generation of toxic by-products) have led to an evaluation of new and more sustainable technologies such as electrochemical reduction. Despite the efforts, no other technology can yet replace reduction with sodium dithionite.

Regarding effluents that contain indigo dye, they are characterized by high conductivity and alkaline pH, both of which are required to reduce the dyes. Generally conventional treatments, such as activated sludge biological plants or/and coagulation–flocculation processes, are applied to treat these effluents. Membrane technology has also shown great potential for treating this type of textile wastewater, because it is especially efficient in retaining the dyes used in the denim industry. In addition, it reuses the concentrated dye and produces permeate of a high quality able to be reused.

From an environmental point of view, future challenges of our society are reflected in the approaches of 'zero discharge' and 'cradle to cradle'. In this sense, membrane techniques are highly promising for the denim industry, to recover and reuse both chemicals and water.

## Sources of further information and advice

1. RedElec Technologie: http://www.redelec.ch
2. DyStar: http://www.dystar.com/
3. Archroma: http://www.archroma.com/
4. Denim Jeans Industry Statistics (Statistics Brain): http://www.statisticbrain.com/denim-jeans-industry-statistics/

## References

[1] Archroma. Denim book: from cotton to fashion. Barcelona. 2014.

[2] Denim jeans industry statistics http://www.statisticbrain.com/denim-jeans-industry-statistics/.

[3] Roshan P, editor. Denim: manufacture, finishing and applications. 2015. Cambridge.

[4] Quintero L. Indigo e indigo carmine: technologies for the decolorization of dyes. Dyna 2010;162:371–86.

[5] Índigo ARO. uma molécula bastante interesante. Rev Quim 2012;106:32–57.

[6] Ghaly A, Ananthashankar R, Alhattab M, Ramakrishnan V. Production, characterization and treatment of textile effluents: a critical review. Chem Eng Process Technol 2014:5.

[7] Buscio V, Crespi M, Gutiérrez-Bouzán C. A critical comparison of methods for the analysis of indigo in dyeing liquors and effluents. Materials (Basel) 2014;7:6184–93.

[8] Meksi N, Ben Ticha M, Kechida M, Mhenni MF. Using of ecofriendly α-hydroxycarbonyls as reducing agents to replace sodium dithionite in indigo dyeing processes. J Clean Prod 2012;24:149–58.
[9] Ben Ticha M, Meksi N, Drira N, Kechida M, Mhenni M. The synergetic effect of α-hydroxycarbonyls mixtures used as green reducing agent on the indigo dyeing process. Chem Ind Chem Eng Q 2014;20:463–70.
[10] Chakraborty J. Dyeing of denim with indigo. Indian J Fibre Text Res 2004;39:100–9.
[11] Božič M, Kokol V. Ecological alternatives to the reduction and oxidation processes in dyeing with vat and sulphur dyes. Dye Pigment 2008;76:299–309.
[12] Blackburn R, Bechtold T, John P. The development of indigo reduction methods and pre-reduced indigo products. Color Technol 2009;125:193–207.
[13] Baig GA. Study on the exhaust dyeing of various synthetic fibres with indigo. Color Technol 2012;128:114–20.
[14] Baig G. Dyeing nylon with indigo in various pH regions. Autex Res J 2010;10:21–5.
[15] Meksi M, Kechida M, Mhenni F. Cotton dyeing by indigo with the borohydride process: effect of some experimental conditions on indigo reduction and dyeing quality. Chem Eng J 2007;131:187–93.
[16] Nicholson S, John P. The mechanism of bacterial indigo reduction. Appl Microb Cell Physiol 2005;68:117–23.
[17] Roessler A, Jin X. State of the art technologies and new electrochemical methods for the reduction of vat dyes. Dye Pigment 2003;59:223–35.
[18] Anbu Kulandainathan M, Patil K, Muthukumaran A, Chavan R. Review of the process development aspects of electrochemical dyeing: its impact and commercial applications. Color Technol 2007;123:143–51.
[19] Chavan R, Chakraborty J. Enhancement in indigo uptake on cotton through metal salt pretreatment. Indian J Fibre Text Res 2000;25:130–7.
[20] Chavan R. Environment-friendly dyeing process for cotton. Indian J Fibre Text Res 2001;26:93–100.
[21] Vuorema A, John P, Keskitalo M, Kulandainathan MA, Marken F. Electrochemical and sonoelectrochemical monitoring of indigo reduction by glucose. Dye Pigment 2008;76:542–9.
[22] Roessler A, Crettenand D, Dossenbach O, Marte W, Rys P. Direct electrochemical reduction of indigo. Electrochim Acta 2002;47:1989–95.
[23] Roessler A. Direct electrochemical reduction of vat dyes in a fixed bed of graphite granules. Dye Pigment 2004;63:29–37.
[24] Anbu Kulandainathan M, Muthukumaran A, Patil K, Chavan R. Potentiostatic studies on indirect electrochemical reduction of vat dyes. Dye Pigment 2007;73:47–54.
[25] Bechtold T, Burtscher E, Bobleter O, Blatt W, Schneider L. Optimization of multi-cathode membrane electrolysers for the indirect electrochemical reduction of indigo. Chem Eng Technol 1998;21:877–80.
[26] Sala M, Gutiérrez-Bouzán MC. Electrochemical techniques in textile processes and wastewater treatment. Int J Photoenergy 2012;2012:1–12.
[27] Hasanbeigi A, Price L. A review of energy use and energy efficiency technologies for the textile industry. Renew Sustain Energy Rev 2012;16:3648–65.
[28] Nawaz MS, Ahsan M. Comparison of physico-chemical, advanced oxidation and biological techniques for the textile wastewater treatment. Alex Eng J 2014;53:717–22.
[29] Arslan-Alaton I, Alaton I. Degradation of xenobiotics originating from the textile preparation, dyeing, and finishing industry using ozonation and advanced oxidation. Ecotoxicol Environ Saf 2007;68:98–107.

[30] Savin I-I, Romen B. Wastewater characteristics in textile finishing mills. Environ Eng Manag J 2008;7:859–64.
[31] Verma AK, Dash RR, Bhunia P. A review on chemical coagulation/flocculation technologies for removal of colour from textile wastewaters. J Environ Manag 2012;93:154–68.
[32] Khandegar V, Saroha AK. Electrocoagulation for the treatment of textile industry effluent–a review. J Environ Manag 2013;128:949–63.
[33] Moreira F, Boaventura R, Brillas E, Vilar V. Electrochemical advanced oxidation processes: a review on their application to synthetic and real wastewaters. Appl Catal B Environ 2017;202:217–61.
[34] Fersi C, Gzara L, Dhahbi M. Treatment of textile effluents by membrane technologies. Desalination 2005;185:399–409.
[35] Körbahti BK, Tanyolaç A. Continuous electrochemical treatment of simulated industrial textile wastewater from industrial components in a tubular reactor. J Hazard Mater 2009;170:771–8.
[36] Uzal N, Levent Yilmaz UY, Uzal N, Yilmaz L, Yetis U. Nanofiltration and reverse osmosis for reuse of indigo dye rinsing waters. Sep Sci Technol 2010;45:331–8.
[37] Riera-Torres M, Gutiérrez-Bouzán C, Crespi M. Combination of coagulation–flocculation and nanofiltration techniques for dye removal and water reuse in textile effluents. Desalination 2010;252:53–9.
[38] De Jager D, Sheldon MS, Edwards W. Colour removal from textile wastewater using a pilot-scale dual-stage MBR and subsequent RO system. Sep Purif Technol 2014;135:135–44.
[39] Danwittayakul S, Jaisai M, Dutta J. Efficient solar photocatalytic degradation of textile wastewater using ZnO/ZTO composites. Appl Catal B Environ 2015;163:1–8.
[40] Sathian S, Rajasimman M, Radha G, Shanmugapriya V, Karthikeyan C. Performance of SBR for the treatment of textile dye wastewater: optimization and kinetic studies. Alex Eng J 2014;53:417–26.
[41] Blanco J, Torrades F, Morón M, Brouta-Agnésa M, García-Montaño J. Photo-Fenton and sequencing batch reactor coupled to photo-Fenton processes for textile wastewater reclamation: feasibility of reuse in dyeing processes. Chem Eng J 2014;240:469–75.
[42] Blanco J, Torrades F, De la Varga M, García-Montaño J. Fenton and biological-Fenton coupled processes for textile wastewater treatment and reuse. Desalination 2012;286:394–9.
[43] Robinson T, McMullan G, Marchant R, Nigam P. Remediation of dyes in textile effluent: a critical review on current treatment technologies with a proposed alternative. Bioresour Technol 2001;77:247–55.
[44] Gosavi VD, Sharma S. A general review on various treatment methods for textile wastewater. J Environ Sci Comput Sci Eng Technol 2014;3:29–39.
[45] Alinsafi A, da Motta M, Le Bonté S, Pons MN, Benhammou A. Effect of variability on the treatment of textile dyeing wastewater by activated sludge. Dye Pigment 2006;69:31–9.
[46] Firmino PIM, da Silva MER, Cervantes FJ, dos Santos AB. Colour removal of dyes from synthetic and real textile wastewaters in one- and two-stage anaerobic systems. Bioresour Technol 2010;101:7773–9.
[47] Amaral FM, Kato MT, Florêncio L, Gavazza S. Color, organic matter and sulfate removal from textile effluents by anaerobic and aerobic processes. Bioresour Technol 2014;163:364–9.
[48] Kapdan IK, Alparslan S. Application of anaerobic–aerobic sequential treatment system to real textile wastewater for color and COD removal. Enzym Microb Technol 2005;36:273–9.
[49] Kapdan IK, Tekol M, Sengul F. Decolorization of simulated textile wastewater in an anaerobic–aerobic sequential treatment system. Process Biochem 2003;38:1031–7.

[50] Shaw C, Carliell C, Wheatley A. Anaerobic/aerobic treatment of coloured textile effluents using sequencing batch reactors. Water Res 2002;36:1993–2001.
[51] Jonstrup M, Kumar N, Murto M, Mattiasson B. Sequential anaerobic–aerobic treatment of azo dyes: decolourisation and amine degradability. Desalination 2011;280:339–46.
[52] O'Neill C. Anaerobic–aerobic biotreatment of simulated textile effluent containing varied ratios of starch and azo dye. Water Res 2000;34:2355–61.
[53] Kokabian B, Bonakdarpour B, Fazel S. The effect of salt on the performance and characteristics of a combined anaerobic–aerobic biological process for the treatment of synthetic wastewaters containing Reactive Black 5. Chem Eng J 2013;221:363–72.
[54] Bonakdarpour B, Vyrides I, Stuckey DC. Comparison of the performance of one stage and two stage sequential anaerobic–aerobic biological processes for the treatment of reactive-azo-dye-containing synthetic wastewaters. Int Biodeterior Biodegrad 2011;65:591–9.
[55] Murali V, Ong S-A, Ho L-N, Wong Y-S. Evaluation of integrated anaerobic-aerobic biofilm reactor for degradation of azo dye methyl orange. Bioresour Technol 2013;143: 104–11.
[56] Farabegoli G, Chiavola A, Rolle E, Naso M. Decolorization of reactive red 195 by a mixed culture in an alternating anaerobic–aerobic sequencing batch reactor. Biochem Eng J 2010;52:220–6.
[57] Martínez-Huitle CA, dos Santos EV, de Araújo DM, Panizza M. Applicability of diamond electrode/anode to the electrochemical treatment of a real textile effluent. J Electroanal Chem 2012;674:103–7.
[58] Aquino JM, Rocha-Filho RC, Ruotolo LAM, Bocchi N, Biaggio SR. Electrochemical degradation of a real textile wastewater using β-$PbO_2$ and DSA® anodes. Chem Eng J 2014;251:138–45.
[59] Riera-Torres M, Gutiérrez M-C. Colour removal of three reactive dyes by UV light exposure after electrochemical treatment. Chem Eng J 2010;156:114–20.
[60] Riera-Torres M, Gutiérrez-Bouzán C. Optimisation of the electrochemical and UV combined treatment to remove colour and organic halogenated compounds of textile effluents. Sep Purif Technol 2012;98:375–82.
[61] Chatzisymeon E, Xekoukoulotakis NP, Coz A, Kalogerakis N, Mantzavinos D. Electrochemical treatment of textile dyes and dyehouse effluents. J Hazard Mater 2006;137:998–1007.
[62] Vlyssides A, Papaioannou D, Loizidoy M, Karlis P, Zorpas A. Testing an electrochemical method for treatment of textile dye wastewater. Waste Manag 2000;20:569–74.
[63] Radha KV, Sridevi V, Kalaivani K. Electrochemical oxidation for the treatment of textile industry wastewater. Bioresour Technol 2009;100:987–90.
[64] Sala M, López-Grimau VV, Gutiérrez-Bouzán C. Photo-electrochemical treatment of reactive dyes in wastewater and reuse of the effluent: method optimization. Materials (Basel) 2014;7:7349–65.
[65] Anjaneyulu Y, Chary NS, Raj DSS. Decolourization of industrial effluents – Available methods and emerging technologies – a review. Rev Environ Sci Bio/Technology 2005;4:245–73.
[66] Ong YK, Li FY, Sun S-P, Zhao B-W, Liang C-Z, Chung T-S. Nanofiltration hollow fiber membranes for textile wastewater treatment: lab-scale and pilot-scale studies. Chem Eng Sci 2014;114:51–7.
[67] Lau W-J, Ismail AF. Polymeric nanofiltration membranes for textile dye wastewater treatment: preparation, performance evaluation, transport modelling, and fouling control—a review. Desalination 2009;245:321–48.
[68] Vergili I, Kaya Y, Sen U, Gönder ZB, Aydiner C. Techno-economic analysis of textile dye bath wastewater treatment by integrated membrane processes under the zero liquid discharge approach. Resour Conserv Recycl 2012;58:25–35.

[69] Chen Q, Yang Y, Zhou M, Liu M, Yu S, Gao C. Comparative study on the treatment of raw and biologically treated textile effluents through submerged nanofiltration. J Hazard Mater 2014;284:121–9.
[70] Chidambaram T, Oren Y, Noel M. Fouling of nanofiltration membranes by dyes during brine recovery from textile dye bath wastewater. Chem Eng J 2015;262:156–68.
[71] Qin J-J, Oo MH, Kekre KA. Nanofiltration for recovering wastewater from a specific dyeing facility. Sep Purif Technol 2007;56:199–203.
[72] Fersi C, Dhahbi M. Treatment of textile plant effluent by ultrafiltration and/or nanofiltration for water reuse. Desalination 2008;222:263–71.
[73] Chakraborty S, Purkait MK, Dasgupta S, Basu JK. Nanofiltration of textile plant effluent for color removal and reduction in COD. Sep Purif Technol 2003;31:141–51.
[74] Kim T-H, Park C, Kim S. Water recycling from desalination and purification process of reactive dye manufacturing industry by combined membrane filtration. J Clean Prod 2005;13:779–86.
[75] Petrinić I, Andersen NPR, Šostar-Turk S, Le Marechal AM. The removal of reactive dye printing compounds using nanofiltration. Dye Pigment 2007;74:512–8.
[76] Buscio V, Marín MJ, Crespi M, Gutiérrez-Bouzán C. Reuse of textile wastewater after homogenization–decantation treatment coupled to PVDF ultrafiltration membranes. Chem Eng J 2015;265:122–8.
[77] Buscio V, Crespi M, Gutiérrez-Bouzán C. Application of PVDF ultrafiltration membranes to treat and reuse textile wastewater. Desalination Water Treat 2016;57:8090–6.
[78] Buscio V, Crespi M, Gutiérrez-Bouzán C. Sustainable dyeing of denim using indigo dye recovered with polyvinylidene difluoride ultrafiltration membranes. J Clean Prod 2015;91:201–7.
[79] Ellouze E, Tahri N, Amar RB. Enhancement of textile wastewater treatment process using nanofiltration. Desalination 2012;286:16–23.
[80] Zheng Y, Yu S, Shuai S, Zhou Q, Cheng Q, Liu M, et al. Color removal and COD reduction of biologically treated textile effluent through submerged filtration using hollow fiber nanofiltration membrane. Desalination 2013;314:89–95.
[81] Żyłła R, Sójka-Ledakowicz J, Stelmach E, Ledakowicz S. Coupling of membrane filtration with biological methods for textile wastewater treatment. Desalination 2006;198:316–25.
[82] Liu M, Lü Z, Chen Z, Yu S, Gao C. Comparison of reverse osmosis and nanofiltration membranes in the treatment of biologically treated textile effluent for water reuse. Desalination 2011;281:372–8.
[83] Debik E, Kaykioglu G, Coban A, Koyuncu I. Reuse of anaerobically and aerobically pretreated textile wastewater by UF and NF membranes. Desalination 2010;256:174–80.
[84] Bes-Piá, Iborra-Clar MI, Iborra-Clar A, Mendoza-Roca JA, Cuartas-Uribe B, Alcaina-Miranda MI. Nanofiltration of textile industry wastewater using a physicochemical process as a pre-treatment. Desalination 2005;178:343–9.
[85] Bes-Piá A, Mendoza-Roca JA, Alcaina-Miranda MI, Iborra-Clar A, Iborra-Clar MI. Combination of physico-chemical treatment and nanofiltration to reuse wastewater of a printing, dyeing and finishing textile industry. Desalination 2003;157:73–80.
[86] Bes-Piá A, Mendoza-Roca J, Alcaina-Miranda MI, Iborra-Clar A, Iborra-Clar MI. Reuse of wastewater of the textile industry after its treatment with a combination of physico-chemical treatment and membrane technologies. Desalination 2002;149:169–74.
[87] Harrelkas F, Azizi A, Yaacoubi A, Benhammou A, Pons MN. Treatment of textile dye effluents using coagulation–flocculation coupled with membrane processes or adsorption on powdered activated carbon. Desalination 2009;235:330–9.

[88] Lee B-B, Choo K-H, Chang D, Choi S-J. Optimizing the coagulant dose to control membrane fouling in combined coagulation/ultrafiltration systems for textile wastewater reclamation. Chem Eng J 2009;155:101–7.
[89] Khelifi E, Gannoun H, Touhami Y, Bouallagui H, Hamdi M. Aerobic decolourization of the indigo dye-containing textile wastewater using continuous combined bioreactors. J Hazard Mater 2008;152:683–9.
[90] Unlu M, Yukseler H, Yetis U. Indigo dyeing wastewater reclamation by membrane-based filtration and coagulation processes. Desalination 2009;240:178–85.
[91] Albuquerque LF, Salgueiro AA, Melo JLDS, Chiavone-Filho O. Coagulation of indigo blue present in dyeing wastewater using a residual bittern. Sep Purif Technol 2013;104:246–9.
[92] Crespi M. Reutilización de los efluentes textiles en Europa. Bol Intexter 1989;96:87–106.
[93] Uzal N, Yilmaz L, Yetis U. Microfiltration/ultrafiltration as pretreatment for reclamation of rinsing waters of indigo dyeing. Desalination 2009;240:198–208.
[94] Porter JJ. Membrane filtration techniques used for recovery of dyes, chemicals and energy. Am Dyest Rep 1990;22:21–6.
[95] Rice EW, Baird RB, Eaton AD, Clesceri LS, editors. Standard methods for the examination of water and wastewater. 22nd ed. 2012. Washington.
[96] Aouni A, Fersi C, Cuartas-Uribe B, Bes-Piá A, Alcaina-Miranda MI, Dhahbi M. Study of membrane fouling using synthetic model solutions in UF and NF processes. Chem Eng J 2011;175:192–200.
[97] Alventosa-deLara E, Barredo-Damas S, Zuriaga-Agustí E, Alcaina-Miranda MI, Iborra-Clar MI. Ultrafiltration ceramic membrane performance during the treatment of model solutions containing dye and salt. Sep Purif Technol 2014;129:96–105.
[98] He Y, Li G, Wang H, Zhao J, Su H, Huang Q. Effect of operating conditions on separation performance of reactive dye solution with membrane process. J Membr Sci 2008;321:183–9.
[99] Forster C. Wastewater treatment and technology. London: Thomas Telford Publishing; 2003.
[100] Qu F, Liang H, Zhou J, Nan J, Shao S, Zhang J, et al. Ultrafiltration membrane fouling caused by extracellular organic matter (EOM) from *Microcystis aeruginosa*: effects of membrane pore size and surface hydrophobicity. J Membr Sci 2014;449:58–66.
[101] Abdelrasoul A, Doan H, Lohi A. Fouling in membrane filtration and remediation methods. In: Nakajima H, editor. Mass transfer – advances in sustainable energy and environment oriented numerical modeling. InTech; 2013. p. 195–218.
[102] Vajnhandl S, Valh JV. The status of water reuse in European textile sector. J Environ Manag 2014;141:29–35.
[103] Chico D, Aldaya MM, Garrido A. A water footprint assessment of a pair of jeans: the influence of agricultural policies on the sustainability of consumer products. J Clean Prod 2013;57:238–48.

# Denim and consumers' phase of life cycle

**10**

*A.P. Periyasamy, J. Militky*
Technical University of Liberec, Liberec, Czech Republic

## 10.1 Introduction

Denim is the oldest, most widely used, most fashionable and most versatile fabric in the world. Although initially the fabric was developed for using as a sailcloth, now denim is mainly used to produce garments, particularly trousers for young people. Denim's origin is as a tough canvas cloth from Nîmes, France (*de Nîmes*: of Nîmes). Genoese sailors used this cloth to make rough-and-ready pants that lasted forever. Called 'Genoese', which in time was corrupted to *geans* or 'jeans', these trousers found favour around the world wherever men had to do work that would ruin most other clothes. In the American Wild West, cowboys practically lived in them, as did railroad workers and farmers.

Today's consumers like to live in a casual, relaxed and yet sophisticated manner. With the emphasis on casual lifestyles, therefore, fabrics must not only provide comfort and beauty but also have the stamina to stand up to the rigors of daily life. Thus denim jeans are becoming one of the fastest-growing apparel products in both the developed and developing world. They are strong and versatile, wear well and become softer and more comfortable with time. Denim has also gone from being a work-wear look to a fashion commodity. Denim garments are a major trendsetter of youth. The spread of denim culture over the world has brought with it the trend of fast-changing fashion. Wet processing of garments imparts an artistic and innovative look, especially to denim fabric. It is considered a value addition to the garments.

Denim is a type of cotton textile known for its use in blue jeans and other clothing. Denim uses a sturdy twill weave with a characteristically diagonal ribbing. Originally used for workers' clothes, denim is now ubiquitous and has even entered the world of high fashion. Nearly everyone has at least one denim garment in the closet. Denim washing is achieved in many ways, using pumice stones or hypochlorite or even enzymes. The reason enzymes are chosen for washing is because of their eco-friendly nature compared with hypochlorite and pumice stones. Enzyme washing is a laundering process which uses enzymes to clean clothing or finish fabric, especially in the case of jeans and other garments with a worn look. Various enzymatic cleaners are available from stores which specialize in laundry supplies; they can also be specially ordered. For regular cleaning, enzymes have numerous economic and environmental benefits. On an industrial scale, enzyme washing has replaced laborious laundering techniques such as stonewashing, saving money and having a better environmental impact for companies [1,2]. Since many decades, cotton was used to produce denim fabrics, but last three decades, different fibres were used to manufacture denim fabrics

**Sustainability in Denim. http://dx.doi.org/10.1016/B978-0-08-102043-2.00010-1**
Copyright © 2017 Elsevier Ltd. All rights reserved.

with improved performance (polyester, lyocell, lycra etc). This has urged the industry to produce innovative denim with different fibres including Tencel, bamboo viscose rayon, polyethylene terephthalate and Lycra (polyurethane) [3–5]. Also, several types of washes have been introduced, such as stonewash and stoneless wash, acid wash, moon wash and monkey wash. With the use of denim, garment export market is concerned that garments of high quality offer greater value for the price and aesthetics. Innovative applications using various types of garment processing bring greater financial profits. The target of the fashion industry is denim. This chapter provides a conceptual and analytical framework for the life-cycle assessment (LCA) of denim in the consumer phase. The chapter illustrates that identifying consumers and their interests, responsibilities and accountability provides an LCA of the consumer phase. It mainly focuses on results regarding the global warming potential during consumer use of one pair of jeans, expressed is $CO_2$ equivalents, and describes water and energy consumption. This study attempts to integrate a more systematic and holistic approach to developing strategies for responding to the environmental aspects of sustainable development in the denim industry with respect to the consumer phase.

## 10.2 The global consumption and market potential of denim

It is estimated that by 2020 the global denim industry will be worth USD 64.1 billion, whereas in 2014 it was USD 56.2 billion (Fig. 10.1). Right now, approximately 1.8 billion pairs of denim trousers are sold annually, which is worth around USD 51 billion. Europe,

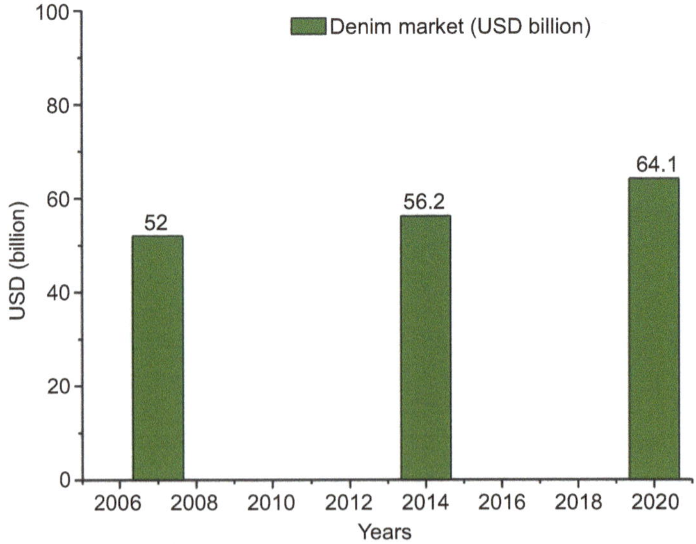


**Figure 10.1** Global denim market potential (expected values for 2020) [6].
Based on Driving demend for denim; (n.d.). http://www.cottoninc.com/corporate/Market-Data/SupplyChainInsights/Driving-Demand-For-Denim-Jeans/.

United States and Japan are the biggest consumers of denim (Fig. 10.2). Younger economies such as China and India are witnessing a steady rise in demand for denim. The global demand is growing 5% and the supply is at 8% [6].

In the production phase, China was the dominant supplier (Figs. 10.3 and 10.4); it makes quality cloth at much lower prices than other suppliers. Meanwhile, China's burgeoning apparel industry has initially fueled the expansion of China's denim capacity; China will also likely carve out a growing share of the global denim market at the expense of other suppliers. Denim is no different from the industry's experience

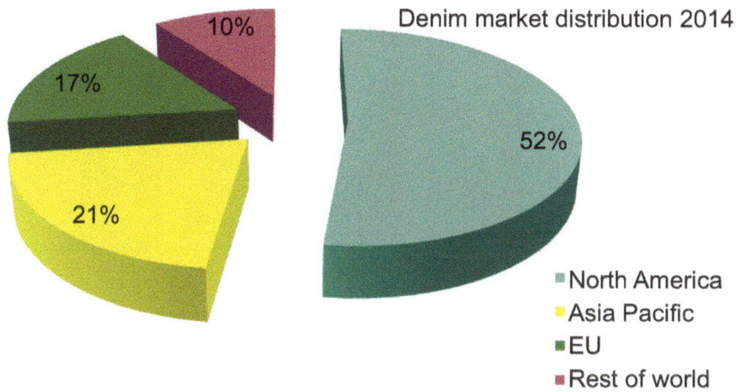


**Figure 10.2** Denim market distribution based on the different continents. *EU*, European Union. Data from: CIA Factbook. 2016. https://www.cia.gov/library/publications/resources/the-world-factbook/.

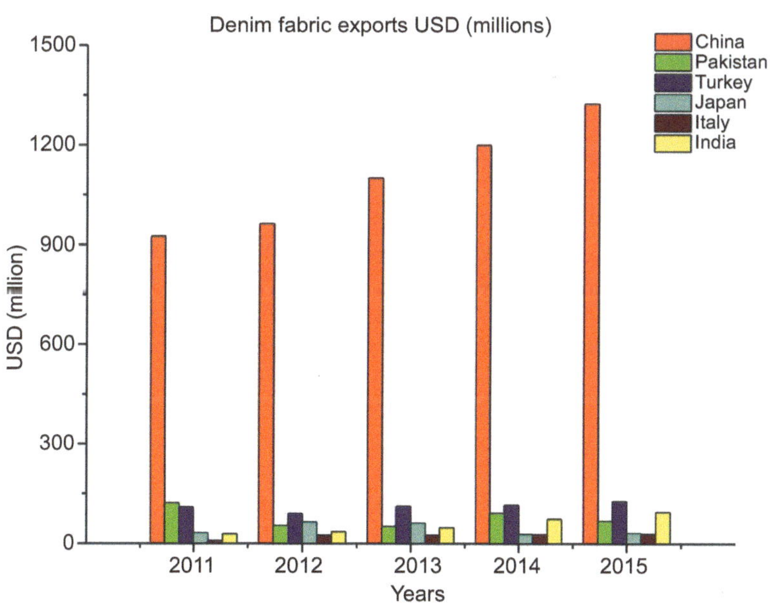


**Figure 10.3** Denim fabric exports (expressed in millions of US dollars) [8].
Based on Newbery M, Global market review of denim and jeanswear–forecasts to 2021. Worcs, UK, 2016. http://www.just-style.com/market-research/.

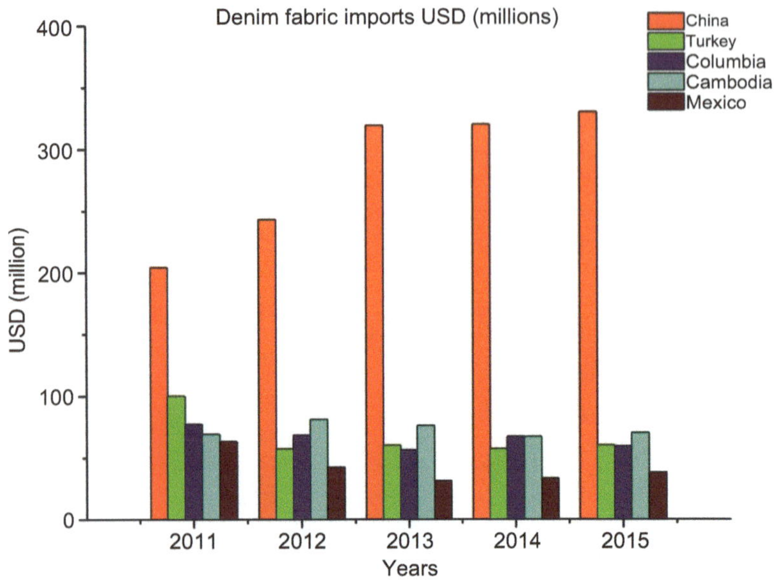


**Figure 10.4** Denim fabric imports (expressed in millions of US dollars) [8].
Based on Newbery M, Global market review of denim and jeanswear–forecasts to 2021. Worcs, UK, 2016.http://www.just-style.com/market-research/.

with other fabrics, because Chinese producers continue to distinguish themselves as low-cost manufacturers. In the past 2 decades the cost of labour has risen and it has had a huge impact on the apparel industry, particularly on exporting apparel goods. Thus the industry has turned to domestic producers owing to the huge potential. There will be strong growth in denim production in India, Turkey, Bangladesh, Pakistan and Vietnam, although Vietnam is still a wild card. Vietnam has the potential to rival China as the dominant denim supplier in the world, although for the moment China continues to be more central to the production of denim than Vietnam. However in China, the cost of labour is increasing every year whereas it is moderate in Vietnam, Bangladesh and Kenya, which has a huge impact on this business. The Chinese denim industry may relocate to these places.

On the other hand, the growing middle class, changes in the retail sector and changes in the private label scenario in the country have worked in favour of manufacturers in the denim market of India. However, the United States has cut down their production of denim as a result of rising labour and energy costs. This has led to significant growth in export demands for the Indian denim market and resulted in increased export demands in the Indian denim market. The country also faced vast competition in the international market along with competition from other Asian countries such as Vietnam, China, Turkey, Bangladesh and Indonesia. Fluctuations in the price of cotton, frequent changes in the central government's policies, the devaluation of the rupee in the world currency market, severe power cuts and the lack of availability of skilled labour were the biggest challenges faced by manufacturers in the denim market of India. Indian denim manufacturing and consumption have grown at a healthy compound annual growth rate of 12%–15% over the past decade and are soon expected to

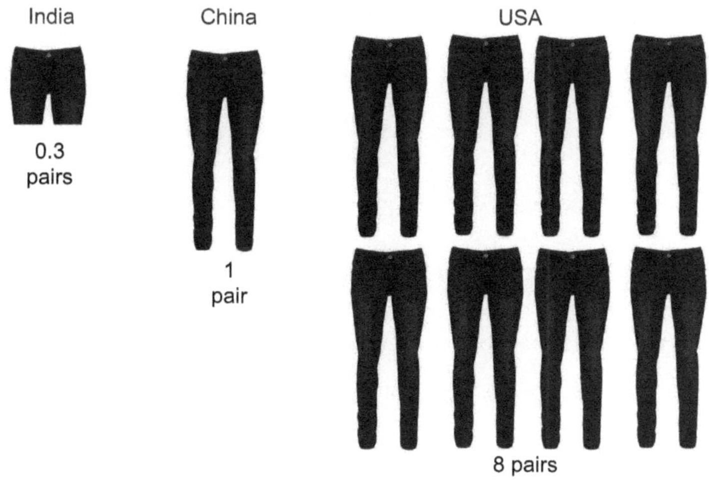


**Figure 10.5** Per capita consumption of denim in India, China, and the United States.
Based on: CIA Factbook. 2016. https://www.cia.gov/library/publications/resources/the-world-factbook/.

grow at similarly compounded rates. The capacity of denim often builds up in India in 'steep steps', i.e., large capacities are added in a short period of time. The current installed capacity of denim fabric production stands at about 1 billion metres per annum. If the capacity continues to grow at even 10% in coming years, India will cross the magical mark of 1.2 billion metres by 2015. The current domestic consumption of Indian denim is approximately 450 million meters and the export of denim fabric from India is approximately 200 million meters, which implies that capacity use is approximately 65%–70%. It was estimated that by 2016 the global denim industry would be worth USD 64.1–70 billion. In 2014, 1.8 billion pairs of denim trousers, amounting to USD 60 billion, were sold annually worldwide. Europe and the United States are the biggest markets for denim fashion even as newer economies such as China and India witnessed a steady rise in demand [6]. According to industry estimates, per capita denim consumption in the United States is eight pairs whereas it is 1 pair in China and 0.3 pairs in India (Fig. 10.5) [7]. However, in the near future, the denim consumption will increase per capita, owing to the young demographic profile of the population in which 50% of the population is in the age group 15–35 years [9].

The European Union (EU) is one of the largest consumers of denim; however, the denim is imported to Italy; for several reasons it can be gotten from Turkey and a few African countries. Fig. 10.6 shows denim suppliers to the EU. African countries are located near to Europe as well as United States in terms of shipping the goods (as compared to Asian countries such as China, India), and the cost of labour is much less even compared with Bangladesh; therefore, African countries are the best option for denim manufactures. As the work in continent grows under the African Growth and Opportunity Act (AGOA) and continues to receive investment from China, India and elsewhere, cutting and sewing operations will take the centre stage and imported fabrics will largely support this production. In part, this will be driven by the fact that the AGOA agreement makes it much more economical to import cloth than to buy it locally [10].

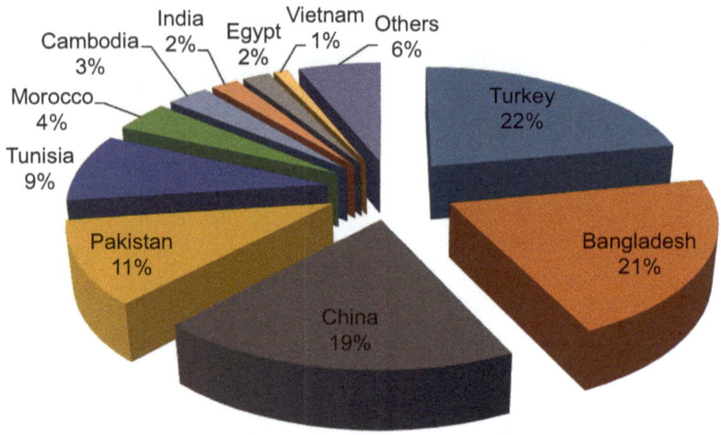


**Figure 10.6** Denim suppliers to the European Union (EU) in 2013 [11].
Based on C. Incorporation, Global lifestyle insights; 2015.http://lifestylemonitor.cottoninc.com/wp-content/uploads/2016/07/MC-GLM_UK_2016-4-4-16.pdf.

## 10.3 Consumer phase for denim

Consumers are not only attuned to the social ramifications of fashion but are becoming increasingly aware of the environmental and social impacts of textile and apparel products [12]. Therefore, reductions in the environmental and social impacts on denim productions have been initiated. Denim manufacturing organizations (brands) have sought ways to implement sustainability strategies while maintaining consumer appeal. Research shows that characteristics of denim such as comfort, fit, price, style and design are the strongest predictors for the purchase of denim rather than social or environmental considerations (Fig. 10.7). Based on a survey conducted by Cotton, Inc., when women in the United States purchase new denim, they concentrated on fit; at 92%, it is mostly attributed factor. The comfort properties of denim are secondary (89%), although they are important. Thereafter, there are many factors such as good looks, price, style, etc. However the truth is that environmental friendliness ranks low and is the least important factor in the survey. It seems that it is not an important factor for American women who purchase denim. Generally, men are most concerned with comfort, and then fit. Men are also more concerned about price than are women. Results show that other factors are not as important, particularly 'multifunctionality'. Environmental factors remain the same and lower than for women: Only 14% of men are concerned about environmentally friendly denim. Environmental consciousness is actually conspicuous in its absence in most markets around the world, where buyers are not aware of actual damage that jeans production causes. However the importance of this factor will increase daily (through the awareness of global warming), and in the near future this factor will come up in the first position. Already many good practices have been initiated, such as organic cotton denim, lyocell denim and Better Cotton Initiative denim. Strict regulations have been enacted to ban harmful production processes such as sandblasting. However, at the consumer level awareness is still low [13].

Thus this survey is important for designers who create denim that appeals to consumers purchasing the product, ensuring a financial goal while pursuing the objectives

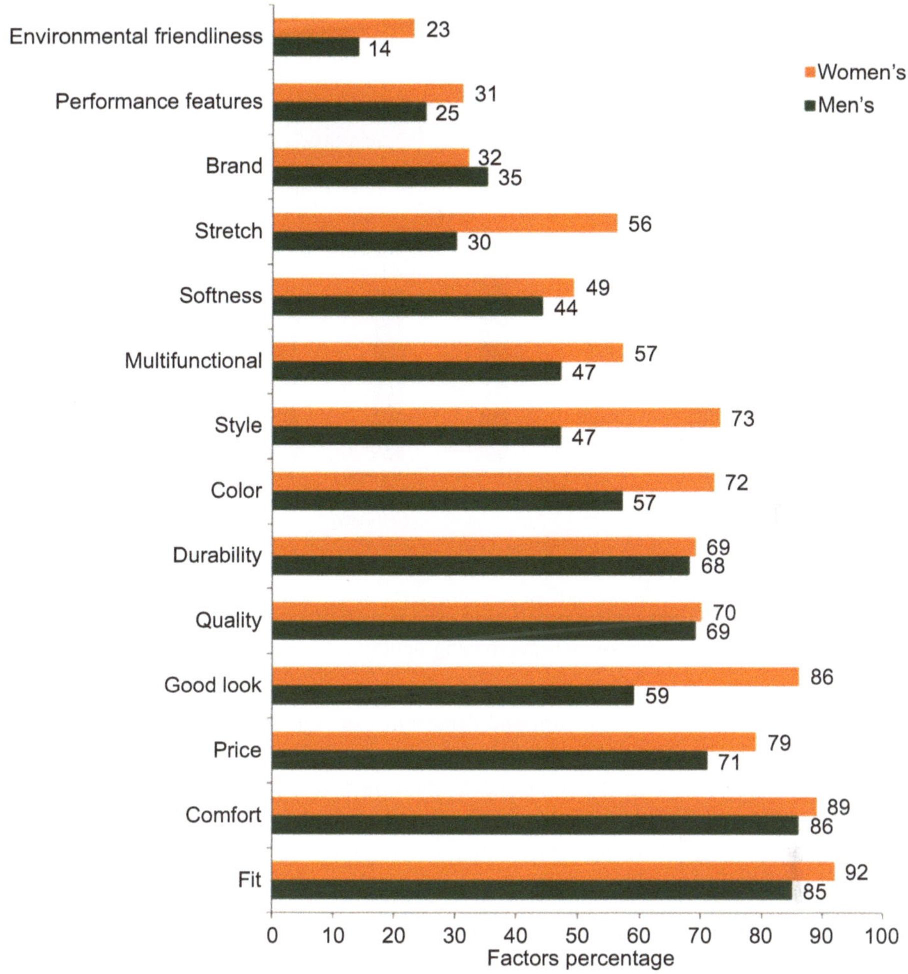


**Figure 10.7** Factors that are important to consumers' purchase of jean (Cotton, Inc. Lifestyle Monitor Survey, 2011) [14].
Based on Consumer concerns when purchasing new jeans; 2012. http://lifestylemonitor.cottoninc.com/consumers-concerns-when-purchasing-denim/.

of enhancing environmental and social sustainability. The design is a key area which interlinks many functions in the denim industry, such as product management and development, marketing, operations and finance; it mutually informs one another. Designers keep some ideas about their products, which depend on aesthetics, cost, quality, function, the business niche and psychological consumer needs; but in addition, they need to make environmental and health hazards prime criteria in the sequence.

### *10.3.1 Influences of premium denim consumers*

Nonpremium consumers wear denim because they like it and they already use it (69%). In this case, premium consumers only comprise 53%. By seeing displays in

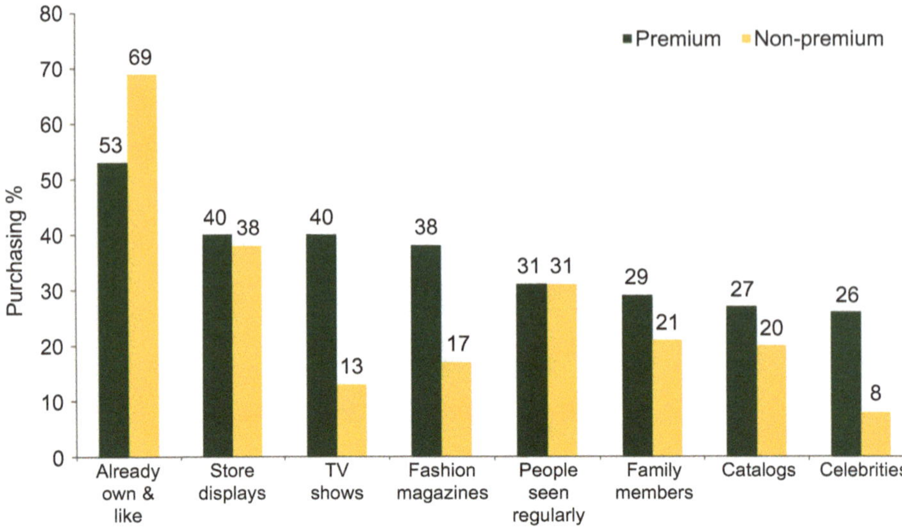


**Figure 10.8** Factors which influence the purchasing percent with respect to the premium and non-premium consumers [14].
Based on Consumer concerns when purchasing new jeans; 2012. http://lifestylemonitor.cottoninc.com/consumers-concerns-when-purchasing-denim/.

stores (visual appearance), premium and nonpremium consumers are attracted to purchase denim at 40% and 38%, respectively, which is not a huge difference for both types of consumer (see Fig. 10.8). However premium consumers are influence by TV shows and fashion magazines. Both types of consumers shared an equal purchasing percentage in the case of seeing people outside and what they wear. Premium consumers more highly influenced by family members, catalogs, and celebrities; nonpremium consumers are not much influenced by this category.

There are many reasons why denim is so popular and has become fashionable wear. Jeans are extremely versatile and provide comfort; therefore many people like to wear them. The colour of indigo is unique and beautiful, and attracts consumers. Temperate zone (tepid latitude) countries such as Europe, Canada, the United States, Australia and China consume more jeans owing to the temperature. The temperatures of these regions are relatively moderate rather than extremely hot or cold as in tropical regions (thicker fabric can sustain the cold weather). Apart from these facts, jeans obtain a better look (vintage) when they become older. Commercial research surveys suggest that the global average for wearing denim is 3.5 days per week (Global Lifestyle Monitor, 2008) and is increasing every year. The highest frequency is in Germany, where the population wears jeans 5.2 times per week. Globally 70% (7 of 10 people) said that they loved and enjoyed wearing jeans (Fig. 10.9) [11].

Consumers have different opinions when they purchase new denim, Cotton, Inc. conducted a survey; Fig. 10.10 shows that good-looking and practical denim are more influencing factors. The highest response was from Brazil (72%) and the lowest rank is India (27%). On average, US consumers spend 73 USD on apparel every month and

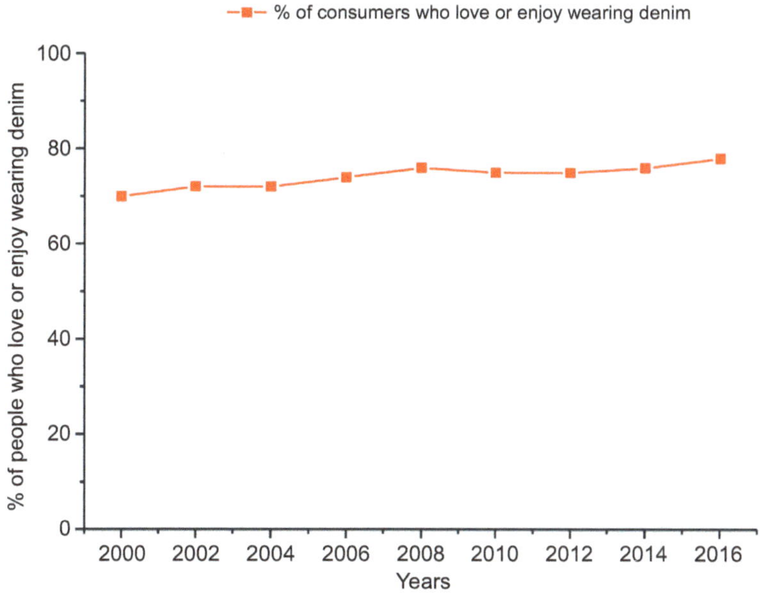


**Figure 10.9** Percentage of consumer who love or enjoy wearing denim [14].
Based on the data from Cotton Incorporated's Lifestyle Monitor™ Survey: 2015; 2016 are expected value and Consumer concerns when purchasing new jeans; 2012. http://lifestylemonitor.cottoninc.com/consumers-concerns-when-purchasing-denim/.

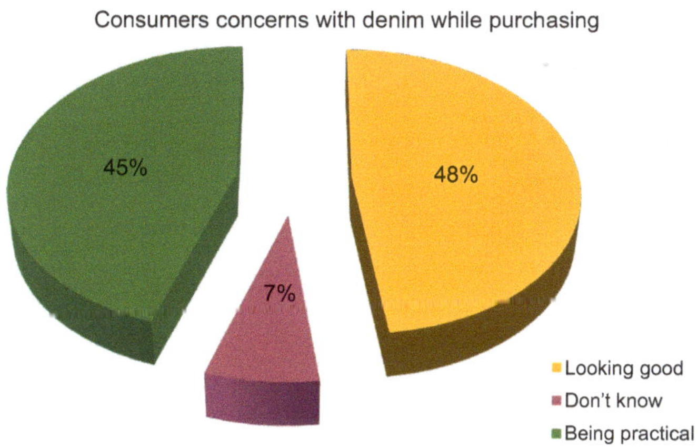


**Figure 10.10** Consumers' concerns with respect to the purchase of new jeans [16].
Modified from Consumer concerns when purchasing new jeans; 2012. http://lifestylemonitor.cottoninc.com/consumers-concerns-when-purchasing-denim/.

denim wear is dominant. In 2015, the global lifestyle monitor reported that apparel purchasing increased (per capita) to 28%, EU, 20%, and for the United Kingdom, 38%; the most important was China. Chinese consumers spent 275 USD per year for apparel and it increased 131% [11,15].

## 10.4 What is a life-cycle assessment?

An LCA is an analytical tool for a systematic and quantitative evaluation of the environmental impacts of a product or service system through all stages of its life. It is an undeniable fact that denim products require the use of raw materials, energy, equipment and labour. Generally these are the key points which determine the design and product development. Therefore an LCA is necessary for an analysis of details about product design by assessing environmental and economic impacts during the product's life cycle with an aim to ensure sustainable development. An LCA studies environmental aspects and potential impacts throughout a product's life (i.e., cradle to grave) from the acquisition of raw materials through production, use and disposal. Life-cycle analyses are critical for understanding the global progress of reducing negative impacts on the environment. Currently, the LCA is the most comprehensive approach to assessing the environmental impacts of textile products; they are graded by the International Organization for Standardization (ISO) 14040-14043 industry standard [17–20]. An LCA not include social or economic impacts of sustainability. An LCA framework has described as having four phases, as shown in Fig. 10.11. This assessment is based on the definition of its goal and scope, inventory analysis, impact assessment and interpretation of results [21–23]. The importance and procedure of an LCA are discussed in Chapter 4 in this book.

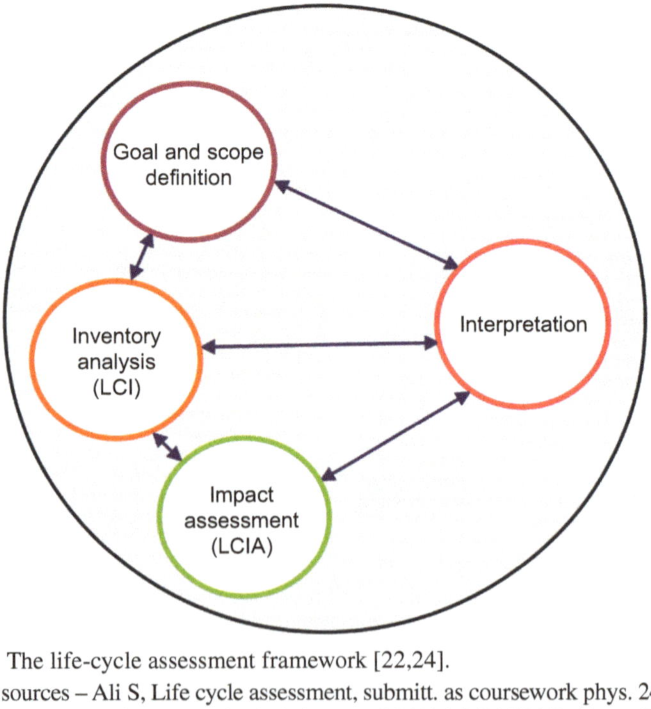


**Figure 10.11** The life-cycle assessment framework [22,24].

Adapted from sources – Ali S, Life cycle assessment, submitt. as coursework phys. 240: Stanford Univ. Fall 2010; 2010. http://large.stanford.edu/courses/2010/ph240/ali2/. Rebitzer G, Ekvall T, Frischknecht R, Hunkeler D, Norris G, Rydberg T, SchmidtW-P, Suh S, Weidema BP, Pennington DW, Life cycle assessment: part 1: framework, goal and scope definition, inventory analysis, and applications, Environ Int 2004;30:701–20. doi:10.1016/j.envint.2003.11.005.

## 10.5 Consumer phase life-cycle assessment of denim

In 2004, Levi Strauss & Co. began a green campaign involving sustainable fashion. They initiated many case studies with respect to their product. In 2007, they conducted an LCA to assess the environmental impact of a pair of Levi's jeans, which are made of cotton. The analysis was made from cradle to grave (seed sowing to landfill) and heralded a new era of awareness about many environmental issues caused by one pair of jeans. From this work, an idea was born to make sustainable fashion in the supply chain (design, production, merchandise and marketing) ensuring the reduction of environmental impacts. The primary objective of this LCA approach was to provide designers and developers with the information they need to produce more sustainable products. A secondary objective was to provide a scientific method to support claims of the environmental improvement of products. Although not an initial objective of the effort, it was later realized that the methodology also provides a rigorous means to communicate environmental performance to suppliers (Levi Strauss & Co.) [25]. The LCA study on denim conducted by Levi Strauss & Co. was based on ISO 14040 standards in which all significant processes from 'cradle to gate' are analysed for the one pair of Levi's 501 jeans. The phases are:

- fibre cultivation (sowing to harvesting)
- yarn formation and fabric formation (ginning and spinning preparatory to weaving and weaving)
- colouration of warp, garment dyeing
- cutting, sewing and finishing (garment making and washing)
- transport and retail (transportation at each stage and distribution to retail stores)
- consumer phase (home laundering, drying and pressing)

The aim of this chapter is to discuss how the consumer phase can have an impact on global climate change. As far as the consumer phase is concerned, three activities cause a huge impact on climatic change: home laundering, drying and pressing.

Much software is available commercially to measure LCA; for this study the Levi's ReCiPe 2008 LCA method was used to compare existing LCA results employing the same parameters [26]. The goal of that assessment was to help Levi's as well as other brands in the apparel sector take more effective and holistic approaches to environmental impacts caused by the industry.

### 10.5.1 Functional unit

Functional units are the number of items to be analysed for the study. Therefore, the functional unit of this study is the production of one pair (women) of Levi's 501 medium stonewash jeans with a weight of 340 g. Details of the functional unit are listed in Table 10.1.

### 10.5.2 System boundaries

This chapter mainly focuses on the consumer phase of the LCA of denim, so system boundaries deal only with home laundering or washing, drying and ironing.

**Table 10.1 Functional unit of life-cycle assessment study [27]**

| | |
|---|---|
| Functional unit | Pair of Levi's 501 women's jeans (one pair) |
| Material | 100% cotton |
| Raw material | Cotton made in United States, Brazil, India, China, Pakistan and Australia |
| Dyeing | Indigo blue |
| Finishing | Medium stone wash |
| Weight of jeans | 340 g |
| Year of production | 2012 |
| LCA conducted | 2013 |
| LCA assessment published | 2015 |
| Consumer phase | United States, United Kingdom, France and China |
| Study conducted by | Industrial Ecology Consultants and Levi Strauss & Co. |
| International Organization for Standardization (ISO) | ISO 14040 |
| Number of studies conducted | Not disclosed |

*LCA*, life-cycle assessment.
Levi Strauss & CO., The life cycle of a jean; 2015:16–50. http://levistrauss.com/wp-content/uploads/2015/03/Full-LCA-Results-Deck-FINAL.pdf.

**Table 10.2 Impact categories and descriptions [27]**

| Environmental impact Categories | | |
|---|---|---|
| **Category** | **Description** | **Units** |
| Climate change | Global warming potential of greenhouse gases released into environment | $CO_2$ emission (kg) |
| Water consumption | Net freshwater taken from environment minus water returned to the same watershed at the same quality or better | Litres |
| Eutrophication potential | Oxygen depletion as a result of nitrogen and phosphorous deposit into freshwater or marine environments | $PO_4$ emission (g) |
| Land occupation | Total land occupied to support product system assessed | $m^2$ per year |
| Abiotic depletion | Measure of depletion of nonrenewable sources including fossil energy, metals and minerals | Sb emission (mg) |

Levi Strauss & CO., The life cycle of a jean; 2015:16–50. http://levistrauss.com/wp-content/uploads/2015/03/Full-LCA-Results-Deck-FINAL.pdf.

### *10.5.3 Environmental impact categories*

The environmental impact potential is expressed at midpoint and end-point levels. Environmental impact categories are listed in Table 10.2. The following categories were adapted from the LCA of Levi's 501 denim jeans conducted by Levi Strauss & Co. based on the ReCiPe 2008 methodology. Essential environmental impact categories are ozone depletion, terrestrial acidification, human toxicity, photochemical oxidant formation, particulate matter formation, terrestrial eco-toxicity, freshwater eco-toxicity, marine eco-toxicity, ionizing radiation, natural land transformation and others which were excluded from this study [27]. To carry out an LCA, these impact categories were investigated (Table 10.2).

## 10.6 Effect of home laundering

Generally climate change results from the emission of greenhouse gases followed by the impacts of global warming. However, the emission of greenhouse gases in textile production is substantial and results in an increase in the temperature of the system and damage to human health, animals and entire ecosystems. To produce one pair of denim jeans, consumer care was considered to have the highest impact on global warming: it measured 37% on the life cycle.

The consumption of nonrenewable energy for one pair denim jeans over its lifetime is shown in Fig. 10.12. In the study, the United States has the highest consumption of nonrenewable energy (60 kW/h for one pair of denim jeans), owing to the highest use of warm water for washing as well as dryers for drying purposes, there is no doubt, which strongly impact on the climatical changes. According to these criteria, China has the lowest consumption of nonrenewable energy because most washing is done with cold water (84%) and it has the lowest use of dryers.

Water consumption in the different countries is shown in Fig. 10.13. The United States has the highest water consumption (1049 L for one pair of denim jeans over its lifetime) compared with other countries; this is related to the highest per capita of denim as well as washing every time for wear. Similarly, the United Kingdom/France

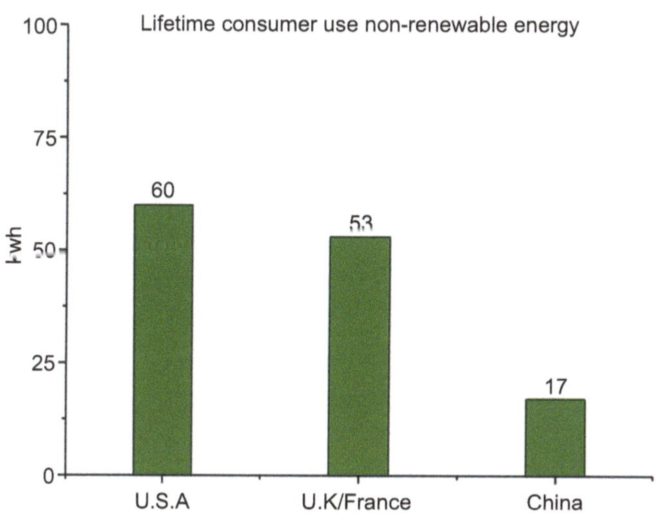


**Figure 10.12** Use of nonrenewable energy for a lifetime of jeans during the consumer phase [28]. *U.K.*, United Kingdom.
Adapted from The life cycle of a jean, by Levi Strauss & Co., 2015; Copyright 2015 by Levi Strauss & Co.

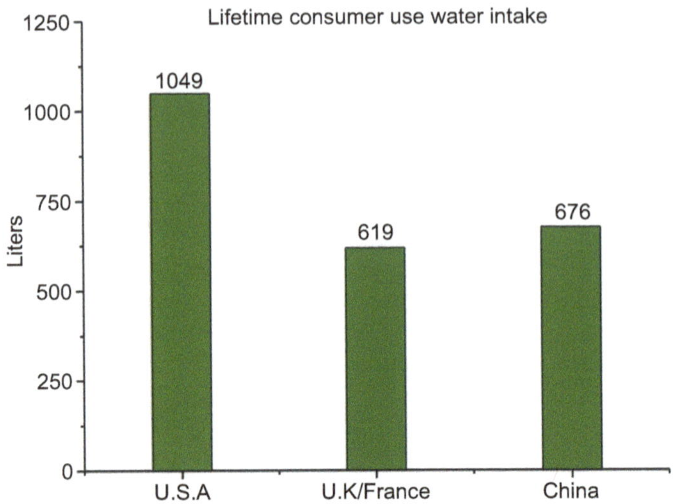


**Figure 10.13** Water intake over the lifetime of jeans during the consumer phase [28]. *U.K.*, United Kingdom.
Adapted from The life cycle of a jean, by Levi Strauss & Co., 2015; Copyright 2015 by Levi Strauss & Co.

and China use 619 and 676L, respectively. Although these figures are low, they could be affecting the environment. Generally, it is known that fresh and clean water are limited resources (the planet is covered in water, but it is saltwater and it is not possible to consume this water without a desalination process). Severe environmental occurrences such as drought and limited access to clean, fresh water occur, which means to take severe action must be taken to reduce water consumption for every product, including textiles and, of course, denim. People who have access to fresh water can take steps to limit the use of water to avoid waste. This study (originally conducted by Levi Strauss & Co.) can give an idea of the water consumption of denim jeans over its lifetime. The reduction of water use is a big thing, but every little thing helps, and it can have a huge impact on the environment.

The average washing frequency between wear for different countries is shown in Fig. 10.14. In these results, Chinese consumers have the highest results, which means that they wash less frequently (wearing jeans 3.9 times for one washing cycle). According to water conservation facts, it is a good practice which reduces the consumer phase of the life cycle in terms of water and it helps to reduce the impact on the entire product life cycle. In the case of the United States, more improvement is needed in the frequency of washing. In general, the washing frequency is lower (i.e., the need to wash clothes every time they are worn) for tropical countries owing to the hot, humid and dusty environment. However, United States, Europe and half of Australia are not located in the tropical regions, therefore it is necessary to reduce the washing frequency, which can show a huge improvement on the environmental concerns.

Fig. 10.15 shows the lifetime water intake for one pair of denim jeans with respect to different washing frequencies. In fact, the water intake for washing every

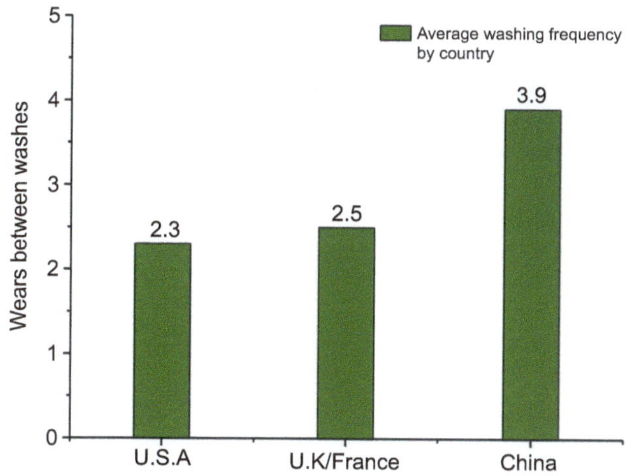


**Figure 10.14** Average washing frequency between wearing [28]. *U.K.*, United Kingdom.
Adapted from The life cycle of a jean, by Levi Strauss & Co., 2015; Copyright 2015 by Levi Strauss & Co.

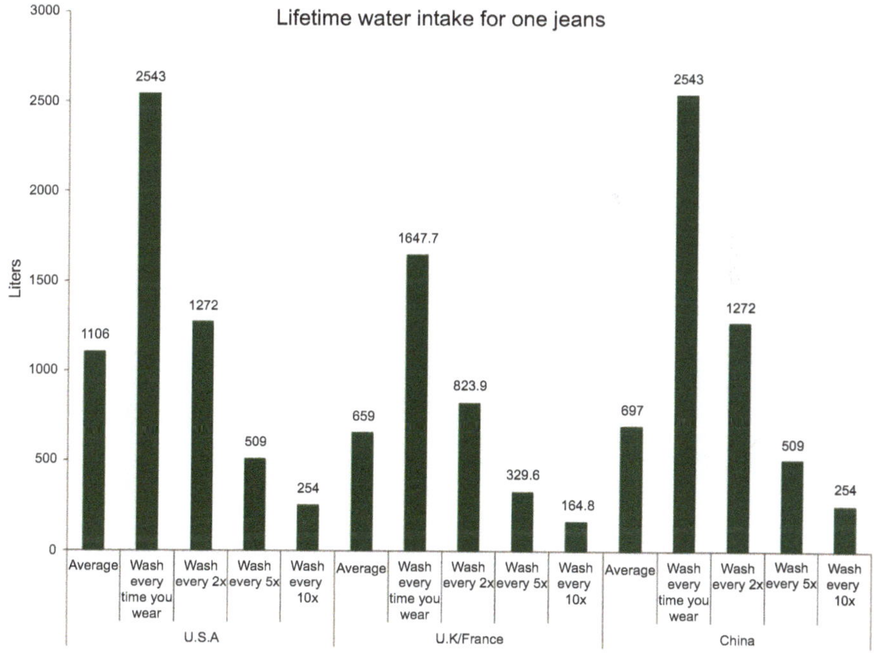


**Figure 10.15** Lifetime water consumption for different frequencies of washing between wear [28]. *U.K.*, United Kingdom.
Adapted from The life cycle of a jean, by Levi Strauss & Co., 2015; Copyright 2015 by Levi Strauss & Co.

time jeans are worn wear is higher than other frequencies. In this, US and Chinese consumers consume more water (2543 L) than do UK/French consumers, who use 1647.7 L of water. Overall results show that US and Chinese consumers are the same in all washing frequencies with respect to the water intake; although China is approximately 40% less than the United States in terms of the average. Among these, Consumers in the United Kingdom and France consumers follow good practice by reducing water intake compared with the United States and China. The graph shows that wearing jeans 10 times before washing them can reduce water use by 77%, 75% and 61% for American, United Kingdom/French and Chinese consumers, respectively.

In general, the lowest number of washing requires higher nonrenewable energy, results prove with numerical values. On average, in the United Kingdom and France, consumers use more energy to wash their jeans than do others, although Americans consume less energy, and compared with Chinese consumers, Americans use more in all washing frequencies. On average, for the energy required for 10 times washing frequency, Chinese consumers use 61%, whereas Americans and people in the United Kingdom/France use 77% and 75%, respectively (Fig. 10.16). The lifetime climate

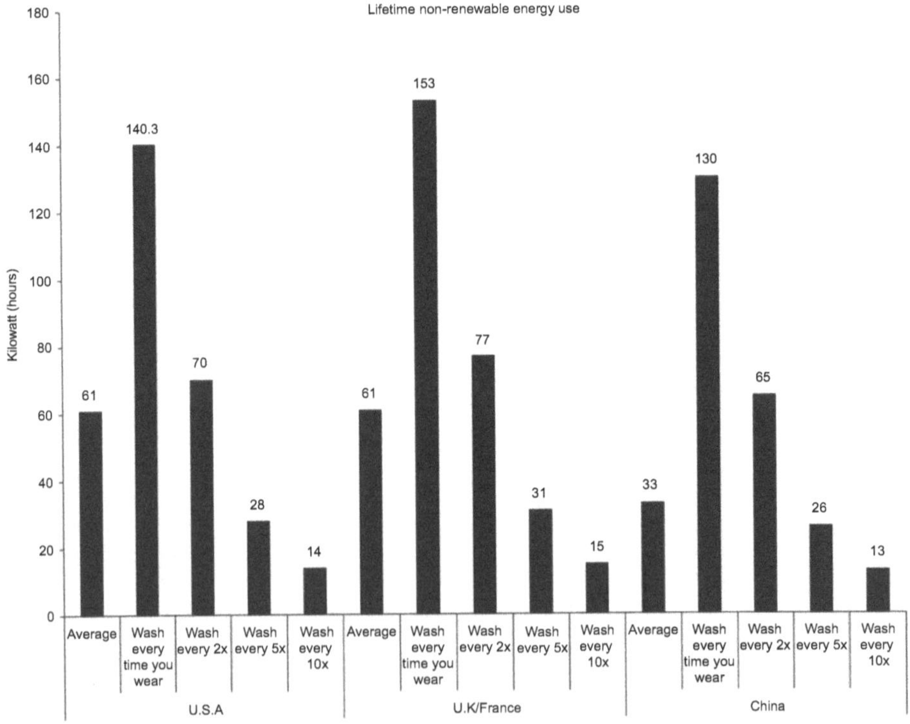


**Figure 10.16** Lifetime nonrenewable energy consumption for different frequencies of washing between wear [28]. *U.K.*, United Kingdom.
Adapted from The life cycle of a jean, by Levi Strauss & Co., 2015; Copyright 2015 by Levi Strauss & Co.

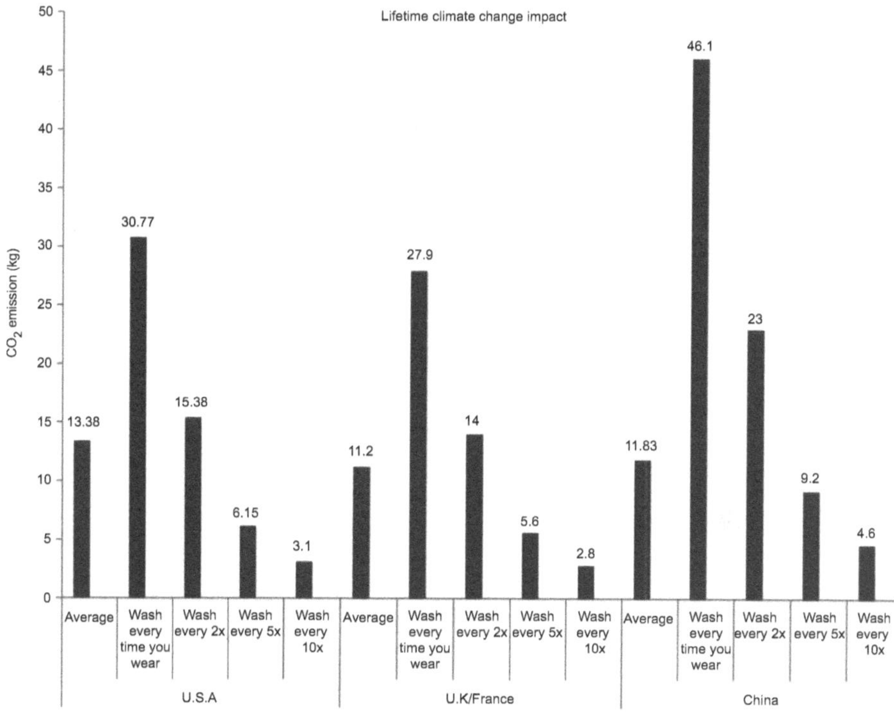


**Figure 10.17** Lifetime climate change impact with different washing frequencies between wear [28]. *U.K.*, United Kingdom.
Adapted from The life cycle of a jean, by Levi Strauss & Co., 2015; Copyright 2015 by Levi Strauss & Co.

change impact variance by washing frequency is shown in Fig. 10.17. On average, frequently washing always shows a higher impact on $CO_2$ emissions. In particular, China has the highest $CO_2$ emission (46.1 per kg) owing to the use of nonrenewable energy, and the impact on climate change always depends on the amount of the use of nonrenewable energy. Similarly, if we see the country average, the United States is higher than other nations, because it has the highest denim per capita. In all washing frequencies, the United Kingdom and France have much lower carbon emissions (in kilograms) than do Americans and Chinese consumers, as the result of to less use of energy and the lifestyle.

## 10.7 Effect of washing with cold and warm water

A total of 84% of Chinese consumers wash their clothes in cold water; only 16% wash them with warm water (Fig. 10.18). In the case of consumers in the United Kingdom/France, around 51% wash denim using cold water. This shows that the lowest percentage of consumers wash with cold water. Of course, this value has a huge impact on the

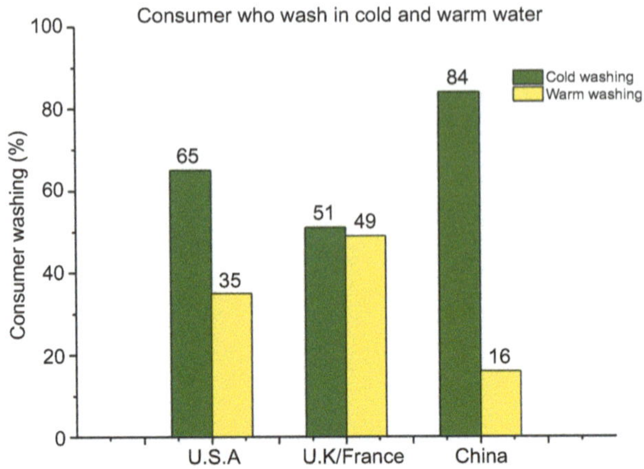


**Figure 10.18** Consumer wash cycles with cold and warm water [28]. *U.K.*, United Kingdom.
Adapted from The life cycle of a jean, by Levi Strauss & Co., 2015; Copyright 2015 by Levi Strauss & Co.

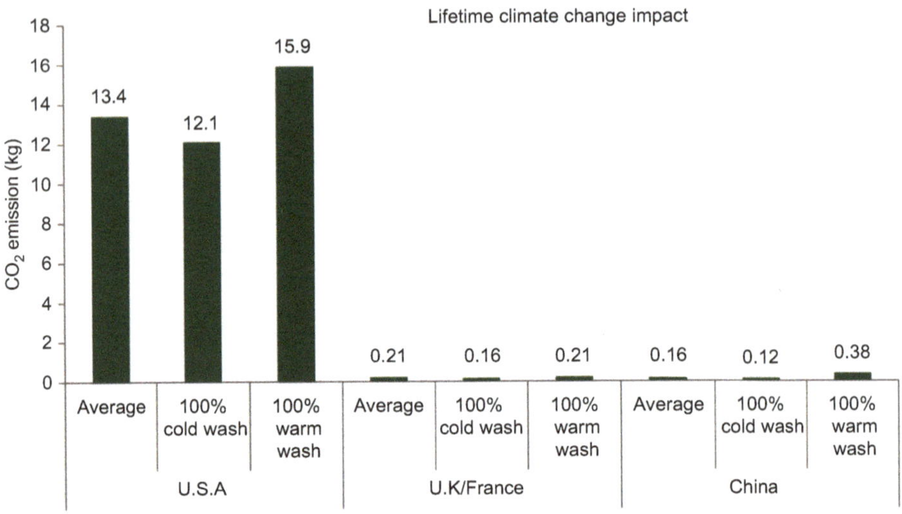


**Figure 10.19** Impact on climate change from various washing techniques [28]. *U.K.*, United Kingdom.
Adapted from The life cycle of a jean, by Levi Strauss & Co., 2015; Copyright 2015 by Levi Strauss & Co.

use of nonrenewable energy as well as climate change. Related to Americans, 65% of people washed with cold water but it is not enough; it needs to improve.

Regarding its lifetime climate change and nonrenewable energy use, the United States has the highest impact. Detailed results are shown in Figs 10.19 and 10.20. On average, the United States emits 13.4 kg of $CO_2$ over the life cycle of one pair of denim jeans, which is much higher than for other countries. The United Kingdom/France and

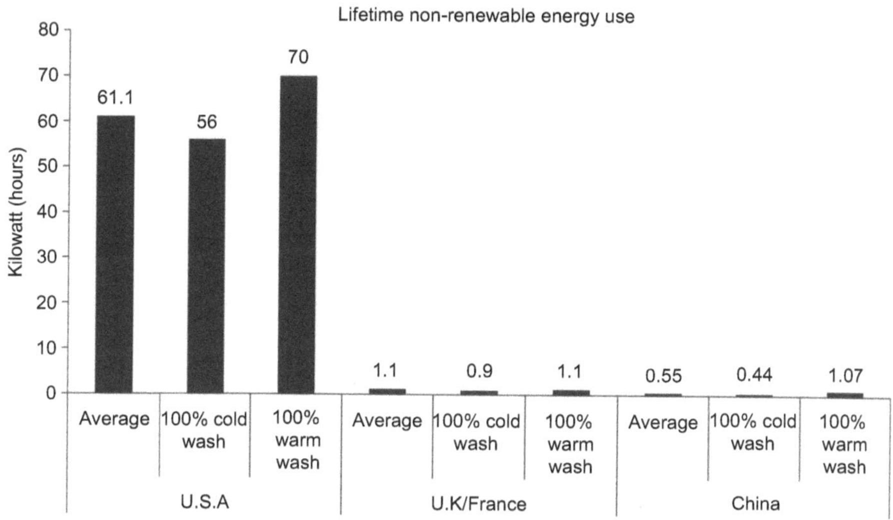


**Figure 10.20** Climate changing impact of nonrenewable energy use for different washing techniques [28]. *U.K.*, United Kingdom.
Adapted from The life cycle of a jean, by Levi Strauss & Co., 2015; Copyright 2015 by Levi Strauss & Co.

China show 0.21 and 0.16 kg of $CO_2$ emissions, respectively. The results demonstrate that the United Kingdom/France and China have a lesser impact on climate change than the United States. Similarly, nonrenewable energy use per denim jeans for the United States is higher than for other countries; doubtless, this has a huge impact on climate change.

## 10.8 Effect of drying techniques

Drying is an important process usually carried out after laundering. Generally there are two methods of drying: line drying and using a dryer. The simplest way to dry wet denim is to vaporize the water molecules. However, denim is made with cotton and is hygroscopic in nature [29,30]. Also it is made with coarse counts, which results in a heavy fabric [31]. Thus denim jeans require more time to drying owing to these properties.

Normally line drying has advantages, such that it is free, no energy is required, it generally leaves your clothes smelling fresh, and it means you do not make your home damp and cold by drying things inside. The most important point is that it is environmentally friendly [32]. The only disadvantage of this process is that it takes more time to dry the jeans (which may be a few hours to a day or sometimes more, particularly in winter). The best condition for drying is warm (summer and spring seasons), and windy days, which have low humidity. Drying jeans outside is usual practice and people in the United Kingdom/France and China do this regularly. Data show that 88% of Chinese consumers drying their jeans on the line, which has a huge impact on reducing global warming by reducing carbon emissions (Fig. 10.21).

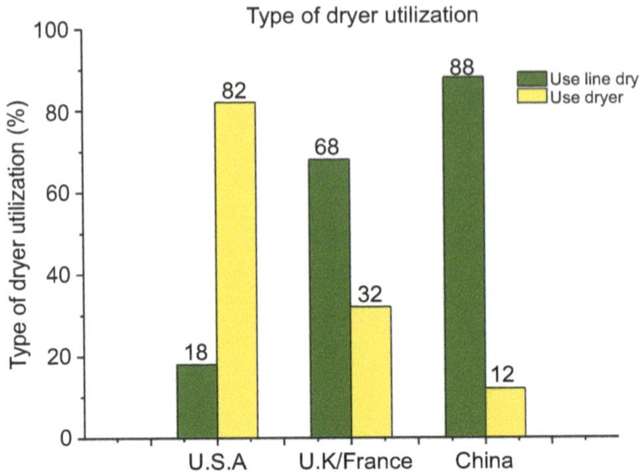


**Figure 10.21** Percentage of consumer drying methods [28]. *U.K.*, United Kingdom.
Adapted from The life cycle of a jean, by Levi Strauss & Co., 2015; Copyright 2015 by Levi Strauss & Co.

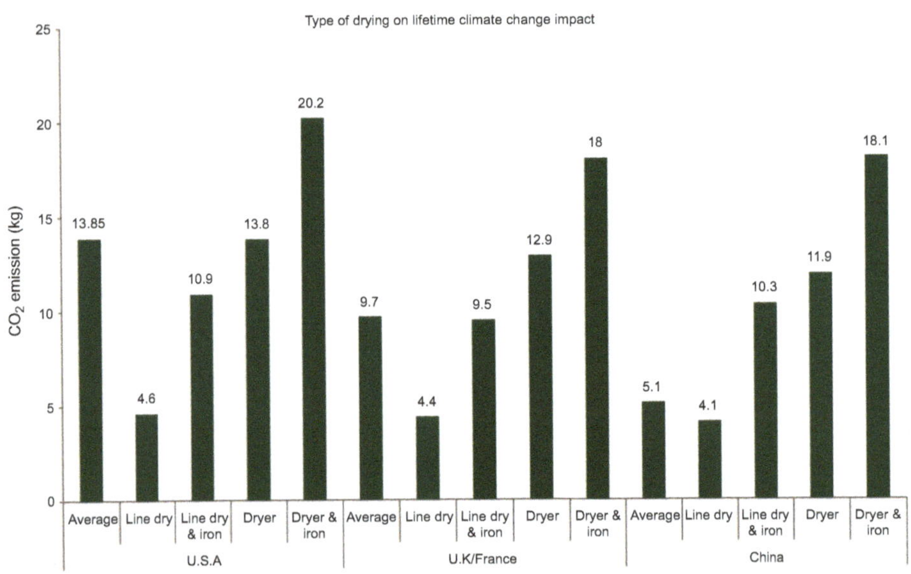


**Figure 10.22** Lifetime climate change impact for the drying process compared with other methods [28]. *U.K.*, United Kingdom.
Adapted from The life cycle of a jean, by Levi Strauss & Co., 2015; Copyright 2015 by Levi Strauss & Co.

Dryers need steam or gas to heating the drum which dries the jeans, but this process requires a huge amount of energy commonly known as the latent heat of vaporization. For example, 2 kg wet jeans (after washing) requires 1.4 kW hours of energy to dry [33–35]. A total of 82% of Americans use the dryer, resulting in approximately 13.85 kg carbon emissions (Fig. 10.22); in the case of China, it

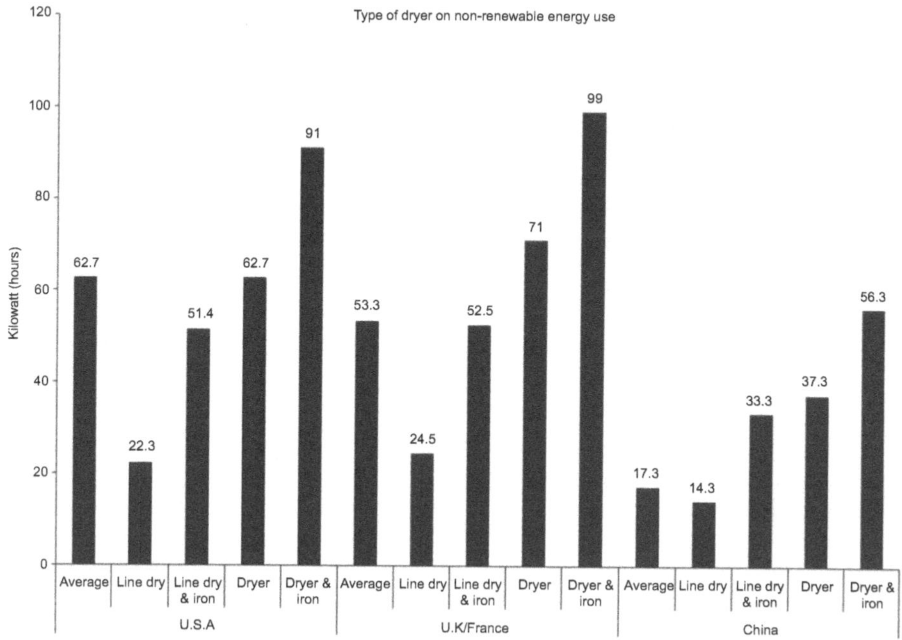


**Figure 10.23** Consumption of nonrenewable energy used for drying and another process [28]. *U.K.*, United Kingdom.
Adapted from The life cycle of a jean, by Levi Strauss & Co., 2015; Copyright 2015 by Levi Strauss & Co.

is 5.1 kg $CO_2$. Therefore line drying can reduce the use of nonrenewable energy by 64% (United States), 65% (United Kingdom/France) and 62% (China); also it reduces the climate impact by 67% (United States) 67% (UK/France) and 65% (China). Line drying alone can help the environment compared with another process. Generally ironing is a process that requires energy and is credited as causing carbon emissions. On average, the consumption of nonrenewable energy for drying followed by ironing is high and it always shows higher carbon emission than for other methods (Fig. 10.23).

## 10.9 Recycling and landfill

Globally, it is estimated that a huge quantity of textiles is thrown away every year. Recycling is a better way that provides important environmental benefits:

- It reduces the need for landfill space. Denim with synthetic fibres can create problems in landfills because it does not decompose.
- It reduces pressure on virgin resources.
- It results in less pollution and energy savings because fibres do not have to be manufactured or transported from abroad.

Denim is made from cotton it is a highly biodegradable polymer that decomposes quickly, from 1 week to 5 months. Blue Jeans Go Green conducted a program and survey, and reported that to wear for many years, on the contrary, it reduces the purchasing of new jeans [36].

Most of American consumers are thinking in terms of the environment, and of recycling. Usually, old denim can be donated to charity and reused it in different ways (turned into bags, pillow covers, as mat in cars or carpet-like materials, etc.). These practices make it possible to reduce landfills and saving the environment.

## 10.10 Interpretation results of consumer phase

*Climate change*: The consumer phase of the life cycle has the highest impact on climate change owing to the number of processes which use resources. On average American consumers are more responsible for the climate change owing to their drying habits and other related practices.

*Water consumption*: By wearing jeans 10 times before washing, American consumers can reduce water and their impact on climate change by 77%, United Kingdom and French consumers by 75% and Chinese consumers by 61%. Therefore, the denim consumers are advised to practice this method, which will have a huge global climate impact.

*Eutrophication*: Eutrophication is largely affected by the production stage, particularly the fibre cultivation stage. There is no significant influence on the consumer phase of the life cycle with respect to the eutrophication process.

*Land occupation*: The highest land occupation for the production of denim jeans is fibre cultivation, which requires a huge area. Apart from cultivation, land occupation may occur in the forest because trees are cut to produce energy. This practice is still carried out in many developing countries.

*Abiotic depletion*: There is no significant impact on the consumer phase of the life cycle with respect to abiotic depletion.

### *10.10.1 Regional highlights*

Average washing, pressing and drying frequencies with respect to different countries are as follows:

- In the United States, consumers typically wear their jeans twice before washing them.
- In the United Kingdom and France, consumers typically wear their jeans 2.5 times before washing them.
- In China, consumers typically wear their jeans four times before washing them.
- Consumer jeans-washing habits vary by region:
- Americans use more water and energy to wash and dry their jeans than do consumers in the other markets.
- Consumers in the United Kingdom and France mostly air-dry their jeans and use more hot water than do others.
- Consumers in China mostly wash their jeans in cold water and air-dry them.

Based on these facts, Levi's and Dockers include importance sentences in the global product tags, encouraging us to 'Wash less, wash cold, line dry, and donate when no longer needed'. Other leading brands were requested to take action on this subject.

According to findings for Levi's 501 Jeans, 37% of energy and 23% of water used during the lifetime of a pair of Levi's 501 jeans occurs during the consumer phase of its life. The study shows that of nearly 3800 L of water used throughout the lifetime of a pair of jeans, cultivation requires 68% and consumer care, 23%. Consumer care, or how we wash and dry our clothes, also adds considerably to the cradle to grave water consumption of a pair of jeans. Consumer care is the second most significant impact on water consumption, which indicates that consumers are responsible for 23% of the water used in the life cycle of a pair of jeans. Consumer care is also responsible for the highest energy use, which is directly proportional to the impact on climate change (37% in 33.4 kg of $CO_2$ emissions); it emits 33.4 kg of $CO_2$ during the life cycle of a pair of jeans. Levi Strauss & Co is working with the Better Cotton Initiative (BCI) to consume less water by 2020. They projected to produce 75% of their denim through BCI techniques (it was 6% in 2013), so there is a possibility of reducing water consumption in the cultivation phase. Similarly it is necessary to reduce the environmental impact during consumer care. The results concluded that Americans use more water and energy to wash their jeans than do consumers in China, France, and the United Kingdom. Consumers in China wear their jeans on average four times before tossing them into the wash, and if American consumers did this, they could reduce water and their impact of climate change 50%. 'It's time to rethink autopilot behaviors like washing your jeans after every wear because in many cases it's simply not necessary', said Chip Bergh, chief executive officer and president of Levi Strauss & Co. 'Our LCA findings have pushed us as a company to rethink how we make our jeans, and we're proud that our water stewardship actions to date have saved 1 billion liters of water'.

## 10.11 Challenges

After this study was published, denim consumers wanted to know the hidden face of the product: how, where and by whom the product was manufactured. Of course, they saw the price, quality and other common criteria. Consumers are driven to an eco-design through their expectations for greener denim. Retailers are in a key position to promote sustainable denim. They can increase customer awareness and their choice of purchasing. However, urban people have more awareness about the environmental impacts of their clothing (including denim) than do consumers in rural places. When we discuss the environmental impacts with denim manufacturers, they suggest that the biggest impact of denim on the environment is consumer care. Therefore, engaging and educating consumers may help them fundamentally, which will change the environmental impact of denim, and, ideally how consumers think about the clothes they wear every day. The main challenges to consumers that arose in this study are to:

- increase the washing frequency between wear (for example, washing 10 times between wear in the United States. Consumers can reduce water consumption by 50%)
- engage consumers in line drying (sun drying)
- reduce the use of pressing. If jeans are dried on the line, they does not need to be pressed, which helps to remove the crease marks

Consumers wash all of their clothes once a week, actually washing, drying and ironing every single item per week. Using this formula revealed a staggering impact on the environmental impact.

## 10.12 Future trends

In the future, the same method can be used to find the consumer phase of the LCA for other denim garments made from other sustainable fibres. It may compare the environmental impacts of similar products with respect to different fibres. LCA research with statistical data is needed yearly to account for changes in fibre production, a company's sustainable initiative, the availability of suppliers, and factors that contribute to harm human health, the ecosystem, diversity, and the availability of resources.

## Acknowledgements

We would like to express our thanks to Levi Strauss & Co. for the useful data, which helped us to write this chapter. The primary author is grateful to the Republic of India for providing the best life and knowledge, and also thanks to the Technical University of Liberec and the Ministry of Education, Youth and Sports of the Czech Republic for the doctoral student scholarship.

## References

[1] Bhala R, Dhandhania V, Periyasamy AP. Bio-finishing of fabrics. Asian Dyer 2012;9.
[2] Periyasamy Aravin Prince. Bio-processing in textiles. Text Mag 2007;48.
[3] Periyasamy AP, Dhurai B, Thangamani K. Salt-free dyeing – a new method of dyeing on lyocell/cotton blended fabrics with reactive dyes. Autex Res J 2011;11.
[4] Periyasamy AP, Dhurai B, Thangamani K. A study on fibrillation properties of Lyocell fiber. Colourage 2011;58.
[5] Periyasamy AP, Dhurai B. Salt free dyeing: a new method of dyeing of Lyocell fabrics with reactive dyes. Pak Text J 2011;60.
[6] Driving demend for denim; n.d. http://www.cottoninc.com/corporate/Market-Data/SupplyChainInsights/Driving-Demand-For-Denim-Jeans/.
[7] Factbook CIA. https://www.cia.gov/library/publications/resources/the-world-factbook/; 2016.

[8] Newbery M. Global market review of denim and jeanswear-forecasts to 2021. U.K.: Worcs; 2016. http://www.just-style.com/market-research/.

[9] Son J. Indian consumer purchase behavior of foreign brand jeans. Oklahoma State University; 2007. http://citeseerx.ist.psu.edu/viewdoc/download?doi=10.1.1.424.9202&rep=rep1&type=pdf.

[10] Wages and working hours in the textiles. Clothing, Leather and Footwear Industries; 2014. http://www.ilo.org/wcmsp5/groups/public/@ed_dialogue/@sector/documents/publication/wcms_300463.pdf.

[11] Incorporation C. Global lifestyle insights. 2015. http://lifestylemonitor.cottoninc.com/wp-content/uploads/2016/07/MC-GLM_UK_2016-4-4-16.pdf.

[12] Allwood JM, Laursen SE, de Rodríguez CM, Bocken N. Well dressed? The present and future sustainability of clothing and textiles in the United Kingdom. Cambridge CB2 1RX, UK: University of Cambridge Institute for Manufacturing; 2006.

[13] Agarwal S. What are consumers looking for when buying denim?. Denim and Jeans; 2012. http://www.denimsandjeans.com/latest-denim-reports/what-are-consumers-looking-for-when-buying-denim/5675.

[14] Cotton incorporated. Cotton incorporated lifestyle monitor survey. 2014. http://lifestylemonitor.cottoninc.com.

[15] Miller D, Woodward S. Blue jeans the art of the ordinary. California, CA, USA: University of California Press; 2012.

[16] Consumer concerns when purchasing new jeans. 2012. http://lifestylemonitor.cottoninc.com/consumers-concerns-when-purchasing-denim/.

[17] ISO 14040. Environmental management: life cycle assessment, principles and framework. 2006.

[18] ISO 14041. Environmental management: life cycle assessment, goal and scope definition and inventory analysis. 1998.

[19] ISO 14042. Environmental management: life cycle assessment, life cycle impact assessment. 2000.

[20] ISO 14043. Environmental management: life cycle assessment, life cycle interpretation. 2000.

[21] Hackett T. A comparative life cycle assessment of denim jeans and a cotton t-Shirt : the production of fast fashion essential items from cradle to gate. College of Agriculture at the University of Kentucky; 2015. http://uknowledge.uky.edu/cgi/viewcontent.cgi?article=1008&context=mat_etds.

[22] Rebitzer G, Ekvall T, Frischknecht R, Hunkeler D, Norris G, Rydberg T, Schmidt W-P, Suh S, Weidema BP, Pennington DW. Life cycle assessment: Part 1: framework, goal and scope definition, inventory analysis, and applications. Environ Int 2004;30.701–20. http://dx.doi.org/10.1016/j.envint.2003.11.005.

[23] Stylianou KS, Heller MC, Fulgoni VL, Ernstoff AS, Keoleian GA, Jolliet O. A life cycle assessment framework combining nutritional and environmental health impacts of diet: a case study on milk. Int J Life Cycle Assess 2016;21:734–46. http://dx.doi.org/10.1007/s11367-015-0961-0.

[24] Ali S. Life cycle assessment, submitt. As coursework Phys. 240. Stanford Univ. Fall 2010; 2010. http://large.stanford.edu/courses/2010/ph240/ali2/.

[25] Schrott W, Paul R. 20-Environmental impacts of denim manufacture BT – Denim. In: Woodhead Publ. Ser. Text. Woodhead Publishing; 2015. p. 563–80. http://dx.doi.org/10.1016/B978-0-85709-843-6.00020-2.

[26] Lamnatou C, Chemisana D. Evaluation of photovoltaic-green and other roofing systems by means of ReCiPe and multiple life cycle–based environmental indicators. Build Environ 2015;93(Part 2):376–84. http://dx.doi.org/10.1016/j.buildenv.2015.06.031.

[27] Levi Strauss & CO. The life cycle of a jean. 2015. p. 16–50. http://levistrauss.com/wp-content/uploads/2015/03/Full-LCA-Results-Deck-FINAL.pdf.
[28] Levis. The life cycle of jean: understanding the environmental impact of a pair of Levi's 501 jeans; n.d. http://levistrauss.com/wp-content/uploads/2015/03/Full-LCA-Results-Deck-FINAL.pdf.
[29] McLoughlin J, Hayes S, Paul R. 2-Cotton fibre for denim manufacture BT – Denim. In: Woodhead Publ. Ser. Text. Woodhead Publishing; 2015. p. 15–36. http://dx.doi.org/10.1016/B978-0-85709-843-6.00002-0.
[30] Hoshino E, Wada Y, Nishizawa K. Improvements in the hygroscopic properties of cotton cellulose by treatment with an endo-type cellulase from *Streptomyces* sp. KSM-26. J Biosci Bioeng 1999;88:519–25. http://dx.doi.org/10.1016/S1389-1723(00)87669-0.
[31] Paul R. 1-Denim and jeans: an overview BT – Denim. In: Woodhead Publ. Ser. Text. Woodhead Publishing; 2015. p. 1–11. http://dx.doi.org/10.1016/B978-0-85709-843-6.00001-9.
[32] Kim J, Yun C, Park Y, Park CH. Post-consumer energy consumption of textile products during "use" phase of the lifecycle. Fibers Polym 2015;16:926–33. http://dx.doi.org/10.1007/s12221-015-0926-8.
[33] California Energy Commission. Clothes dryers. Consumer Energy Center; 2016. http://www.consumerenergycenter.org/residential/appliances/dryers.html.
[34] Lee J, Hoelle N, Rogers D, Musnier S, Salustri FA. An empirical study of energy efficiency of clothes dryers. In: Int. Conf. Eng. Des. ICED 2009. Stanford, CA, USA: Stanford University; 2009. p. 24–7. http://citeseerx.ist.psu.edu/viewdoc/download?doi=10.1.1.538.7116&rep=rep1&type=pdf.
[35] What's the carbon footprint of a load of laundry? Guard. 2010. https://www.theguardian.com/environment/green-living-blog/2010/nov/25/carbon-footprint-load-laundry.
[36] Recycle denim; 2013. http://bluejeansgogreen.org/Recycle-Denim/(viděno 21. říjen 2016).

# Social aspects in denim production

# 11

*I. Braga*[1], *M.J. Abreu*[2]
[1]Federal University of Piauí, Teresina, Brazil; [2]University of Minho, Guimarães, Portugal

## 11.1 Introduction

The popular market in Brazil, especially the denim market, has been showing great economic movements in several Brazilian regions, mainly in the northeast [13].

The denim industry in Brazil has increased on the order of 3.9% per year [61]. It has expanded in production, reaching 365.1 million units of denim jeans [31], and has seen prominence growth in the popular trading market [14,56].

According to data presented in 2016 by the Brazilian Association of Textile and Garment Industry [1], concerning performance in 2015, Brazil is the second biggest producer and the third biggest consumer of denim in the world. In northeast Brazil, the informal market of popular fashion has been demonstrated to be important to the economy according to indexes of economic, production and consumption development in different sectors; in 2011, 120,000 jobs were directly generated as a result. The states in the Brazilian northeast that stand out owing to the development of the activities of the popular production of denim clothing are Paraíba, Pernambuco and Ceará [60].

To consider such information in a relevant light, this chapter presents a case study concerning aspects of the production of denim practiced in the northeast, in particular in Toritama, a city in the region of Agreste in the state of Pernambuco.

This research, it was applied a combination of the following methods [59]: bibliographic research and field observation, by visiting popular markets that trade denim in the popular fashion centres in the cities of Toritama and Caruaru, in the region of Agreste, Pernambuco. The main aim of the research was to collect information that describes the production of the popular segment of denim that is practiced in that region.

The chapter is organized according to topic. The first section is the introduction, which presents the following points: the initial information that justifies the commitment and importance in carrying out the case study on the popular fashion market of Agreste, Pernambuco; the method used to collect the data and the enumeration of how information is arranged throughout the text.

The second section, global perspectives on social aspects of informal production, deals with a bibliographical survey of reports on the development of the informal market from a global perspective, highlighting how informality has developed in the following formats: in less developed countries, developed and postcommunist countries. It presents different contexts of establishing these business relations.

**Sustainability in Denim. http://dx.doi.org/10.1016/B978-0-08-102043-2.00011-3**
Copyright © 2017 Elsevier Ltd. All rights reserved.

The topic on denim in Brazilian popular fashion consists of a bibliographical survey and scientific publications that report on the Brazilian socioeconomic scene that resulted in the establishment of the market of informal fashion within the national territory. It presents the rates of development of the market of denim production and the products made, and describes the main commercial spaces of popular fashion in the Brazilian territory.

In the fourth section, the denim market: informal trade in northeastern Brazil, describes the northeastern region of Brazil, and its environmental, social and economic conditions in the creation of a scenario conducive to the development of the informal market in services, as a survival strategy that favours the formation of the informal market of denim clothing production and marketing.

The fifth section, on aspects of denim production in Agreste, Pernambuco, presents a historical survey of the denim market in the Agreste region of Pernambuco, which emerged as artisanal activity to supplement income, and which has been considered one of the largest production and marketing markets for denim trousers in Brazil.

The sixth section is a case study on the modes of production practiced by the informal enterprises of jeans popular of Agreste, Pernambuco. It emphasizes the process of sewing, that owing to the flexible strategies of production, have a large part of the piecework carried out by female labour, in domestic confections and in rural areas of the municipalities of Toritama and Caruaru. In addition to allowing the growth of production, this constitutes an important source of income for the survival and permanence of families in the countryside.

The seventh section presents conclusions obtained from the work, and the eighth and last section provides references of scientific studies used in this investigation.

This theme is considered relevant for an understanding of the social aspects of methods of producing denim in the popular market, so that Brazilian governors can be sensitized to strengthen the labour laws and to direct investment policies for popular enterprises.

## 11.2 Global perspectives on social aspects of informal production

The concept of an informal economy first appeared in 1973 in an International Labour Organization report; it was defined as an economic activity developed in the informal sector, not regulated by law but governed by custom and/or personal ties [32].

Popular and informal enterprises, according to some studies [11,23,25,40], have been important sources of generating income in different developing countries.

Hart [25,26] was the first to investigate the subject. In his early investigations he confirms that informal activities are economic practices commonly found in underdeveloped countries. In reporting on the reality of economic activities in Ghana, he states that informal enterprises are more easily developed to carry out processes in line with the reality of local communities and regional skills, and with the ease of absorption of the economically active population, which defines the development of existing skills. Hart further states that research on the informal economy in the underdeveloped countries should be more frequent because of the importance of this segment to local economic growth [26].

Gërxhani [23] confirms that the first studies on the informal sector began through anthropological and sociological studies in Third World countries through the investigations of Hart [25,26], and that only more recently there has been investigations about the informal sector in developed countries and former socialist countries in transition. Gërxhani [23] points out that this segment is considered a central aspect of the economic and social dynamics of any country, but it has greater importance in the economy of less developed countries. Gërxhani [23] further emphasizes that to analyse the importance of the informal segment it is essential to take into account that the nature of the informal activity differs significantly among countries with different structures.

Hart [26] describes the insertion of the rural migratory population from southeastern Ghana to urban centres. This population starts to live in conglomerates in the periphery of big cities, in the formation of *favelas*. A large part of this population cannot be inserted into formal economic activities, and begins to carry out activities such as domestic production, street vending and handicraft production, called workers. Also, this movement is reflected by low wages and the nonabsorption of labour by the formal market.

Beneria [11] observed the movements of the informal economy in both poor and developed countries. Starting from the microlevel of the company, the emphasis is on how economic restructuring and globalization generated the growth of informal activities, resulting in the economic insecurity and poverty of a significant portion of the population, especially female labour.

Based on the analysis of data presented on the participation of women in informal activities, mainly in small confectionery companies in North and South Africa, Latin America and Asia, Beneria [11] noted that small-scale and microenterprises such as garment production of the Asian Tigers cost reduction, productivity increase and low wages, detected the existence of gender discrimination in the division of labour and the value of wages.

Beneria [11] also points out that from the movements of the fronts for the defence of women's rights, she observed a greater insertion of women in productive activities and an increase in the rates of the female population in education programs. However, Beneria's [11] research concludes that informalization of work is a problem that contributes to a scenario of poor working and living conditions because of poverty; the author emphasizes the need for programs to generate employment and income. She states that we should be concerned about generating decent jobs, mechanisms for redistributing income and social protection programs so that real poverty eradication can take place.

Regarding the development of the informal economy in countries in transition, Katalin's [32] text reports on the context of the development of an informal economy in Romania that emerged with the end of communism after 1989, and on the process of restructuring the country's economy. According to data presented by Katalin [32], most informal activities in Romania are concentrated in the retail trade, transport, construction, repair and maintenance, agriculture, hotels and restaurants, health and education. The reasons for informality are as follows: there are people who work informally for survival and there are those who deliberately avoid taxes and social security contributions.

About informal activities in the economics of developing countries, Brown et al. [15] reports that part of its insertion process occurred through the facilities of multinationals, implementing outsourcing strategies and subcontracting the workforce, impelling the formation of a scenario conducive to the development of these activities. Based on data collected from the investigation of footwear poles in Mexico, to evaluate empirical evidence on the effects of multinational production on wages and working conditions in developing countries.

Brown et al. [15] found several conditions that allow the informalization of labour: the lack of protection of the internal market with laws that establish labour rules, and also because most multinationals seek strategies that do not respect the labour laws of the country, and pay low wages. In the case of this investigation, informality deals mainly with labour relations. Over the past few years, because of the pressure of the media and consumer groups that oversee modes of production, especially clothing, multinationals have implemented changes. Therefore the study concludes that foreign firms raise wages by increasing labour productivity. Expanding the scale of production improves working conditions. In addition, there appears to be evidence that foreign-owned firms use aspects of labour organizations and democratic institutions that improve the efficiency of their manufacturing operations.

What these investigations observed is that in the different scenarios, the formation of the informal market happens because of a search for survival, an escape from bureaucracy or payment of taxes and or the desire for low-cost workmanship.

This is in line with the conclusions of the Gërxhani [23] survey, which reports that a comparison of studies on the informal sector in developed and less developed countries showed that they converge on some basic criteria such as the use of declared work, evasion tax, unregulated or unlimited companies, illegality or criminality. The essential divergence is related to the use of the criterion of 'survival'. As a result, studies in developed countries show that the informal sector offers possibilities for growth, whereas research in less developed countries proves that survival is the main feature of the informal sector. Finally, it confirms that survival always 'legitimates' the evasion of the law in extreme situations, where a conflict arises between morality and individual rationality: agents justify their actions for lack of choice.

On the other hand, the context of the informal market in Brazil, especially in the production of clothing, can be found in the search for tax evasion, but economic activities are developed in the search for income and for survival. The activity that has grown the most in Brazil is the activity of clothing industry in denim in the region of the Brazilian northeast, which will be presented subsequently.

## 11.3 Denim in Brazilian popular fashion

According to data presented by the Brazilian Association of Textile and Garment Industry [1] in 2016 concerning performance in 2015, Brazil is the second biggest producer and the third biggest consumer of denim in the world. According to a survey conducted by the Brazilian Institute of Public Opinion and Statistics [30], 46% of Brazilians use denim clothes on a daily basis; among the most commonly used clothes, denim jeans are worn by 85% of the population interviewed and 54% of this sample wear denim to go to work.

The growth of the Brazilian popular denim market in past years is a consequence of the countless transformation processes reflected in the expansion of the markets. Among the transformations that occurred in the country, Rocha and Silva [54] emphasize improvements in the quality of life, especially in better care of basic needs such as health, education and feeding; and improvement in the working conditions and better wages; and, as a consequence, access to information, which has generated changes in social and cultural behaviour, consumption behaviour, as well as changes in social values.

Castilhos [17] reports that such transformations have generated changes that can be better observed in the life of the working class; these changes have reached their behaviour deeply, as well as their conceptions of the world and of themselves. Families have access to the markets to purchase not only food but also home appliances, mobile phones, laptops, automobiles and clothes, among others.

Because the working class composes most of the population in every country in the world, Paz and Leão [51] stated that improving life conditions and amplifying access to information and consumption by this portion of the population generated growth in the market in different sectors.

The importance of understanding social changes and consequently the viewpoint of the markets directed the study of North American researchers such as Prahalad and Hart, who in 2002 conducted investigations into business actions directed to poor clients in countries such as India. The authors observed that there were new market opportunities and claimed that the population that constitutes the base of the social pyramid of developing countries, such as Brazil, Russia, India, China and South Africa showed great opportunities for profitable trade.

In Brazil, to better understand what the population that constitutes the base of the social pyramid represents, is important to know the criteria of classification of the social strata of the country. In 2014, the Brazilian Association of Enterprise and Research (ABEP), following the Brazil criteria presented the following data of social classification, as demonstrated in Table 11.1, which presents the household income of each Brazilian social stratum.

Based on this socioeconomic classification of social strata, Classes C1, C2, D and E are considered social layers that are part of the popular categories of the Brazilian society; according to ABEP [2], they correspond to 74.1% of the country's consumers, and are the poorest (41.3)%.

Because of the economic importance of this population group, Barreto and Bochi [9] reported that consumption by the low-income population has caught the attention

**Table 11.1 Average household income for each social stratum in Brazil**

| Social economic stratum | Household income median (R$) | Household income median (€) |
|---|---|---|
| Class A | 20,272.56 | 6,199.55 |
| Class B1 | 8,695.88 | 2,659.30 |
| Class B2 | 4,427.36 | 1,353.94 |
| Class C1 | 2,409.01 | 73,670 |
| Classes D and E | 63978 | 19,566 |

Author's own source, 2015.

of the media and small domestic companies, mainly in the fields of foods and beverages, personal care and cleaning, causing them to stop producing for multinationals and start investing in this new segment of local consumers.

Giovinazzo [24] points out that not paying attention to the low-income population means losing a large segment of consumers because, according to data published by the Boston Consulting Group, in 2002 in Brazil, the annual consumption of families belonging to the popular classes corresponded to 389 billion reais (Brazilian currency), equivalent to about 106 billion Euros.

This economic representation encouraged the research of Barreto and Bochi [9] to characterize the popular market, with the aim of collecting information about the economic profitability of Brazilian companies that produced for the popular segment, to become acquainted with the habits of consumption for low-income customers.

In that research, Barreto and Bochi [9] pointed out that the growth of domestic companies that invested in the popular segment created interest in the international market, especially multinational companies. Companies such as Coca-Cola and Unilever began to invest in the Brazilian popular market and conduct research to determine the idiosyncrasies and complexities of this segment and accordingly develop products within the consumer capabilities of the poor population in Brazil.

Montenegro [42] states that although the major companies have an interest in offering products to low-income consumers, their efforts are not enough. The products of these companies do not reach the popular public primarily because there are difficulties in distributing and selling within the housing areas and circulation of this poor and peripheral population. In his observance of the dynamics of supply for this population, the author observed the existence of a market movement for the proper supply of peripheral residents.

In the discussion of this market movement, Montenegro [42] states that as far as the supply of products and services is concerned, because the poor populations that live in poor neighbourhoods and peripheral regions of Brazil were not assisted by the services and trade of the formal higher economy, they started to develop informal economic activities through production and repair workshops in small food trading, clothing, shoes and accessories. The author also states that besides supplying the marginal spaces of the Brazilian society, these activities of informal economy also become important sources of income for poor Brazilian families.

This dynamic of the lower-economy popular markets is also mentioned in the investigations of Sarti [58] and Barros [10], who studied working-class housemaids. The authors found that to support their families, it is necessary for housemaids to seek a complementary income which usually does not come from so-called formal sources of employment.

According to the Brazilian Institute of Geography and Statistics (IBGE) [29], formal employment is defined as having an employee who has an employment contract and/or signed working papers in the form of a formal contract, which serves as proof of the possession of employment, ensuring to the employee all of the benefits of Brazilian labour laws.

Therefore, regarding the lower economy previously mentioned by Montenegro [42], Ramos and Jorge [52] explain that its execution is made through informal work activities: that is, an employment relationship without a formal contract and without security or social protection, termed by IBGE [29] as an independent worker and home worker.

Montenegro [43] reports that because of the growth of consumption by the popular classes and the transformation of urban areas of larger cities, popular enterprises left the suburbs and moved to the city centres. The author also points out that such shifts add to the territorial expansion of informal businesses, the occupation of the centres of large cities in Brazil by production workshops, popular fairs and street trading as the main selling points of products manufactured by means of peripheral informal production, thus forming what is known as the popular market.

Nakano et al. [46] state that this movement of occupation of the spaces of city centres, as in the case of the city of São Paulo, where this movement has been happening since the 1990s, is related to the construction of collective public transport networks, which has facilitated the displacement of the peripheral population to the city centre, and which has since been frequented mostly by people with lower purchasing power.

Nakano et al. [46] also point out that this occupation happened mainly as a result of the survival strategies of impoverished segments such as homeless people, unemployed persons and other sectors excluded from the formal productive circuits.

These survival strategies of the population favour the occupation of public spaces by activities in the fields of the lower economy, such as those of informal production and trade, and the production of clothing, which has moved from the restricted area of the suburbs to the city centre, setting up in abandoned buildings.

The establishment of denim manufacturing centres in informal companies also was related to the process of implementing the popular clothing production market. According to Abreu et al. [50] and Araújo and Amorim [6], the implementation of informal companies was driven by the context of the economic instability of the unemployed working class. In the quest for survival, the unemployed workers in the garment industries organized and started working informally to manufacture pieces in denim with characteristics similar to those of the big brands, but more accessible owing to the lower prices [6].

Araújo and Amorim [6] also state that the production of denim practiced by popular enterprises is characterized as an informal, home and family production mode because the unemployed workers of the garment industry form precarious structures in their sewing workshops in small houses, involving the whole family and neighbours, with labour relations characterized by not having contracts and without paying taxes.

According to Silva [63], production was initially directed toward selling in the workers' own neighbourhoods; then it started to receive orders from multibrand companies that supplied large department stores. However owing to the instability and uncertainty of the payment of those orders, workers started to sell only to traders of the Feira da Madrugada held in the centres of the cities.

Braga and Abreu [13] state that since then, businesses have been progressing with significant growth in production and the expansion of the market, giving rise to great popular fashion marketing centres in Brazil, with a circulation of thousands of buyers. As an example, there is the Feirinha da Madrugada do Brás in the centre of São Paulo; the Sociedade de Amigos das Adjacências da Rua da Alfândega [Friends and Adjacencies of Alfandega Street Society (SAARA)] in the centre of Rio de Janeiro, as illustrated in Fig. 11.1; the hippie market in the city of Goiânia; the popular shopping area of Porto in Porto Alegre; the Feira Sulanca in the Agreste, Pernambuco; and Feira da Madruga da José Avelino in Fortaleza, Ceará.

Besides being the main centres of clothing market, such cities are considered to be the largest denim producers in the country.

Research by SEBRAE [61] and the European Institute of International Management (IEMI) [31] points out that in these regions of denim industry, the market has shown growth on the order of 3.9% per year [61] and expansion in production and marketing, highlighting the popular markets [48].

In Brazil, according to IEMI [31], in 2012 349.8 million pieces were produced, an increase of 27% compared with 2008.

Publications by Braga et al. [14] and Braga and Abreu [13] focusing on the popular denim trade practiced in the centres of the cities of Sao Paulo, Rio de Janeiro, Toritama and Fortaleza, identified that the displacement of popular fashion enterprises happened in a similar way in all of these cities. They moved from peripheral regions towards the centres of the cities, and have been progressing over the years through significant growth

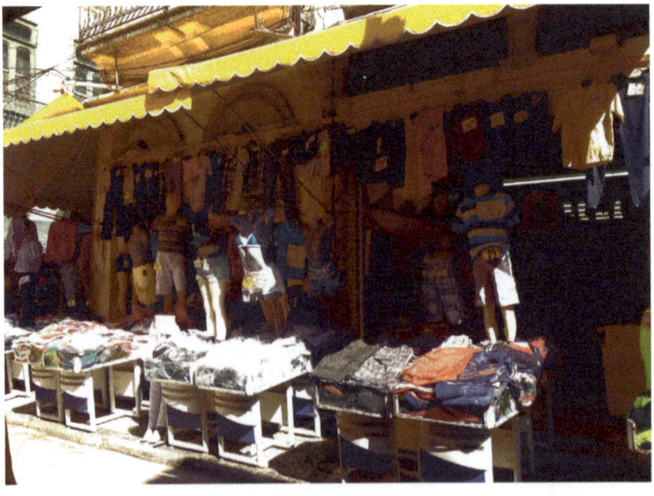

**Figure 11.1** Popular clothing trade in SAARA, Rio de Janeiro.
Author's own source, 2015.

in production and the expansion of trading fields, in the occupation of streets and old buildings with free fairs of clothing and with the building of shopping centres and popular malls.

An example is the Feirinha da Madrugada Brás, a free fair that takes place every day in the city centre of Sao Paulo, which, according to Sobrinho [66], receives about 15,000 shoppers per day. Braga and Abreu [13] found that among these buyers, 80% are retailers from cities in the interior of the state of São Paulo and other states such as Brasilia, Minas Gerais and Goiás; 20% are popular consumers of the city of São Paulo.

As far as Rio de Janeiro is concerned, Braga and Abreu [13] reported that among the marketing spaces in the city centre that remain strong and active, SAARA, an association formed by traders of one of the oldest shopping areas of the city, occupies the streets of Andrades, Buenos Aires, Alfândega and Praça da República, forming a trading conglomerate that has become considered the main centre of popular commercialization in the city.

In comparing the popular denim markets in São Paulo and Rio de Janeiro, Braga and Abreu [13] sensed that they differed as far as the buyers are concerned, because in Rio de Janeiro 75% are popular consumers who reside on the periphery and in the metropolitan area, whereas 25% of buyers are retailers in cities in the interior of Rio de Janeiro and other states such as Minas Gerais.

Concerning goods made of denim that are sold in São Paulo and in Rio de Janeiro, Braga and Abreu [13] found jeans of similar models in both cities, as shown in Figs 11.2 and 11.3. In São Paulo, there was a greater variety of models, as shown in Fig. 11.4, which had different washing techniques to the denim; in Rio de Janeiro, the models of jeans were repeated in their variations of pieces, in a skinny high waistband.

As for the origin of the pieces for sale in these markets, Braga and Abreu [13] report that they are all made in the same production centres, coming from Brás factories in São Paulo and enterprises in the Brazilian northeast.

Concerning the denim market in the northeastern region of Brazil, Braga et al. [14] state that popular denim projects have been highlighted concerning production indexes and commercial expansion. Therefore, they has attracted the interest of the government, private institutions (SEBRAE 2013) and the media, resulting in research, investigations and the spread of the popular universe created, produced and marketed by this region, and, as previously mentioned, bought not only by consumers of the region but also throughout the country.

**Figure 11.2** Jeans models in the market of São Paulo.
Author's own source, 2016.

**Figure 11.3** Jeans models in the streets of Rio de Janeiro.
Author's own source, 2016.

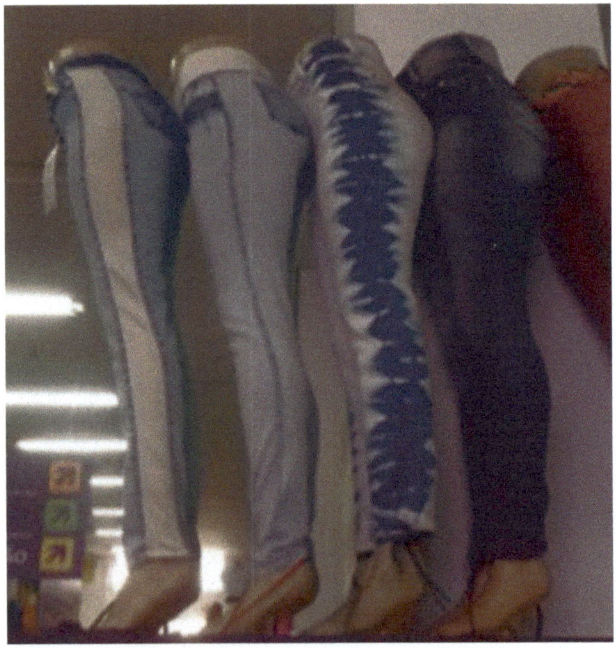

**Figure 11.4** Variations in models and different techniques of denim washes in the market of São Paulo.
Author's own source, 2016.

## 11.4 The denim market: informal trade in northeastern Brazil

The northeastern region of Brazil consists of nine states: Maranhão, Piauí, Ceará, Rio Grande do Norte, Paraiba, Pernambuco, Alagoas, Sergipe and Bahia. It has a land area of 600,102 square miles and is the third largest regional grouping of Brazil, occupying 18.2% of the country's area [29].

According to IBGE [29], the northeast is the second most populous region, with 27.7% of the population, totalling 53,081,950 inhabitants; 73% of the population is concentrated in urban areas located on the coast [21]. It is a region considered to be one of the poorest in Brazil, where 47.2% of the population belongs to the poorest sectors of the Brazilian society, social Classes D and E [2].

This information highlighted the northeast, for many years a topic of interest to academia, government, national and international institutions and nongovernmental organizations, and in areas of research of humanities and social sciences. Its social problems are the main focus of research.

Publications concerning northeastern problems were related to the following approaches: unemployment issues [36], the high levels of social inequality [27,55], drought and lack of water supply [20], the constant presence of hunger, lack of food and malnutrition [41] and the situation of poverty, among other issues.

Such statements are emphasized by Ribeiro [53], who confirmed the conditions of poverty in which the population live and the worst rates of economic and social development of the country, which made the northeast the most investigated region of Brazil and the one with the most published articles until current days.

Ribeiro [53] also points out that for many years, the conditions of poverty in the region were disseminated by politicians, the media and academia under the focus of theories of geographical determinism, blaming drought and the scarcity of water in the wilderness for being largely responsible for the misery that hits the region.

Research by Duarte [20] clarifies that the droughts, loss of agricultural production, weakening, or in many cases, destruction of livestock and emptiness of water tanks are not the main causes of the depletion of the northeastern population. Duarte [20] presents data that show that the real cause of poverty in the northeast is directly related to the unequal distribution of income, land ownership inequality, intermittent work and lack of fair compensation for rural workers.

Duarte [20] also points out that drought conditions in the semiarid region of northeastern back lands leads to calamitous proportions owing to high levels of vulnerability of the population of that region. Ribeiro [53] points out that this condition is also related to the lack of political initiative that may really solve 'the question of the northeast'.

Garcia [22] says that despite the problems of the population affected by drought being the subject of discussion since the arrival of the Portuguese royal family to Brazil, five centuries later, the political pronouncements remain the same and continue to be mitigated with the same remedial actions.

The only option the population has had to survive is to plunder, surround the building of the city hall (*prefeitura*, in Brazil), beg, wait for welfare actions such as food distribution (basic food baskets), migrate to urban centres in search of work, often in low-level jobs and/or join waiting lists in labour fronts. Labour fronts were created by the government to temporarily employ people affected by drought, to help in the construction of new water reservoirs and dams [22].

The actions of welfare programs were involved mostly in the development of projects supposed to generate income and reduce poverty in the northeast. Theodoro [67] made a historical survey of the actions done with the aim of reducing poverty in the northeast, and notes that from the early 1980s until 1991, numerous social programs attempted to generate employment and income for the poor northeastern populations.

Carvalho [49] confirms that over the years, various government measures have been taken in the northeast. Highlights of these initiatives are projects to strengthen craft activities, training courses, the creation of production groups and incentives for installing and modernizing industry in the metropolitan areas of Salvador, Recife and Fortaleza. These actions, including tax incentives by the Superintendence of the Development of Northeast, were not able to resolve the issues of unemployment and reduction in poverty rates, because they could not reach and absorb all of the manual labour of the working population from the northeast. Theodoro [67] points out that despite implementing and establishing industry in metropolitan areas and economic growth, there was also an increase in income, an expansion in underemployment and the strengthening of informality. It was observed that the formal market was not able to absorb all of the manual labour of the northeast, and that the informal sector occupied half of the economically active population, not counting the farm workers, in simple activities such as crafts and activities.

Based on data generated by the informal sector and its capability of including the poor people of the northeast in the labour market, government initiatives are performed to encourage programs and vocational training, job creation and income actions. Macambira and Carleial [36] argue that investment in the informal sector was an option defined by the government as a means of support and guidance for unemployed people in activities capable of generating income for subsistence and self-support.

Theodoro [67] reports that government actions directed to the informal sector received nearly \$80million from the federal government in investments from 1980 to 1991. Programs for the informal sector were divided into two fronts of action: projects for the creation or strengthening of groups that produce handicrafts, quilting, blacksmithing, fishing, food and cleaning products, among others; and professional training projects of central services, such as tinsmiths, cooks, nannies, masons and painters, among others.

Theodoro [67] concludes that during this period, in addition to investments from governmental programs, the informal sector of the northeast received incentives from national and international private institutions, nongovernmental organizations and the Catholic church, favouring the strengthening of informality in this region.

Incentives for activities such as the formation of groups of informal production and popular enterprises boosted the occupation of workers in activities on their own, in production activities, as in the process of selling the products, with initial emphasis on rural areas of small towns in the interior, and subsequently in urban centres [49,67].

Coutinho and Neves [19] report that the free markets are the structures most commonly used to carry out the flow of this informal production. The fairs are considered to be market, consumption and supply environments for the northeastern population, and are the main trading routes of production of family farming, trading goods such as handicrafts, household items and clothing at low prices and easy negotiation, as shown in Figs 11.5 and 11.6. On the other hand, they are places of social interaction and cultural strengthening and are important for local economic development.

Carvalho [49] reports that in 2004, after the implementation of public policies of distribution of monthly allowances for the population below the poverty line, the northeast started to present new levels of social and economic development, expressed by access to consumption by a large part of the population. It was watched with interest by private investment and resulted in the implementation of small and medium enterprises and, particularly, strengthening of the informal sector.

**Figure 11.5** Free fair in the northeast: family agriculture products, Fortaleza/Ceará. Author's own source, 2016.

**Figure 11.6** Free fair of Fortaleza: marketing of informal household production. Author's own source, 2016.

Because of these changes, Carvalho [49] points out that for the first time in the history of the northeast, research started reporting on positive indicators and the results of economic growth coupled with a reduction in social inequality in the indices of unemployment and poverty.

Because of this, the northeastern population has become interesting to researchers, because of the number of families who benefit from the programs of direct monthly income transfer and the reflection of this on economic growth in the region. As registered by IBGE, in 2013, this was the region with the highest growth; it reached 3%, higher than the country's growth rates, which were 2%.

Assunção [8] reports that owing to new consumer practices in the northeast, the northeastern popular class has started to generate interest in research institutes and different academic areas such as administration, management and marketing, evaluating changes and an increase in food consumption, changes in family income and the impact on education levels [18] and forms of leisure of this new consumer class [62].

Based on data collected from these investigations [8,18]; found that the main area of occupation of the northeastern poor population is the informal market, with 39.7% of the labour-active work. According to data from 2013 presented by the IBGE [29], 17.6% of workers in the northeast are involved in activities on their own in the informal trade sector.

These data reflect the significant growth of the free markets of popular clothing and the emergence of shopping centres and popular shopping malls as the main development of informal street trading activities. As pointed out by Mascarenhas and Dolzani [37], the popular market centres have grown, and this phenomenon has been the focus of many investigations aiming to understand the production process [5,34] whether in the marketing environment [35], the construction of relationships during the consumer act of the popular classes of the northeast [47,57]. Thus, the structure of the northeastern popular fashion market, as well as many popular informal enterprises, is part of a positive economic representation of new conditions of consumption for the popular classes.

The popular fashion market in northeastern Brazil has become prominent owing to the indices of economic development, production and consumption in different sectors, such as shoes and accessories and various types of clothing. In 2011, the denim market directly generated 120,000 jobs. The states in the region that stand out regarding the development of clothing production activities are Paraiba, Pernambuco and Ceará [61].

Ceará has an important garment centre which makes the state an example for a fashion market with statewide, regional, national and even international coverage. Over time, it has been highlighted because of the diversity, economic growth and income from the activities of business and informal popular enterprises [34,38,39,45].

According to the reports of Mota and Barbosa [44], Fortaleza, the capital of Ceará, has a popular fashion shopping area with low prices, which attracts many buyers from peripheral regions, other states and even other countries. In the traditional city centre is the Beco da Poeira mall, shown in Fig. 11.7, the José Avelino street fair, shown in Fig. 11.8 and Buraco da Gia, shown in Fig. 11.9, the main trading points of a popular fashion centre created, produced and traded by the popular classes, mostly under house production methods or micro and small enterprises.

Mota and Barbosa [44] also state that the popular trade began to develop in the 1980s and continues to date, facing government attempts to move it and to discipline

**Figure 11.7** Popular fashion trade in the Beco da Poeira mall.
Author's own source, 2016.

**Figure 11.8** Popular fashion fair in Rua José Avelino in Fortaleza, Ceará, Brazil.
Author's own source, 2016.

**Figure 11.9** Trading centre of popular fashion in Buraco da Gia in Fortaleza, Ceará, Brazil. Author's own source, 2016.

it, to submit it to the urbanization policies of the city centre. However, even with the conflicts, the marketers, vendors, hawkers and traders that compose this marginal market, informal or semiformal, have maintained themselves there and even expanded.

As in Ceará, the popular fashion trade of the state of Pernambuco has fairs as the main way of trading. Lyra [35] points out that development and strengthening of the units of production and the informal clothing trade in the region of Agreste, Pernambuco have transformed the economic environment and living conditions of the population.

Because the popular market of denim in Agreste, Pernambuco is the main focus of this research, its socioeconomic framework will be described in detail, especially the popular context of the city of Toritama.

## 11.5 Aspects of denim production in Agreste, Pernambuco

Agreste is one of five mesoregions of Pernambuco, extending over an area of approximately 9420 square miles, inserted between the Zona da Mata (region of the woods) and the Sertão (wilderness). It represents 24.7% of the territory of Pernambuco;

according to data from the demographic census of the 2010, the population is about 1,800,000 inhabitants [28], which represents 25% of the population of the state [7].

As far as this mesoregion is concerned, specifically in the municipalities of Caruaru, Toritama and Santa Cruz do Capibaribe, as well as the entire rural surroundings of these cities, Bezerra [12] describes this conglomerate as comprising an important production unit, commonly called the pole of textile and garment production of Agreste of Pernambuco.

The regional nature of production and marketing of clothing, according to reports by Burnett [16], began when the cotton crisis in the region intensified, reinforcing a migration process to the southeast/south of the country in the 1950s and 1960s. As an means of survival, the local population started quilting using scraps from the textile industry of the city of Recife, and later combined them with scraps from the textile industry in São Paulo.

These scraps served as raw material for rural women who initially produced pieces of clothing and blankets for their own use; later they sold the textiles in open fairs. With the sale of these pieces of cloth, the women obtained income in addition to what they received from agricultural activities. The production was a craft and domestic activity, with the labour force of the family.

Owing to the historical and circumstantial context, Arruda et al. [7] state that the method of popular clothing production developed in Santa Cruz de Capibaribe and this region became known as *Sulanca*, because they were produced with scraps, and therefore were of lower quality and low cost, a subsistence activity of local families produced by seamstresses, by hand and with domestic machines.

Burnett [16] describes that over the years, the production of *Sulanca* has turned into productive units; they expanded their businesses and moved to urban areas. With the inclusion of new machinery, production lines were created, because the scraps of denim coming from the southeast demanded more complex equipment for production and production became semiindustrial. This phenomenon became established with the formation of a network of working relationships among relatives and friends; it expanded and joined other conglomerates, forming a pole of textile and garment production of Agreste, Pernambuco.

Studies [60] on the structure of the pole Agreste report that it originated in Santa Cruz do Capibaribe. In 2002 it was firmly established in Caruaru and Toritama, and in 2010 production had significantly expanded to other municipalities, up to 18 cities.

In Santa Cruz and Toritama, the economy is largely dominated by the clothing sector, which includes trade and directly involving a network of related activities such as technical assistance, lodging, food and other services. Therefore, the significant growth of the gross domestic product (GDP) of these two cities can be attributed to the expansion of the production of clothing. In part, this also applies to Caruaru, where according to SEBRAE reports (2013a), it was estimated that in 2010, the clothing industry employed 9% of the working-age population.

Between 2000 and 2010, the growth in clothing production in the municipalities of Agreste, especially the three most important ones, led to expansion of the economy, with growth rates much higher than those recorded in Brazil, in the northeast, and even the state of Pernambuco. Between 1996 and 2009, Toritama was the city with second greatest growth in the state, recording a rate of 12% per year; Santa Cruz do Capibaribe had 9%, whereas the average for annual growth of Caruaru was high, reaching 6% per year.

In Caruaru, economic growth is due not only production, but to the marketing process. The city houses the largest open-air free fair in the country, the main selling point for food production and crafts, where jeans and knitting products are marketed from clothing enterprises across the region. The products are cheap and of low quality. Buyers circulating the fair look for low-priced garments. They may be resellers, 'baggers', who buy in huge amounts to resell in their cities, or direct consumers. All come from different regions of the country, such as Brasilia, Goiania and Belo Horizonte. Every Monday, the city of Caruaru receives 60,000 people, and in June, November and December the total reaches 100,000 people, with negotiations that yield R$30 million reais (€9.37 million) (SEBRAE, 2013).

Regarding this event, which is also known as the Sulanca Fair, Burnett [16] states that according to data presented by the Board of Trade and Markets of the City of Caruaru, in December 2015, in just 1 day, buyers from several states in the northeast and Brazil paid R$50 million reais (€13 million).

The formation of the popular denim production market in Toritama resulted from the economic crisis of 1980, which subsequently hit the production of manual sandal and shoe manufacturing in leather and affected large footwear industries in the region in the 1990s. Workers in the city of Toritama left the leather industries and began to adapt their skills and equipment to work with denim. They took denim leftovers from factories in Sao Paulo and produced garments to be sold in popular fairs at low prices [4].

According to studies carried out in 2012 by SEBRAE [61], Agreste is one of the main centres of production and commercialization of denim in Brazil, owing to its representation in the segment, with about 20,000 enterprises. It generates more than 100,000 jobs for the whole region and produces 900 million pieces per year. In the Sulanca Fair alone, there are 60,000 people, moving around R$1 million (€226,000) [14].

Toritama has been responsible over the past decade for 16% of denim garment manufacturing produced in the country. It is one of the largest producers and hosts in its territory the *Parque das feiras*, a centre for production flow in the city, which, besides having a covered area in the park, includes a free fair held on the outskirts of the commercial centre.

Data concerning the number of informal enterprises involved in popular fashion production in the state of Pernambuco show that in 2011, they comprised 90% of fashion companies in the region, corresponding to 17,000 units, generating 2.8% of the GDP of the state and producing more than one billion pieces per year [61].

The development of the popular fashion market in the region has attracted the interest of the media, the government, private institutions [60] and the academic community, which has conducted numerous investigations such as those of Alves [3], Oliveira and Braga [48] Silva [65], Silva and Gehlen [64] and Lima [33], who direct their research toward the popular universe created, produced and marketed by this region and consumed not only in Pernambuco but throughout the country.

Therefore, because of the economic and social representation of the popular market of denim in Toritama, it is an important place that useful for describing production practiced by popular enterprises to produce denim clothing. The following section explains how the denim manufacturing process has taken place in Toritama, from its creation to production, and from laundry to commercialization.

## 11.6 Domestic denim production: employment and sustainability of the families in the countryside

The town of Toritama has an area of 13.44 square miles and a population of 26.895 inhabitants. Located in the northern region of Agreste in the state of Pernambuco, it belongs to the microregion of Alto Capibaribe. The city is 103,77 miles from Recife, the capital of the state, and 22,37 miles from Caruaru.

The town shows growth without urban planning. When entering the main access roads it is possible to observe that it has problems concerning urban paving, as shown in Fig. 11.10, a polluted river that runs through the city (Capibaribe River), shown in Fig. 11.11, irregular roads and heavy traffic owing to the intensive movement of people and vehicles. The main streets and avenues are home to numerous large, medium and small shops, which deal in locally produced brands and are spread throughout the city, in addition to clothing stalls spread along the sidewalks. These present urban problems common to all small towns in the interior of Brazil, which are growing inordinately.

Following along the road, on the state highway, BR-104, at mile 18.64 is the *Parque das Feiras* (Park of the Fairs), built by entrepreneurs who live or work in town. It opened in 2001 and is a place where denim is the main product marketed. The Park of the Fairs is considered to be one a leading centre of fashion marketing of the pole of textile and garment production of the Agreste, Pernambuco.

**Figure 11.10** Paving of streets of Toritama.
Author's own source, 2016.

**Figure 11.11** Pollution of banks of Capibaribe River.
Author's own source, 2016.

On every street, one can clearly observe that the economy of the town develops around activities that are directly and indirectly related to the production of denim clothing. The balconies and yards of almost all of the houses in the town are transformed into small factories, stores of cloth and sewing pieces, like small shops and laundries. All of this creates an intense movement of negotiations, with unusual indexes for a provincial town. To better understand the reality of the denim market in Toritama, the Park of the Fairs was visited to observe these commercial movements and to talk to some vendors.

### 11.6.1 Script of observation of the field of research: park of the fairs

#### 11.6.1.1 Observation of the street and adjacent areas of the park of the fairs

The Park of the Fairs is located near the road that allows access to the cities of Caruaru and Santa Cruz de Capibaribe, as presented in Fig. 11.12: more specifically, at mile 18.64 of state highway BR-104.

On the outer area of the Park of the Fairs, to the right, is 'Denim Street' (*Rua do Jeans*), presented in Fig. 11.13, a place where the fair of denim of Toritama occurs beginning at dawn. In addition to being carried out within the space limited by road markings protected with metal, it extends to the outskirts of the mall of Park of the Fairs with the strong movement of people and business.

Across the road are large stores selling local brands from local denim factories, as shown in Fig. 11.14. These stores sell in large quantities to retailers, who come from other cities and states.

#### 11.6.1.2 Physical structure of the shopping centre

The shopping centre Park of the Fairs has 12,355 acres of covered area and 700 stalls/stores for clothing sales. The organization is shown in Fig. 11.15. The Park of the

**Figure 11.12** State highway BR-104, in front of the Park of the Fairs.
Author's own source, 2016.

**Figure 11.13** *Rua do Jeans* (Denim Street).
Author's own source, 2016.

**Figure 11.14** Large stores reselling local brand jeans adjacent to the Park of the Fairs.
Author's own source, 2016.

**Figure 11.15** Design of the shopping centre Park of the Fairs.
Author's own source, 2016.

Fairs is considered one of the main business centres of fashion production in Agreste, Pernambuco.

Because it is a place for large-quantity sales, the shopping centre offers customers shopping carts at the large entrance door, like the ones shown in Figs 11.16 and 11.17.

A large popular shopping centre with ample circulation space, as shown in Fig 11.18, where the corridors between the stores have not good lighting and ventilation conditions. Inside the shopping centre, besides stores, there is a food court, as shown in Fig. 11.19, bathrooms, service areas, like banks, and a parking lot with capacity for 2000 cars.

### *11.6.1.3 Structure of places of sale*

The selling places have different sizes. There are small shops fitted with metal structures which function as a reception counter; at the bottom are shelves for product

**Figure 11.16** Entrance door of the shopping centre Park of the Fairs.
Author's own source, 2016.

**Figure 11.17** Shopping carts of the shopping centre Park of the Fairs.
Author's own source, 2016.

display, as illustrated in Fig. 11.20. Medium and large stores have air conditioning, showcases and spaces to store goods (Fig. 11.21).

### *11.6.1.4 Buyers*

The main customers circulating in the halls of the shopping centre Park of the Fairs are buyers and resellers coming from neighbouring towns and dealers from cities in the northeast and other states of the southeast and midwest. Generally, buyers from other states arrive to Toritama in groups, in caravans of buses. Tour companies organize

**Figure 11.18** Corridors in the shopping centre Park of the Fairs.
Author's own source, 2016.

**Figure 11.19** Food court in the shopping centre Park of the Fairs.
Author's own source, 2016.

**Figure 11.20** Small store of the Park of the Fairs.
Author's own source, 2016.

**Figure 11.21** Large store of the Park of the Fairs.
Author's own source, 2016.

weekly trips for traders who are interested in shopping in Toritama and in the pole of production of the Agreste, Pernambuco.

#### 11.6.1.5 Observation of traded pieces

The pieces in denim traded in the Park of the Fairs present a high variety of models, as shown in Figs 11.22–11.25. The most common models are pants with a destroyed wash, ripped effects in the front, tight modelling, skinny legs, a high waist, with button and zipper closures, and a simple wash in the most varied shades from the darkest to lightest indigo blue.

#### 11.6.1.6 Source of fabrication of pieces

All pieces are manufactured in local factories in the city of Toritama, and in its surrounding areas and neighbouring towns.

#### 11.6.1.7 Methods of production

Products traded in the Park of the Fairs come from the denim factories of in Toritama, which are mostly informal and follow a fragmented mode of production in which each step of manufacturing is performed in a different factory.

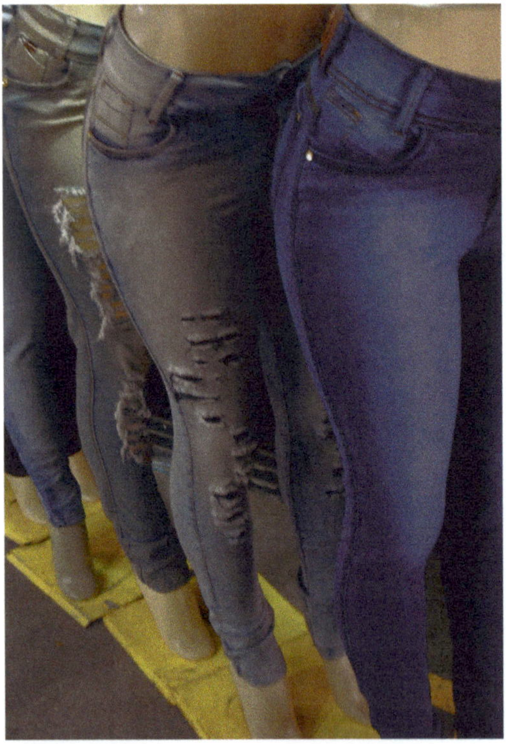

**Figure 11.22** Model of denim pants with destroyed wash.
Author's own source, 2016.

**Figure 11.23** Model of denim pants with different washing techniques.
Author's own source, 2016.

**Figure 11.24** Jeans with different models and washing effects.
Author's own source, 2016.

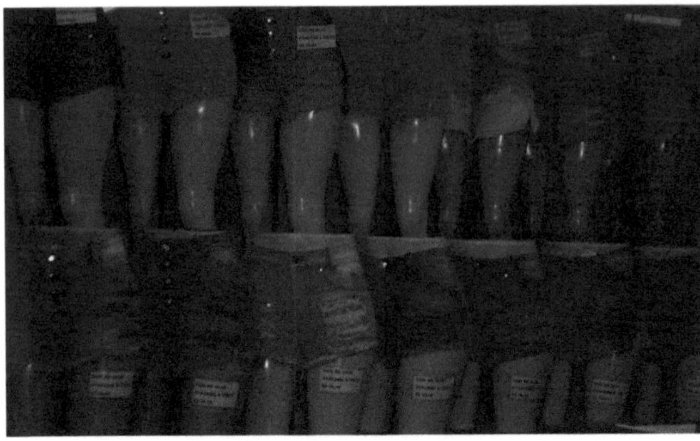

**Figure 11.25** Model of pants shorts traded in the Park of the Fairs.
Author's own source, 2016.

### 11.6.2 Interviews with sellers in shops of the park of the fairs

A conversation with the sellers in the stalls and shops provided an opportunity to talk with them informally about their activities.

In the initial conversation with the subjects, some of them reported that they produced the pieces that were selling. Generally, the pieces were produced in the neighbourhood or in their own homes with the support of their families. Others said they were already producing pieces in spaces exclusively reserved for the production of the pieces, a place detached from the home environment, using both relatives and nonrelatives as the labour force.

Another portion of the interviewees said that they were only resellers of denim products that were produced in some local factories and workshops. These first contacts provided some important information about how the division of labour is performed.

Traders/sellers reported that a large part of the production in the pole of textile and garment production of denim is performed by both men and women, and that among the activities developed, there is a division of labour. Therefore, in this context, the proportion of female labour in factories has been a key to the progress of that economic activity both in the city and throughout the rural area of the municipality, and it is an important source of income for the family.

The manufacture of denim pieces follows a fragmented model, considering the flexibility of home production. Women, who have as their main activity the sewing of pieces, divide their time between the family care and the production of pieces for the companies.

Besides this juxtaposition of production and domestic activities, working conditions offered by informal companies take place mostly in precarious environments because traditionally sewing activities are carried out at home, in facilities that do not possess adequate an physical environment for the development of the activity. Because they work at home, seamstresses spend more time involved in production activities than they should, and are exposed to exhausting work for indefinite periods of time because the more they produce, the higher the profits will be.

**Figure 11.26** Man working on finishing pieces of denim.
Author's own source, 2016.

Such working conditions offered by denim factories in the region are easily accepted and adhered to by people because they are a working alternative as well as a way of survival, because much of the manual labour involved in this type of productions is from former rural workers of the semiarid region of Pernambuco, which underwent numerous restrictions of survival in periods of drought.

Concerning production in the city centre of Toritama, the daily routine overlaps with the production process. Men and women, adults and children are included in the denim production process. As illustrated in Fig. 11.26, men sit in garages finishing denim pieces and teenagers walk on the streets carrying cut pieces of denim on their shoulders, taking the products from one place of manufacture/factory to another.

Across town, houses function as production workshops to process denim. Some homes have entirely become productive workshops; others have had only one room adapted to execute the work, and are commonly extended to the sidewalks, as the example of some women that, sited on the sidewalk in front of their house, were cutting pieces of denim that were already half cut but needed a more delicate handwork.

Many of these factories and workshops provide inadequate conditions, such as insufficient lighting and weather conditions for the activities that are performed, electrical installations in hazardous conditions, and a lack of personal protective equipment. On the streets near a factory that specialized in cutting denim pieces, there was an accumulation of waste coming from the door of a factory and occupying the sidewalk.

The small production factories acquire textile materials such as fabrics from distributors and shops in town. The haberdashery stores are family business and have considerable sales figures, especially for products such as buttons and rivets used to assemble denim pieces. Similarly, the fabrics stores, embroidery production and laundries are integrated into the local dynamics.

**Figure 11.27** Adornments applied in jeans at the Park of the Fairs.
Author's own source, 2016.

Most production workshops and factories operate informally, even though the presence and installation of technological equipment, such as the use of small and large industrial machines used to produce and embroider denim clothes, speeds the production process.

Embroidery factories apply traditional embroidery with thread, gems and sequins, attending to demands from the smallest workshops to the largest factories. Embroideries with lines are made before the pieces are produced; gems are applied after their production.

In these factories, there are several trend books, with models of embroidery for the customer's choice. Still, according to one interviewee, many customers prefer to bring a design sketch to be applied to denim. This drawing is set for reproduction in scale.

For the production of more elaborate embroidery, some factories have mobilized the population of neighbouring communities of the city of Toritama, especially women in rural areas and agricultural exproducers, to form embroidery production groups, owing to their natural handmade and craftsmanship skills. Their manual labour is used for finishing activities and the application of adornments to the pieces, such as stones and ornaments, mainly in denim pieces directed for the female segment, such as pants and shorts, as illustrated in Figs 11.27 and 11.28.

The laundry facility is a major challenge because Toritama is located in the region with the lowest rainfall index in the state of Pernambuco. Laundries, which are a necessary part of the process of improving denim pieces, must have a lot of creativity and adaptability in existing techniques, because during the dry season, laundries are supplied with 'kite cars', water transport trucks. Because of these limitations, operations are performed in such a way that they use a lower amount of water, and have to apply measures to reuse the water.

In many cases, laundries have rudimentary structures, not just the ones located in the city but also those in rural areas, which are run by workers with little training and

**Figure 11.28** Adornments and stones applied to denim pieces at the Park of the Fairs. Author's own source, 2016.

often without the necessary technical knowledge to develop and implement the procedures. When these structures were first installed, this caused numerous problems, especially environmental ones, such as pollution to rivers and streams. Although there was a restructuring process, the problem of the disposal of pollutants has not been fully resolved.

The application of seams for finishing purposes, such as rivets, in pieces of denim and in more basic pieces, as in the case of basic shorts with a low waist, is usually carried in housing spaces; it has become an activity that integrates a familiar scenario.

Because interviews have reported Toritama to be an informal 'factory town' demonstrating a rapid pace in launching new products, the process of creating the collections launched by the factories of Toritama are seen as models of pieces that started from an image constructed from aesthetic standards of the popular classes, the result of numerous connections and combinations of information arising from catalogues and books' trends, combined with models seen in the media, in soap operas and in social networks, and interwoven in popular demonstrations expressed in the city streets, all of which has created products.

Micro, small and medium entrepreneurs, to have access to the book's trends and lectures about fashion trends, as well to buy raw materials and equipment, have joined associations such as American Society of Composers, Authors and Publishers, the Business Association of Santa Cruz do Capibaribe and the Association of Shopkeepers of the Park of the Fairs in Toritama, among others, as a consortium and partnership so that the acquisition may have a lower cost value for each company.

In addition to the production workshops, home production and informal production, it is known through the respondents that in Toritama some offices have a strong relationship with the production of denim, which is primarily engaged in opening businesses and other establishments.

## 11.7 Conclusion

With the realization of this work it was possible to find that the concepts of a lower economy advocated by Montenegro in 2007 and 2011 can also be applied in Toritama and in all cities belonging to the pole of textile and garment production of Agreste, Pernambuco, because the practice of sewing clothing pieces emerged from the search for a productive activity capable of generating income for poor families in cities and in the rural areas of the municipalities.

Through visits to the city of Toritama and the shopping centre Park of the Fairs, and through what subjects said who were interviewed, it was possible to confirm the reports of Alves [3], when he states that the enterprises of the city of Toritama are informal popular businesses, because the data collected show that these enterprises have manufacturing practices based on production flexibility; the home environments are the main area of development of manufacturing activities. It was also identified that the methods of manufacturing applied are rudimentary, based on empirical knowledge, even though investing in technological equipment with industrial machinery, the physical structures are precarious and unhealthy.

Despite the social aspects of unfavourable work, the informal mode of denim production, the popular factories of Toritama and the pole of textile and garment production of Agreste, Pernambuco has shown positive economic returns, growth in production and expansion of trade. It is the Fair of the Dawn and the commercial centre Park of the Fairs, the main routes of negotiations for the popular denim fashion in this region.

In Toritama, the seams of denim pieces are made in large numbers by seamstresses who live in rural areas near the city and neighbouring cities, and who the struggle for survival and income to support their families. They submit to work in inadequate conditions and without establishing a work schedule. Because it is an activity in which remuneration depends on the index of production, these workers, who search for higher income, spend many hours involved in production. Although the activities leave little or no time for leisure, these working men and women do not grieve or complain about their conditions.

Another important factor to be mentioned is that in the Toritama market, even with the contingent workshops, factories and formal and informal companies, what prevails throughout the city is not competition between those involved, but a network of business relationships that rally together, share information and knowledge and are increasingly strengthened in the struggle for existence in the daily time and the survival of the population in the arid Agreste, Pernambuco.

By analyzing the general context of the market, it can be understood that the work prevails as the basis of society throughout the city of Toritama and countryside, setting the own rhythm for the city and its own dynamics, which are expressed in the ethics and aesthetics of existence, are noticeable in the city's daily life and are its greatest means of expression, the production of denim clothing.

Finally, to know denim production in Agreste, Pernambuco is an opportunity to understand the importance of the continuing and deepening investigations into and discussions about the social aspects of the methods of production of the popular denim market. In this way we can learn that in popular fashion enterprises, development strategies have subsistence as their principle, and therefore the intention of an organization is to generate income for individuals, families and businesses within a framework of solidarity. Finally, this chapter presents data to sensitize the Brazilian government to strengthen laws to regulate labour issues and direct investment policies for popular enterprises.

# References


[1] ABIT. Perfil do Setor Têxtil Brasileiro. 2016. p. 1. Available at: http://www.abit.org.br/cont/perfil-do-setor.

[2] Associação brasileira de empresas de pesquisa (ABEP). Critério de Classificação Econômica Brasil. 2014. p. 1–6. Available at: www.abep.org.

[3] Alves RP. Moda e desenvolvimento local: reconversões culturais na criação do jeans em Toritama – Pernambuco. Universidade Federal Rural de Pernambuco; 2009. Available at: http://200.17.137.108/tde_arquivos/13/TDE-2002-01-01T230644Z-814/Publico/Rosiane Pereira Alves.pdf.

[4] Alves RP. Moda e desenvolvimento local: reconversões culturais na criação e confecção de jeans em Toritama – Pernambuco. 2009b. Available at: https://scholar.google.com/scholar?q=toritama+Alves&btnG=&hl=pt-BR&as_sdt=0%2C5#0.

[5] Alves RP. Referências culturais na moda: diferenças e semelhanças na configuração do jeans das marcas Adágio e Zagnetron. Oikos Rev Bras Econ Domést 2012;23(2):191–204.

[6] Araújo AMC, Amorim ERA. Redes de subcontratação e trabalho a domicílio na indústria de confecção: um estudo na região de Campinas. Cadernos Pagu; 2002. p. 267–310.

[7] de Arruda AMCT, et al. Avaliação da qualidade microbiológica das águas subterrâneas em municípios da mesorregião do Agreste Pernambucano – Brasil. In: XVI Congresso brasileiro de águas sbterrâneas e XVII Encontro nacional de perfuradores de poços. São Luís, MA. 2010. p. 1–15.

[8] de Assunção V. Comer Mais e Melhor: Os Impactos do Programa Bolsa Família na Alimentação de Famílias de Pescadoras Artesanais de Pernambuco. Amazônica-Revista de; 2012. Available at: http://www.periodicos.ufpa.br/index.php/amazonica/article/viewArticle/961.

[9] Barreto F, Bochi R. Mercados pouco explorados: descobrindo a classe C. The Boston Consulting Group; 2002. Available at: https://scholar.google.com/scholar?hl=pt-BR&q=Mercados+pouco+explorados%253A+descobrindo+a+classe+C&btnG=&lr=#1.

[10] Barros C. A 'Invisibilidade' do Mercado de Baixa Renda nas Pesquisas de Marketing: as Camadas Populares Consomem ou Sobrevivem?. In: II EMA – Encontro de Marketing da ANPAD. 2006. p. 1–11.

[11] Beneria L. Shifting the risk: new employment patterns, informalization, and women's work. Int J Pol Cult Soc 2001;15(1).

[12] Bezerra E. O Trabalho a domicílio das mulheres do Cariri paraibano no polo de confecções do Agreste de Pernambuco. Universidade Federal de Campina Grande; 2011. Available at: http://www.ufcg.edu.br/~ppgcs/wp-content/uploads/2012/09/Disserta%C3%A7%C3%A3o-ELAINE-BEZERRA.pdf.

[13] Braga I, Abreu MJ. The jeans in the popular Brazilian panorama. In: TIWC 2016. Poznan. 2016. p. 534–40.

[14] Braga I, Abreu MJ, Oliveira M. O mercado de moda popular brasileira: os centros de comercialização de vestuário popular na região Nordeste do Brasil. In: 3º Congresso Internacional de Negócios de Moda. Porto, Portugal. 2015.

[15] Brown DK, Deardorff A, Stern R. The effects of multinational production on wages and working conditions in developing countries. University of Chicago Press; February 2004. Available at: http://www.nber.org/books/bald04-1.
[16] Burnett A. Os 'estabelecidos e os outsiders' da Sulanca no agreste Pernambucano. Ciências Sociais 2016;46(1):199–220.
[17] Castilhos R. Subindo o morro: consumo, posição social e distinção entre famílias de classes populares. 2007. Available at: http://www.lume.ufrgs.br/handle/10183/8933.
[18] Cavalcanti DM, Costa EM, da Silva JLM. Programa bolsa família e o nordeste: impactos na renda e na educação, nos anos de 2004 e 2006. Rev Econ Contemp 2013;17(1):99–128. Available at: http://www.scielo.br/scielo.php?script=sci_arttext&pid=S1415-98482013000100004&lng=en&nrm=iso&tlng=pt.
[19] Coutinho E, Neves H. Feiras livres do brejo paraibano: crise e perspectivas. In: XLIV Congresso da Sober:'Questões Agrárias, Educação no Campo e Desenvolvimento'. 2006. Available at: http://www.sober.org.br/palestra/5/663.pdf.
[20] Duarte R. Seca, pobreza e políticas públicas no nordeste do Brasil. 2001. Available at: https://scholar.google.com/scholar?q=desenvolvimento+do+Nordeste%2C+popula%C3%A7%C3%A3o+pobre&btnG=&hl=pt-BR&as_sdt=0%2C5#2.
[21] Francisco de WC. A Região Nordeste. Região Nordeste. A região Nordeste do Brasil – Brasil Escola. 2015. Available at: http://www.brasilescola.com/brasil/a-regiao-nordeste.htm.
[22] Garcia MF. Nordeste: o reverso da medalha. Lua Nova 1984;1(1):67–71. (online) Available at: http://www.scielo.br/scielo.php?script=sci_arttext&pid=S0102-64451984000100015&lng=en&nrm=iso.
[23] Gërxhani K. The informal sector in developed and less developed countries. Public Choice 2003;114:295–318.
[24] Giovinazzo RA. Um estudo sobre o desempenho e a estratégia das empresas que atuam no mercado de bens populares no Brasil. 2003. Available at: http://www.teses.usp.br/teses/disponiveis/12/12139/tde-13112003-114140/.
[25] Hart K. Small-scale entrepreneurs in Ghana and development planning. J Dev Plan July 1971:104–20.
[26] Hart K. Informal income opportunities and urban employment in Ghana. J Mod Afr Stud 1973;11(1):61–89. Available at: http://www.jstor.org/page/info/about/policies/terms.jsp.
[27] Hoffmann R. Mensuração da desigualdade e da pobreza no Brasil. Desigualdade e pobreza no Brasil. 2000. Rio de Janeiro Available at: http://empreende.org.br/pdf/Programas e Pol%C3%ADticas Sociais/Mensura%C3%A7%C3%A3o da desigualdade e da poobreza no Brasil.pdf.
[28] IBGE. IBGE | Cidades | Ceará | Fortaleza. Sistema Nacional de Informação de Gênero - Uma análise dos resultados do Censo Demográfico - 2010. Censo Demográfico 2010. Available at: http://www.cidades.ibge.gov.br/xtras/temas.php?lang=&codmun=230440&idtema=132&search=ceara%7Cfortaleza%7Csistema-nacional-de-informacao-de-genero-uma-analise-dos-resultados-do-censo-demografico-2010.
[29] IBGE-Diretoria de Pesquisas. Estudo e Pesquisa: informação demográfica e socioeconômica – uma análise das consições de vida da população brasileira. 34th ed. 2014.
[30] IBOPE. Pesquisa identifica hábitos de compra e consumo de Jeanswear no Brasil. 2013. Available at: http://www.ibope.com.br/pt-br/noticias/paginas/pesquisa-identifica-habitos-de-compra-e-consumo-de-jeanswear-no-brasil.aspx.
[31] IEMI. Jeans cresce mais que total de vestuário. Inteligência de mercado; 2013. p. 15–7. Available at: http://www.iemi.com.br/jeans-cresce-mais-que-total-de-vestuario/.
[32] Katalin G. The informal economy: conceptual background and theoretical framework. In: ARHIPELAG XXI PRESS. 2015. p. 57–71.

[33] Lima AS. "Empreendendo" a Sulanca: o SEBRAE e o polo de confecções do Agreste de Pernambuco. Universidade Federal de Campina Grande; 2011.
[34] Lopes HP, de Matos CHS. Aspectos sociais dos Objetos copiados do mercado informal de Fortaleza: as feiras livres das proximidades da Igreja da Sé. 2013. p. 250–8. fav.ufg.br.
[35] Lyra M. Sulanca X muamba: rede social que alimenta a migração de retorno. São Paulo em Perspec 2005. Available at: http://www.scielo.br/scielo.php?pid=S0102-88392005000400010&script=sci_arttext&tlng=pt.
[36] Macambira J, Carleial LMda F. Emprego, Trabalho e Políticas Públicas. 2009.
[37] Mascarenhas G, Dolzani MCS. Feira Livre: Territorialidade popular E Cultura Na Metrópole Contemporânea. Ateliê Geográfico 2008;2(2):72–87.
[38] Matos ALB, Mota de B, Dolores M. Caminhando pelas ruas do Beco, para democrático encontro com a Moda. 2006. coloquiomoda.com.br.
[39] Matos T, Maciel R, Maia L. Feiras de roupas em Fortaleza: relação entre redes sociais e trabalho informal. 2011. Available at: xiconlab.eventos.dype.com.brhttp://www.xiconlab.eventos.dype.com.br/resources/anais/3/1308288096_ARQUIVO_MatosMaciel_TrabCompleto.pdf.
[40] Meghir C, Narita R, Robin J-M. Wages and informality in developing countries. Am Econ Rev 2015;105(4):1509–46. Available at: http://dx.doi.org/10.1257/aer.20121110.
[41] Monteiro C. A dimensão da pobreza, da desnutrição e da fome no Brasil. Estudos avançados; 2003. Available at: http://www.scielo.br/scielo.php?pid=S0103-40142003000200002&script=sci_arttext.
[42] Montenegro M. O circuito inferior da economia urbana na cidade de São Paulo no período da globalização. 2006. Available at: http://www.teses.usp.br/teses/disponiveis/8/8136/tde-25062007-140750/en.php.
[43] Montenegro MR. Globalização, trabalho e pobreza no Brasil metropolitano. O circuito inferior da economia urbana em São Paulo, Brasília, Fortaleza e Belém. Universidade de São Paulo; 2011.
[44] Mota de B, Dolores M, Barbosa RCA. Comércio popular de moda -o centro faz a moda, a moda faz o centro na cidade de Fortaleza-ce. In: 5º ENP Moda: Encontro nacional de Pesquisa em Moda. Novo Hamburgo/RS. 2015. p. 15.
[45] Muniz A. Tributação e comércio internacional informal – estudo das relações Cabo Verde/Ceará. 2008. Available at: http://uol01.unifor.br/oul/conteudosite/F1066347188/Dissertacao.pdf.
[46] Nakano K, Campos C, Rolnik R. Dinâmicas dos subespaços da área central de São Paulo. São Paulo: Caminhos para o Centro; 2004. Available at: http://www.fflch.usp.br/centrodametropole/antigo/v1/divercidade/numero2/caminhos/11Nakano_Malta_Rolnik.pdf.
[47] Neco I. A transferência da feira livre de Sumé-PB e as redifinições territoriais da cidade. 2014. Available at: http://dspace.bc.uepb.edu.br:8080/xmlui/handle/123456789/5526.
[48] de Oliveira R, Braga B. Território comercial de Toritama: persistência e metamorfoses da informalidade. Revista Política Trabalho 2015;(41):193–225. Available at: http://www.ies.ufpb.br/ojs2/index.php/politicaetrabalho/article/view/21820.
[49] de Oliveira Carvalho CP. Nordeste: sinais de um novo padrão de crescimento (2000/2008). Rev Econ Política do Desenvolv 2008;1(2):7–40. Available at: http://www.ufal.br/seer/index.php/repd/article/view/94.
[50] de Paiva Abreu AR, Jorge A, Sorj B. Projetos de geração de renda para mulheres de baixa renda. Estud Fem 1996:4.
[51] da Paz AS, Leão ALMDS. Práticas Discursivas de uma Publicação de Moda Feminina de Classe Popular: Uma Análise Crítica do Discurso de Informação de Moda. Comunicação, Mídia Consumo 2012;9(25):221.

[52] Ramos C, Jorge R. Classe C, ascenção e alguns reflexos mercado consumidor. Revista de Gestão; 2013. Available at: http://www.revista.unisal.br/lo/index.php/reget/article/view/55.

[53] Ribeiro R. Seca e determinismo: a gênese do discurso do semi-árido nordestino. Anuário do Instituto de Geociências; 1999. Available at: http://ppegeo.igc.usp.br/scielo.php?pid=S0101-97591999000100004&script=sci_arttext&tlng=es.

[54] da Rocha A, da Silva JF. Inclusão social e marketing na base da pirâmide: uma agenda de pesquisa. RAE Eletrôn 2008;7(2).

[55] Rocha S. Pobreza e desigualdade no Brasil: o esgotamento dos efeitos distributivos do Plano Real. 2000. Available at: http://repositorio.ipea.gov.br/handle/11058/2365.

[56] Rodrigues A, Barbosa C, Braga I. Um olhar sobre o negócio de moda popular de Teresina. In: 9°Colóquio de Moda. Rio de Janeiro/RJ. 2012.

[57] dos Santos C. Produção e repetição: uma reflexão a partir do espaço de comércio e consumo do shopping center popular do Feiraguai em Feira de Santana-BA. 2013. p. 19. Available at: http://www.simpurb2013.com.br/wp-content/uploads/2013/11/GT04_1260_Claudio.pdf.

[58] Sarti C. A família como espelho: um estudo sobre a moral dos pobres. 2003. Available at: https://scholar.google.com/scholar?lr=lang_pt-BR&q=SARTI,+Cynthia.+A+fam%C3%ADlia+como+espelho:+um+estudo+sobre+a+moral+dos+pobres.+Campinas,+S%C3%A3o+Paulo:+Autores+Associados,+1996.&hl=pt-BR&as_sdt=0,5#0.

[59] Saunders M, et al. Research methods for business students. 5th ed. Edinburgh: Pearson Education Limited; 2011.

[60] Sebrae. Estudo econômico do arranjo produtivo local de confecções do Agreste. 2013.

[61] SEBRAE. Moda Jeans: mercado cresce reinventando tradição. SEBRAE; 2013. Available at: http://softcomunicacao.com.br/oficina/sebrae/blog-mercados/category/consumidor/page/11/.

[62] de Sena J, Alves M. Lazer em famílias de classes populares: breve perfil de dois bairrosde um município da região metropolitana do Recife. 2013. Available at: http://www.eventosufrpe.com.br/2013/cd/resumos/R0079-3.pdf.

[63] Silva C. Trabalho informal e redes de subcontratação: dinâmicas urbanas da indústria de confecções em São Paulo. 2008. Available at: http://www.teses.usp.br/teses/disponiveis/8/8132/tde-24112009-113627/en.php.

[64] de Silva J, Gehlen V. As construções sociais de gênero e raça no espaço de produção em Toritama e sua relação com o meio ambiente. 2010. Available at: http://www.contabeis.ufpe.br/propesq/images/conic/2010/conic/pibic/60/Resumo_CONIC_10060420PP.pdf.

[65] Silva WCB. Processo de desenvolvimento de calça jeans nas indústrias de Toritama. 2009. Available at: http://coloquiomoda.com.br/anais/anais/8-Coloquio-de-Moda_2012/GT10/COMUNICACAO-ORAL/103614_Processo_de_desenvolvimento_de_calca_jeans_nas_industrias_de_Toritama.pdf.

[66] Sobrinho WP. Feira da Madrugada completa 12 anos com polêmicas e lucro de 400% em SP. IG; 2013. p. 1–8. Available at: http://ultimosegundo.ig.com.br/brasil/sp/2013-01-24/feira-da-madrugada-completa-12-anos-com-polemicas-e-lucro-de-400-em-sp.html.

[67] Theodoro ML. Os programas de apoio ao setor informal no nordeste. Planej Políticas Públicas 2009;12:147–66.

# Environmentally friendly denim processing using water-free technologies

**12**

*K.K. Samanta*[1], *S. Basak*[2], *S.K. Chattopadhyay*[2]
[1]ICAR-National Institute of Research on Jute and Allied Fibre Technology, Kolkata, India;
[2]ICAR-Central Institute for Research on Cotton Technology, Mumbai, India

## 12.1 Introduction

The chemical processing of textiles is an important operation that can impart the highest value to textiles in terms of fashion and functional utility, and ensuring the requisite aesthetics and comfort. Among the various types of wet-chemical processing of textiles, chemical and chemomechanical processing of denim fabrics are important. However, because of heightened environmental awareness among citizens and the enforcement of stringent effluent laws, along with constant efforts to cut costs of production, a number of new technologies have been developed, validated and/or implemented in textile preparatory processing, dyeing/printing and finishing. Some of these are: (1) low-material to liquor processing, (2) spray and foam finishing, (3) the use of enzymes, (4) the application of natural dyes, (5) digital printing, (6) infrared dyeing and drying, (7) radio-frequency (RF) drying, (8) ultrasound methods for dyeing and dispersion, (9) supercritical carbon dioxide dyeing, (10) the application of plasma for water-free processing and (11) UV and laser-assisted processing and finishing [1]. However, textile chemical processing industries have yet to implement and adopt such developments in a massive way so as to remove their stigma of being a water-intensive industry, particularly in developing countries. Nevertheless, current research and development in the textile field is evolving toward the waterless or less–water based technologies.

Denim, which is made from a 100% cotton fabric, is an important and demanding casual dress material, especially for the tech-savvy young generation. Denim, also popularly known as blue jeans, can provide a much smarter casual look to the wearer compared with other dress materials. Both the production and consumption of denim are increasing daily. Worldwide, there are about 300 denim mills producing around 6.4 billion metres of fabric, which is likely to reach 8.2 billion metres by 2021 [2]. The current market value of jeans is over US$66 billion. Consequently, denim fabric and its processing is an important operation for added value to final products in the apparel business. Therefore, it is necessary to take a look at the denim processing sector, and whether it can satisfy the demands of modern eco-concerned young customers, both physiologically and psychologically. No doubt, faded textile garments are popular among the younger generation. Numerous technological advancements

**Sustainability in Denim. http://dx.doi.org/10.1016/B978-0-08-102043-2.00012-5**
Copyright © 2017 Elsevier Ltd. All rights reserved.

are improving the faded look of the jeans, such as sanding, sandblasting, brushing, prewashing, rinsing, stonewashing, sand washing, snow washing, stonewashing with enzymes and bleaching [3]. During denim processing, a part of the surface colour (mostly indigo dye) is removed with a pumice stone. Although it has some technical merits, the process has several adverse effects on the final product; for example, it creates an unnecessary harsh surface feeling, causes a loss in fabric mechanical strength and imparts a dull, used and degraded look. Some of these disadvantages could be overcome by biotechnological interventions, such as through the application of a cellulase enzyme. However, compared with stonewashing, such an enzymatic process is always more costly and needs precise control and intervention by skilled technical operators. Denim processing using either stone (pumice) or enzymes consumes a high amount of water. Although the desired fading effect could be achieved by such traditional processes, the following problems must be overcome to produce a classy denim quality: (1) difficulty in application with time-consuming work flow; (2) a decrease in the wear resistance of the jeans; (3) nonstandard and nonreproducible designs; (4) designs that are not applicable for all kinds of textile surfaces; (5) limited variations in shading; (6) the inability to produce identical fading effects on both sides of the fabric and (7) a loss of product quality.

Plasma, an ionized gas composed of ions, electrons, excited molecules and energetic photons, can be used to modify the surface of the textile fibre in a dry state. The plasma treatment of textiles is a new concept in textile processing and is being explored to impart different value-added functionalities, such as water/stain and oil repellents, hydrophilic, hygienic, antimicrobial and flame retardant properties, and also to improve dyeing [1,4–6]. Plasma treatment for denim processing, e.g., discolouration or fading of indigo dyes, is an emerging science that has been explored for the surface discolouration of denim by breaking the bonds of the dye molecules, dye–fibre interaction, and/or by etching the surface molecules of the textile. Reported results indicated that the decolourization was highly affected by the plasma parameters and the desired ‘worn-look’ effect could be imparted by adequate control of the plasma processing parameters, while paying due attention to limit plasma-induced damage. Low-pressure plasma and corona treatments can be considered cost-effective yet viable alternatives to the conventional bio-stoning process to obtain the ‘worn’ look of indigo-dyed denim fabric. It has been postulated that finishing with lasers and ozone are reduced water treatment processes and water jet fading systems, although they consume water, but by assembling a water recycling system this makes it more ecological and economic. $CO_2$ laser treatment has been used in different areas of the textile industry because it involves a shorter duration for designing surface patterns with good precision, and desirable effects, sizes and intensity, yet it does not cause much damage to the bulk properties of the textile materials. When laser fading was compared with manual dry processes, it was noted that (1) the effect produced in the manual system is slightly finer than the laser system, (2) the hand-feel is relatively finer in the manual system than the laser system, (3) the hairiness is comparatively less in the laser system, and (4) the production cost is higher in the laser system. Laser treatment has been found to be a useful tool for colour fading and the evenness of denim textiles. It is a much faster process, yet it provides an accurate reproduction of design by controlling the laser process parameters. The laser action on the coloured denim textile is mainly a surface interaction with indigo colour by a thermal effect. There are three steps in the process: (1) the generation of a computerized map of the fabric area that will be irradiated with different frequencies; (2) control of the laser, scanning and conveyor systems; and (3) the process of laser fading. Colour fading is affected not only by the laser power density but also by the laser resolution and pixel time.

Therefore, to maintain sustainable growth in the denim production and processing industries, it is important to develop sustainable process technologies which are much more economical and ecological in terms of water, energy, cost and effluent, yet meet the demands of modern consumers. The current chapter discusses the application of emerging water-free technologies such as plasma, laser, ozone and sandblasting and their effectiveness in decolouring fabrics. It also briefly summarizes the different methods of denim finishing and technological advancements in traditional processes to make them environmentally friendly, such as water jet-based finishing, biotechnological interventions, pumice stone and spot bleaching. The chapter also highlights some important finishing methods to improve the aesthetics, comfort and much desired worn look of denim fabric.

## 12.2 Sustainable textile processing

The chemical processing of textile goods includes a preparatory process, colouration and functionalization, which are highly essential to remove oil, wax, fatty matter, dust, dirt, colouring matter, spin finish, etc., from the fibres, thus making them amenable to dyeing and finishing, and leading to improved aesthetics and functional and fashionable value in the final product. Wet-chemical processing in today's textile industry, including small-scale processing plants, consumes a large amount of water, various chemicals and others. Several such processes generate a large quantum of liquid effluent, which is finally discharged into the water stream contaminated with residual dyes, pigments, acid, alkali, sizing material, bleaching agents, oxidative and reducing agents, and other such chemicals. Some of the chemicals used in textile preparatory processing, dyeing and finishing may cause environmental pollution [7]. The cost of the final textile products also usually remains high owing to multiple wet-processing operations and drying of the wet textiles.

Sustainability is the current hot topic of consideration in all spheres of science. Sustainability is often talked about in the subdomains of fibre, fabric, fashion, economics, food and agricultural production. The most common definition of 'sustainability' is the 'processes or products' that can meet the needs of society without compromising the resources needed by future generations [8]. Indeed, it is said that for something to be sustainable, it needs to survive for a long time without damaging the environment or society, or becoming too expensive to continue, so that one day it will have to die naturally. To be sustainable within the given definition, a fibre or textile or a fashion design needs to be produced without harming the environment and the people of the entire production value chain, and without costing too much, so that one day it will no longer be economically viable to produce. More often, it is assumed that a 'sustainable fibre' means an organic fibre or a natural one; but this is not correct. It may also be noted that some synthetic or man-made fibres may be similar or more sustainable than natural fibres, because they do not use as many resources as those that are used by 'natural fibres' [8]. Nevertheless, some natural fibres that are sustainable, green, ethical and eco-friendly are at the helm of the current fashion movement [9]. Naturally coloured/pigmented cotton, organic cotton, organic wool, wild silk, and even flax and hemp fibres are important fibrous materials as far as the sustainable textiles are concerned. For the production of sustainable textiles with appropriate chemical processing, the development of the product should also be sustainable in terms of energy, water, cost, raw material, chemicals used and biodegradability, and from the point of their impact on the environment. The other dimension of sustainability is the concern regarding employees' working conditions in the textile and garment factories, which are often associated with long working hours, exposure to hazardous chemicals used in bleaching and dyeing processes, and the scourge of child labour [8]. To process 1 kg of cotton textiles, starting from its preparatory processing to finishing, approximately, 100 L of clean water is required that is either partially recycled or discharged as an effluent contaminated with residual colouring materials, textile chemicals and leftovers. The shortage of water associated with issues relating to climate change and the carbon footprint will have a significant impact on research, development and the adoption of sustainable processing technologies in the textile and allied industries in coming decades. Owing to the adoption of stringent pollution laws and government legislation in past decades in this direction, a few technologies have been developed and validated in the textile production and processing value chain to ensure a reduction in the consumption of water, energy, processing costs, and/or effluent load, such as low material to liquor processing, foam and spray finishing, use of natural dyes/pigments, supercritical fluid–based dyeing, enzyme application, water-free digital printing, water-free plasma processing, water-free excimer and laser applications, infrared-assisted dyeing and drying, drying with RF, and dyeing and dispersion with ultrasound. All of these developed technologies are associated with either pre-or-postprocessing of textiles or their associated machineries. Despite such notable technological advancements, textile chemical processes remains water and energy intensive and uses some non–eco-friendly chemicals and auxiliaries routinely.

The chemicals and other products used in textile processing are caustic soda, acid, hypochlorite, sizing material, stain removers (carbon tetrachloride) and unfixed dyes. Some textile chemicals as well as auxiliaries possess suspended solids and produce a significant amount of effluent with high biological oxygen demand and chemical oxygen demand (COD) [7]. Similarly, human skin is sensitive to some synthetic dyes (for example, azo dyes) and they have a carcinogenic effect. In polyester dyeing, a toxic chlorinated compound is used as a carrier, and formaldehyde-based chemicals are used in crease-recovery finishing. In antimicrobial, rot resistance and mothproof finishing, chlorinated compounds such as hexamethylene biguanide, quaternary ammonium compounds and pentachlorobenzene are used. As far as UV protection of textile material is concerned, benzophenone, phenyl salicylate and benzotriazole-based chemicals are used. A few chemicals and auxiliaries used in textile processing have an adverse effect on the environment. However, serious efforts have also been made to develop various environmentally friendly synthetic dyes, chemicals and add-ons such as 1,2,3,4-butanetetracarboxylic acid and dimethylol dihydroxyethyleneurea for crease-recovery finishing; halogen-free flame retardants such as sodium-meta-silicate-nonahydrate and ammonium sulphamate for flame-retardant finishing; enzyme-based applications for desizing, biopolishing and denim processing; application of natural dyes for colouring and/or antimicrobial finishing; use of hydrogen peroxide ($H_2O_2$) for bleaching; the application of laser technology for denim fading; and polycarboxylic acids (maleic, citric and itaconic) for functional finishing [10–12]. In addition, for waterless operation plasma treatment, UV excimer and laser treatment have been explored as pre- and posttreatments, in situ reactions and postpolymerization for surface activation, oxidation, etching, polymerization, coating and deposition to impart value-added functionality such as water and oil absorbency, water and oil repellency, UV protection, flame retardance, antistatic quality and crease-resistance, and to improve the dyeing and printing of textiles [4,13,14].

Because of the global awareness of environmental pollution, climate change, carbon footprint, global warming, health and hygiene, the demand for organic materials such as organic fruit, vegetables, fertilizer, crops and pulses, organic cotton and naturally pigmented cotton are growing exponentially and have garnered a significant market. Natural fibres (nonorganic to organic) are biodegradable and are produced from renewable sources with requisite moisture to keep their value, soft feel, fair strength, and good appearance after chemical treatments. They are commonly used to making apparel and home textiles. Furthermore, for the sustainable development of high-end textile products, specialty apparel and home furnishing application, these should be made of natural fibres and be finished with eco-friendly chemicals and auxiliaries, preferably those derived from natural resources to ensure appropriate added value to the natural products, simultaneously preserving the natural resources. In this regard, a number of plant extracts, biomaterials, biopolymers and biomolecules such as enzymes, natural dyes, aromatic and medicinal plant extracts, chitosan, aloe vera, neem, lignin, silk-sericin, grape and mulberry fruit extracts, banana pseudostem and peel sap, spinach juice, citrus oil and such have been explored in the production of sustainable health and hygiene, well-being, skin care, and comfortable, and UV-protective textiles [15,16]. Similarly, deoxyribonucleic acid biomacromolecules, whey proteins, casein, chicken feather protein, banana pseudostem sap, spinach juice, and green coconut shell extracts have been used effectively to improve the thermal stability of cellulosic, lingocellulosic and protein textiles, and paper substrates in the context of sustainability [17–23].

Some estimates have been reported that to produce a pair of jeans requires more than 2500 gallons of water, nearly a pound of chemicals and vast amounts of energy [24]. If it is multiplied by 2 billion, the number of jeans produced worldwide every year, one can get a snapshot of the share of wastewater and greenhouse gases pumped into the environment. In this context, a process called Advanced Denim was developed that can produce a pair of jeans using up to 92% less water and up to 30% less energy compared with conventional denim manufacturing methods, according to Clariant, a specialty chemical company that developed Advanced Denim. In addition, it generates up to 87% less cotton waste (which is often burned, adding carbon dioxide and other greenhouse gases to the atmosphere) and virtually no waste water. It offers new possibilities for enlarging the number of tones and effects currently achievable, making the production more simple and efficient, and all this with minimal use of resources. The 'Holy Grail' of denim, the Levi's brand, researched a life-cycle assessment of a pair of their Levi's 501 jeans [25]. It was found that 58% of the climate change impact happens at the consumer-use phase. Together with Goodwill International, Levi Strauss & Co. launched the 'Care Tag for Our Planet' campaign. This particular collaborative project shares data from the life-cycle assessment, educates consumers about the benefits of cold washing and line drying and insists they donate used clothes to keep them out of landfills.

## 12.3 Denim fabric

The story of jeans begins in the mid-19th century, simultaneously in two European cities: Nîmes, in France, where the fabric itself was created with the name *serge de Nîmes*, which led to the name 'denim', and Genoa, in Italy, from where the colour of the fabric came, which was called in French *bleu de gênes*. Both sources jointly led to the expression 'blue jeans'. The fabric and the colour together came to be the denim and jeans. Denim trousers, commonly known as 'blue jeans', have maintained its popularity over the years [26]. Like denim, perhaps no other piece of clothing caused so much status, confusion and ambiguity since the onset of Western fashion about 700 years ago. Similarly, no other apparel or garment other than denim has received such wide acceptability by all categories of people irrespective of their age, class and gender [27,28]. With the passage of time, denim/jeans have become symbols for cowboys and icons for women, youth and people of varying economic status, and they have evolved from work wear to casual wear, and then to premium wear and functional wear. Denim trousers was introduced by a Jewish vendor named Morris Levi Strauss from Bavaria, who had settled in San Francisco (California) and started his firm, Levi Strauss & Co. There, in 1853, he started producing jeans from strong cotton fabric dyed with indigo, more or less what is done today as well [29]. Denim is basically a cotton twill-weave fabric that uses coloured warps and white/grey weft yarn [30]. It is also one of the world's oldest fabrics, yet it remains eternally young, influencing the lives of the consumers since its inception in many ways. It is traditionally coloured blue with indigo dyes to make blue jeans, although jeans denote a different lighter cotton textile [31]. As per the Global Market Report (2012), the global denim fabric production capacity is more than 7 billion yards and is estimated to increase to 9 billion yards by the year 2021 [32].

In textiles, fashion and apparel trends have been dominated by the introduction of faded and torn looks, in which one imperative examples is faded blue jeans [33]. For the purpose of supporting customers' purchasing behaviour and to fulfil their aesthetic taste textile companies have tried to develop and introduce various techniques to improve the visual aspect of denim fabrics. Such techniques mainly include printing on fabrics, and embroidering and washing the final product. Specially, fraying of certain areas of the fabric by sanding and stonewashing to create designs is a popular technique [29]. As far as garment finishing is concerned, finishing methods using new processing technologies and equipment to ensure the desired and novel effects are getting attention [34]. This is because a freshly assembled denim garment is uncomfortable to wear as a result of its weaving and dyeing effects, which are required to be finished or modified by a suitable washing technique and the introduction of a new look and fashion design [30]. For instance, a range of treatments are available in the finishing of denim, such as enzymes, bleaching, acid, silicone, ozone, plasma and laser, sandblasting, and many more [34]. In fact, blue jeans were a good idea for promoting the faded and torn look that was a way to express poverty [29]. Consequently, within a short time, jeans manufacturers started imparting the effects of fading and tearing. Such a type of blue faded jeans provided a poor-guy look and those with torn legs and knees became popular among young customers over the bright blue jeans that looked brand new, even though they could fetch a higher price in certain markets. Consumers evaluate a jeans garment based on its style, brand, country of origin and ethics of the company producing the product. The challenges faced by denim apparel manufacturers and fashion designers include the need to reinvent some products desired by niche markets, and also to meet consumer demands for better apparel sizing [27].

Nearly 80% of denim washing industries use cellulase enzymes alone or combined with pumice stone [32]. During the denim manufacturing process, the warp yarns are first sized and then desized in successive wet chemical processing. Today, most jeans are stonewashed. The technique was first popularized in 1970 to achieve a soft handle as well as the desired worn look [31]. Later, enzymes were also used on the cotton fabric in addition to pumice stones, to create the stonewashed look. The typical aged look of the denim is commonly obtained as a result of nonhomogeneous colour removal, trapped inside the fabric by the cooperative action of enzyme and mechanical factors such as beating and friction. Although the finishing of denim is important owing to its effects on appearance and comfort, with the increasing awareness of and concern for environmental issues, such as the production of large amounts of effluents and high consumption of water and energy, wet processes related to denim washing, which are considered not to be environment friendly, are getting a lot of attention. To address these environmental issues and also because of certain inconveniences caused by traditional procedures, and to meet the growing demands for denim from today's consumers, research is being conducted to obtain a similar appearance by creating a better quality and/or more advantageous manufacturing conditions. Emerging techniques such as laser, plasma, ozone and water jet have also been explored as alternative approaches to conventional wet processing, as discussed in the following sections [27]. To enhance the physical and comfort properties, finishing of cotton denim with herbal nanoencapsulation was also attempted; it was found to be durable even after 30 washing cycles [35]. Since the 1970s, the denim wash is mainly done using a pumice stone or superstone for a long time, or by sandblasting [36]. In 1980s, acid-washed denim was introduced using a stonewash with a weak chlorine. The fabric was used to tumble in a dry washing machine with chlorine-wetted pumice stones; afterward, potassium permanganate was used instead of chlorine. Later, in 2003, it was mentioned that an enzyme wash was commonly used as a wet process technique in the textile industry. Laser fading ($CO_2$) of denim fabric is an important process because it is not a process in a wet condition, but one that can be used for cleaner production in controlled faded patterns. Plasma treatment of textile is an environmentally friendly, inexpensive process compared with other denim-fading methods. In denim processing, the superstonewash can longer up to 6h, whereas enzyme fading of denim takes about 55–85 min. A sample of 20×20 cm would take only 45 min on a 70-cm/s $CO_2$ laser machine and the same area could be faded within 6–10 min using a 10×14-cm plasma head.

## 12.4 Water-based denim processing

### 12.4.1 *Pumice stonewashing and other methods*

Denim finishing, especially to improve the visual appearance by imparting a 'faded look', has become an important value-imparting operation in textile production. Several methods are available to fade denim, such as sanding, sandblasting, sand washing, snow washing, stonewashing with enzymes, prewashing, brushing, rinsing and bleaching. During denim processing, a part of the surface colour (mostly indigo dye) is removed by pumice stones. The desired fading effect could be achieved by traditional processes, but the industry reportedly encounters many problems, such as (1) application difficulty, in addition to the process being time-consuming owing to poor work flow; (2) loss of quality; (3) the inability to create standard and reproducible designs, the required nuances in shading; (4) identical fading effects on both sides of the fabric; (5) a decrease in wear resistance of the product; (6) the impossibility of designs on all textile surfaces and (7) the inability to apply the original forms, writings and design on products and to create the shade difference at the desired levels [29,37]. In addition, in the production of faded attributes in the fabric, the use of pumice stones has specific drawbacks such as difficulty in removing residual stone from the fabric; damage to the equipment owing to overloading of tumbling stones, stones and particulate materials; clogging of the drainage passage of the machine; and the use of a large amount of water, most of which is highly contaminated by chemicals [31]. Indeed, time-consuming and old-fashioned processes are not suitable for mass production, because ultimately such processes increase the cost of production [37].

To minimize unwanted loss of strength in the enzyme-treated sample and provide a distressed, worn-out fashion look to the denim garment, stone-bleach washing is carried out using a fixed amount of pumice stone mixed with various concentrations of bleaching powder [30]. A stone-bleach wash of denim garment is popular among today's youth; however, sufficient scientific literature on the process is not available, although the process has been in vogue since 1970. Various kinds of natural and synthetic stones are available on the market for denim washing. Of these, volcanic rocks and pumice stones are most commonly used for denim washing. Concerning the stonewashing process, the fabric is washed with oval or round lightweight pumice stones that have a rough surface structure. A stone-based wash is carried out to discolour the dye on the fabric surface and achieve a broken, worn effect on the denim garment [38,39]. The pumice stones abrade the superficially coloured fibres and remove them from the fabric's surface. In such a process, the denim garments as well as the stones are loaded into the machine (0.5–3 parts stone and 1 part of garment by weight). The garment is washed and tumbled for 15 min with alpha enzyme and detergent by maintaining a material to liquor ratio of 1:10 at 55°C. Thereafter, the treated denim fabrics are tumbled with calcium hypochlorite for 15 min at 55°C, followed by anticolouring with sodium bisulphite or hydrogen peroxide. The finished fabrics are then collected after a softening treatment. However, such a type of chlorinated treatment has been abandoned owing to the ecological hazards. In stonewashing, the duration of treatment is one of the most important parameters, because it can control the colour and colour uniformity of the washed jeans [39].

Khan and Mondal examined the effect of various concentrations of bleaching powder with a fixed amount of pumice stone, temperature and time in a stone-bleach wash to produce the most preferred fashionable and durable denim garment [30]. The effects of pumice stone (30% on weight of fibre) with various concentrations of bleaching powder ranging from 5.0 to 15.0 g/L on the properties of denim garments were investigated. Treatment of denim trousers with 5.0 g/L bleaching powder combined with 30% pumice stone causes a decrease in tensile strength, and the reduction was higher with a higher bleach concentration. The sample with 15.0 g/L bleaching powder concentration exhibited 19% loss in the tensile strength when the bleaching was carried out in the presence of soda. Similarly, with increasing bleaching powder concentration, the elongation at break in both warp and weft directions decreased. The mechanical action of the pumice stone removes the dye rapidly from the denim surface. The bleach in combination with pumice stone helps to achieve a light shade. It also accelerates colour fading, affecting the fabric properties that have a major positive role in producing the faded fashion look. Organo-montmorillonite was introduced as a new material for washing of denim garments in a conventional washing machine to produce old-looking garments before desizing and chemical softening [40]. In this method, pumice stone was replaced with organo-montmorillonite in denim washing to obtain the desired properties. Such a new and quick nanomineral-based washing/finishing process of denim is considered to be economical and simple.

Many of these challenges associated with stonewashing could be overcome by biotechnological interventions such as applying enzymes, as discussed subsequently. Cellolytic enzymes have been introduced into the denim processing, converting stone washing into a bio-stoning process. However, the process is more costly and requires a longer period (90 min), and it needs precise control and a skilled technical operator compared with the stonewashing process [31]. In this context, low-temperature plasma as well as laser technologies have also been explored as alternative methods for sustainable denim processing, as discusses in Section 12.5.

### 12.4.2 Biofinishing of denim

Textile industries use various chemicals in different processes such as sizing, scouring, bleaching, softening and denim washing; these sometimes reportedly cause environmental pollution [41]. Applications of biotechnology in the wet chemical processing of textiles are an example of a more environmentally friendly approach because the accelerated reactions are operated under mild conditions, acting on a specific substrate and providing ease of control, biodegradability and wider industrial applicability [38]. The use of enzymes in textile processing dates to the middle of the 19th century, when malt extract was reportedly used to remove amylaceous sizes from some textile substrates before the printing process. Enzymatic processes have been developed for the wet chemical processing of textiles directly from cleaning to finishing operations. Natural sizes (starch) are most frequently removed by enzymes, because they do not have an adverse effect on the cellulosic polymer component. Different kinds of enzymes such as cellulase, amylase, proteases, lipases, pectinases, peroxidases, lignases, and such have been explored for added value to textile substrates [41]. Of these, cellulase enzyme is increasingly being used in the textile industry. The most successful end application is in denim washing to produce the stonewashed look [42]. Besides catalase and peroxidase enzymes have also been used in denim processing. In a study on the enzyme washing process, denim washing was carried out at 60°C using alpha-amylase enzyme and a detergent as a desizing agent, and 0.5–2 g/L alkaline trisodium phosphate to accelerate the washing process for 20–60 min, depending on the desired effect. After washing, the fabric was treated with a softener to obtain the requisite comfort properties [38,39,43]. The effect of Bio prip, Cellusoft L and Denilite enzymes on cotton fabric properties were examined at different stages of textile wet processing, including scouring, singeing, bleaching and finishing, to obtain a cotton fabric with good physicomechanical properties, along with high wettability and dye absorbance. When applied to textiles, cellulase enzymes caused a controlled superficial attack on the fibre, leading to uneven decolourization but without causing excessive loss of fabric strength. Furthermore, the researchers attempted to improve the fabric's appearance by removing the fuzz fibres and pills, along with deriving the softening benefits using traditional softeners [38,44]. Cellulase enzymes have also been incorporated into domestic fabric washing products, where they have been claimed to aid the detergent, so as to clean the fibre's surface and thus improve the fabric's appearance and colour brightness. The effluents of stonewashing usually contain indigo dye coming from the garment's surface [45]. Many researchers have investigated both enzyme washing and bleach washing for denim garments [30]. To improve the fabric's handle and other important properties, softeners are regularly used in finishing operations. In denim processing, biotechnological intervention could reduce the abrasive action of stones and chlorine while ensuring the desired attractive worn look. Like stonewashing, cellulase enzyme-based washing is a popular method for washing a denim fabric. Cellulase has been obtained from *Trichoderma* and *Humicola* fungi. The working mechanisms of the cellulase and amylase enzymes are different from each other. Cellulase enzyme is effective only on the cellulosic fibrous surface at a particular temperature and pH; it attacks the 1-4-β-glucosidic linkage of cellulose polymer chain, where cellulose acts as a food ingredient for the enzyme microorganisms. Cellulases are a multicomponent enzyme system with (1) endoglucanases (EGs) that hydrolyze the cellulose chains randomly, (2) cellobiohydrolases that hydrolyze the cellobiose from the polymer ends, and (3) cellobiases that hydrolyze the cellobiose to glucose. These components act in a synergistic mode during degradation of the cellulose [42,43].

In the cellulase enzyme washing process, 0.5–2.0 parts weight of the stones and 1 part weight of the denim garment are loaded in a machine, where the alpha-amylase enzyme and detergent can also be added. An adequate quantity of the cellulase enzyme is added at a particular temperature and pH. The optimum working pH of the cellulase enzyme is 4.5–6 (*Trichoderma*-based cellulases) and 6–6.5 (*Humicola*-based cellulases). The reaction temperature is also important because the reaction rate slows at a lower temperature and the enzyme action gets deactivated by the destruction of its molecular alignments at a higher temperature. The garment and enzyme mixture are tumbled and washed for 30–90 min, and unloaded from the machine after application of a softener. The mechanical agitation inside the machine has an important role because as it controls the efficiency of the hydrolysis process. At the end of the process, the enzyme action is deactivated by raising either the bath temperature beyond 70°C or the pH above 10. Actually, the enzyme process mitigates the use of stones and accelerates the superficial degradation of the fabric. In a few areas, the enzyme has also been used with a certain quantity of pumice stones to enhance the worn looks of the denim garment [39,43]. Montazer and Maryan used laccases along with cellulase to investigate their effects on the garments and the possibility of reusing the biowashed effluent to save water, enzymes and energy [45]. Under the optimum concentration, laccases and celulases produce the same biowashing effect on the garment, even after three repeated biowashes. In such a process, laccases discolour the effluents and the enzyme residuals of both cellulases and laccases can be used for repeated processing. Laccases (benzenediol:oxygen oxidoreductase, EC 1.10.3.2) are copper-containing oxidases that accept a wide range of aromatic alcohol and amines as substrates [45]. Laccases have been reported to bleach indigo-dyed fabrics and degrade indigo dye present both in solution and on the denim fabric. Spectroscopic results exhibited that samples treated with a mixture of cellulases and laccases, have a higher L* than treated samples with cellulase alone. The mixture of enzymes and laccases-treated samples possess a lower crystallinity.

Cellulase enzyme treatment of cotton denim fabrics is an environmentally friendly approach to improving the properties of the fabric. In such a treatment, the removed indigo dye may also be redeposited on the white yarn, which is known as back-staining and ultimately impairs the look of the denim garment. There have been studies on back staining that mainly cover the mechanism of indigo redeposition on the denim fabric; nevertheless, it is still a problem for the manufacturer that needs to be addressed [42,46]. Efforts have been made to find the efficacy of immobilized acid cellulases on the removal of indigo dye from denim fabrics and its action in reducing back staining. Although the enzymatic method is eco-friendly, enzyme treatment on cellulosic garments degrades cellulose at the ends or in the middle of the cellulose chains, yielding a shorter chain of cellulosic polymers and thus severely reducing the fabric mechanical strength [30]. The first cellulase binding process has a major role during back-staining, which is reduced by the use of proteases to prevent enzyme binding during washing.

### 12.4.3 Water jet fading of denim

Hydrojet treatment has been developed for pattern making and/or enhancing the surface finish, texture, durability and other characteristics of a denim garment [27]. Hydrojet treatment generally involves exposing one or both surfaces of the garment through hydrojet nozzles. The degree of colour washout, clarity of patterns and softness of the resulting fabric are related to the type of dye, its amount present in the fabric, and the manner by which the impact energy of the fluid is applied to the fabric. A good result is especially obtained with blue indigo-dyed denim. Because no other chemical is involved in the process, it is a pollution-free process. Indeed, by incorporating a suitable water recycling system, the technique can be used for economical, environmentally friendly denim processing. The colour washout of the dye in the striped areas produces a faded effect with no blurring, loss of fabric strength, durability or excessive warp shrinkage.

### 12.4.4 Local tint/bleached spots

To obtain a bleached effect on the denim fabric, an oxidizing agent such as permanganate solution or ozone gas is sprayed onto the denim fabric at a controlled rate. These chemicals can also be applied by sponge to get a vintage look with the denim fabric.

### 12.4.5 Acid wash effect

During the 1980s, the acid wash process was also popular for washing denim fabrics. For stonewashing, pumice stones to be used were soaked with various oxidizing chemicals. The treatment used to provide an uneven visual effect on the fabric's surface. The colour contrast was obtained with a discharge oxidative treatment by combining a stable sulphur dye or indigo with sodium hypochlorite. Potassium permanganate has been used as an oxidizing agent instead of chlorinated chemicals to take care of environmental concerns [39].

## 12.5 Sustainable denim processing using water-free technologies

### 12.5.1 Sandblast finishing of denim

As reported earlier, to achieve a special look as well change the fashion, requisite washing methods of denim are required, including stonewashing, sand washing and bleach washing [44]. In the case of stonewashing, there are difficulties in removing residual pumice from the processed cloth items and in ensure a reduction in damage to equipment by an overload of the tumbling stones, as discussed previously. Pumice stones and particulate materials can also clog machine drainage passages, sewer and drainage lines at the machine site. On the other hand, sandblasting does not lead to these types of problems. Sandblasting is a modern technique applicable to specific portions of the indigo-dyed denim using special types of sand. The sandblast fading process for denim has been the most commonly used process in the industry to achieve a worn vintage effect. Actually, it is a mechanical process used to create a fading effect on heavy fabrics such as twill, denim, canvas, etc. Very fine granular powder and hard aluminium oxide has been used to sandblast denim. In the process, aluminium oxide sand mixed with dry air has been passed through the nozzle at very high speed to effect the sandblasting action. The sand removes colour from specific portions of the apparel and thus creates the fashionable effects. After sandblasting, the apparel is washed with enzymes and silicon softener to increase the softness of the fabric [44]. Because of the abrasion damage, the ends per inch and picks per inch of the denim fabric were found to increase, resulting in a decrease in the fabric's surface density. Furthermore, the fabric tensile strength and seam strength were found to decrease owing to enzyme washing with sandblasting. Nevertheless, the stiffness of the denim fabric decreased after silicon washing, which indirectly ensured an increase in fabric softness. Here, the denim garments are placed in a closed chamber and are instantly faded by the frictional effect of the sandblast. The flow angle of the sandblast, an important parameter, is kept somewhere between 10 and 20 degrees on the surface. A properly set blowing angle is vital because a process with a higher blowing angle will result in a higher fading effect, in addition to higher fabric damage. After mechanical sandblasting of the denim fabric, a chemical wash such as stonewashing or a combination of stone and enzyme washing is required to meet the customer's expectations for denim quality [47]. Numerous operations exist to create the fading effect of the fabric, but in the context of sustainable and environmentally friendly denim processing using water-free or low water–based processing technologies, the use of laser, ozone and plasma are mentioned, as discussed subsequently.

### 12.5.2 Ozone fading of denim

Like plasma-enhanced fading of denim fabric, ozone gas–induced fading is an approach in sync with low water–based denim processing. Ozone is released in the presence of sunlight as the result of the interaction of hydrocarbon, oxides of nitrogen, oxygen, etc. Indigo-dyed denim fabric has been faded to yellow because of the ozone reaction after UV exposure. Although the ozone finishing of denim fabrics does not completely eliminate the use of water, it could noticeably reduce the consumption of water, energy, chemicals, enzymes and stones. The fading rate of indigo dye depends on the portion of dye in the fibre and its solubility, and the diffusion of ozone in the fibre. Besides, the absorption of ozone on the fibre surface is an aspect of ozone-fading in the denim fabric [31,37]. For denim washing by ozone, the garment is bleached with dissolved ozone in water in a washing machine. Denim garments can also be bleached or faded using ozone gas in a closed chamber. Some techno-environmental advantages of ozone-based denim processing are: (1) minimal loss of the fabric's strength, (2) a straightforward process, (3) involvement of low water and chemicals, and (4) economy of process. The method is simple and environmentally friendly, because after laundering the ozonized water can easily be deozonized by applying UV radiation. Ozone characteristically acts as a mild bleaching and sterilizing agent. Ozone fading can also be accomplished by plasma processing [27]. During plasma treatment, high-energy electrons react with moisture in the air, generating a mixture of radicals. In the process of generating ozone plasma, a combination of charged particles, free radicals and UV light is also generated. UV light in the plasma zone can also contribute to the production of •OH radicals, which are the most oxidative radicals among radicals generated in the plasma and are mainly responsible for the degradation of indigo dye in textiles. Compared with untreated samples, the

Table 12.1 *K/S* value in differently processed denim fabrics [27]

| Sl. No. | Different processed treated denim fabric | *K/S value* |
|---|---|---|
| 1 | Untreated | 430 |
| 2 | Plasma-induced ozone-treated | 382 |
| 3 | Enzyme desized | 425 |
| 4 | Plasma-induced ozone-treated, followed by cellulase-treated | 365 |
| 5 | Enzyme-desized, followed by cellulase-treated | 379 |

Khalil E, Sustainable and ecological finishing technology for Denim jeans. AASCIT Commun 2015;2(5):159–63.

*K/S* values of differently modified denim fabrics using various technologies are shown in Table 12.1. In the *K/S* value the symbol *K* is the absorption coefficient at a specific wavelength and *S* is the scattering coefficient.

Plasma-assisted, ozone-treated denim fabric exhibits a paler shade than enzyme-desized denim fabric, because ozone in the plasma oxidizes indigo dye present on the fabric surface [27]. During enzyme desizing, the enzyme reacts only with the sizing material of the fibres. It has no effect on indigo dye molecules; thus no significant shade change of the fabric takes place. However, in the case of cellulase enzyme treatment, it can penetrate effectively into the denim fabric in an aqueous medium. Enzymatic hydrolysis induced by cellulase in plasma-assisted, ozone-treated denim is more severe than that of the enzyme-desized denim fabric. Khan et al. described the optimization of process conditions such as dye concentration, pH and the oxidant dose for the UV, $O_3$, $H_2O_2$/UV, $O_3$/UV, $H_2O_2/O_3$ and $H_2O_2/O_3$/UV by applying Red Cl-5B azo dye with varying concentrations (e.g., 100, 300, and 500 mg/L) [48]. The ozone treatment resulted in more than 90% colour removal, whereas $H_2O_2/O_3$ showed no advantage over $O_3$ treatment alone. Similarly, $H_2O_2$/UV was found to be suitable because it could give almost 100% decolourization in a relatively shorter time period. It was also found that decolourization rates for all of the processes were reduced to half when the dye concentration was enhanced from 100 to 300 mg/L. The rate constants indicated that $H_2O_2$/UV treatment is four times faster than that of UV treatment alone and $O_3/H_2O_2$ treatment was three to four times slower than $O_3$ treatment alone. The ozonation process was found to be promising for decolourization of Red Cl-5B azo dye solution, which resulted in more than 90% colour removal and 80% of COD removal at all selected concentrations.

### 12.5.3 Denim processing using water-free plasma technology

#### 12.5.3.1 Plasma

Plasma technology has been used to treat textiles over the past decades, but a generally available and easily implemented technology has yet to be developed [49]. At the same time, advances in the modeling and diagnostics of the plasma process to produce integrated circuits, nanotechnology and biomedical applications raise hopes for better control and tailor-made properties for the treatment of textiles that can be accomplished based on knowledge gained in other fields of research [49]. Plasma is a partially ionized gas composed of both positive and negative ions, electrons, neutrals, excited molecules, photons and UV light. It can be used for nano-level surface modification for both natural and synthetic textile substrates without changing the inherent bulk properties of the fibrous material [1,50,51]. Other major advantages of plasma processing over traditional wet-chemical processing of textiles are its surface modification capability, liquid-free and eco-friendly dry operation, single-step faster operation, less need for chemicals, cost-effectiveness in terms of processing time and temperature, and imparted functionality independent of substrate chemistry [1,50–52]. The superficial modification of fibres induced by various plasma processes such as polymerization, reaction, activation, cleaning, oxidation, changes in surface energy, increases in surface roughness/area, etching, coating and deposition ultimately could lead to improvement in aesthetics and functional value of the textiles, such as improvement in water and oil absorbency, water and oil repellency, UV protection, antimicrobial results, flame retardance, crease resistance, antifelting, antistatic properties, dyeing, printing, desizing, adhesion strength and biocompatibility [1,4–6,13,50–55].

### *12.5.3.2 Plasma processing of denim fabric*

Plasma generation with the help of an electrical discharge using polymerizing to non-polymerizing gaseous molecules was used for denim processing. However, efficacy of the process depends largely on the nature of gases used [31]. The production of a faded and torn looks in the fabric using conventional technologies involves the consumption of large amounts of water and chemicals, which leads to environmental pollution. Dry finishing or processing is a promising solution to tackle these problems. Laser and plasma technologies have shown a potential for creating different surface effects on fabric without using much water and chemicals. Chemically, a reactive plasma discharge has been used in various industries to modify the specific surface properties of the materials; in the past two decades, it has also been amply applied in textile processing [1]. Washing of denim garments is a polluting activity in the textile industry; however, if plasma is applied to denim jeans, it minimizes or altogether avoids the use of water. So far, plasma exposure has mainly been used to fade indigo and sulphur-dyed denim garments. Mostly, it is the oxygen cold plasma that has been tried for desizing and colour fading of denim cloth.

Cheug et al. reported the fading efficacy of a reactive dyed fabric surface with plasma exposure [37]. In 2016 Gabr et al. reported the use of oxygen and water vapour on cotton/polyester-based twill indigo-dyed denim fabric [36]. As per the research findings, plasma application etched the surface of the indigo-dyed warp yarn of the denim fabric and caused the yarn to stretch and lose weight. In addition, the extent of etching was more for the oxygen plasma–treated denim than that noted with the water vapour plasma–treated sample. It was revealed that the exposure of plasma treatment time on fabric fading was limited to only 3–5 min, which was significantly shorter compared with the stone and enzyme-washed samples. By increasing plasma exposure beyond 5 min, the weight of the denim fabric was found to increase, possibly owing to redeposition of the indigo-dyed layer on the fabric's surface. On the other hand, the weight and the thickness of the water vapour plasma–treated denim were found to be more, probably owing to the absorption of water molecules and swelling of the fabric. Indeed, the plasma-treated denim fabric turned to a greenish tone with an increase in the ΔE value. The treated fabric also had a yellowish tone as a result of the thermooxidative effect of the plasma. From these research findings, it was postulated that plasma treatment for 3 min is appropriate for observing the fading effect in indigo-dyed denim fabric. After plasma treatment in both of the atmospheres (e.g., oxygen and water vapour), the stiffness and bending length of the fabric increased, depending on the plasma exposure time and the magnitude of current used for the treatment process [36,56]. Ghoranneviss et al. reported the efficacy of DC plasma and corona treatment on the decolourization of indigo-dyed denim fabrics without using water. It was commented that plasma treatment could replace the less eco-friendly conventional stonewashing or enzymatic bio-stoning process that is used to generate the popular 'worn look' of denim [31,57]. Cheung et al. reported the application of an atmospheric pressure plasma jet on colour fading of 100% cotton single jersey-knitted fabric dyed with C.I. Reactive Blue 19 under industrial conditions [37]. Glow plasma was generated by RF of 13.56 kHz in the presence of helium (carrier gas) and oxygen (reactive gas). Plasma parameters were as follows: jet distance 3 mm, treatment time 5 s/mm, oxygen concentration 1% and plasma power 160 W. It was revealed that atmospheric pressure plasma (APP) treatment could achieve a similar colour-fading effect with a shorter treatment time compared with the conventional enzymatic colour-fading process without a significant loss in fabric weight. In the neutral enzyme (activity 130 EGU/g)-treated sample, owing to colour fading and the action of enzymes, there was a 10% loss in weight, whereas it was only 5% in the plasma-treated sample. Unlike enzymatic hydrolysis, in plasma process the percent loss of fabric weight remained lower, possibly owing to localized ablation and only surface etching. In general, APP treatment induced lower weight loss than did enzyme treatment, because plasma treatment affected only the surface of the material while keeping the bulk properties unaltered [37]. The *K/S* sum value was found to be reduced from 436 (untreated sample) to 387 and 384 in the plasma- and enzyme-treated samples, respectively. This resulted from the oxygen plasma oxidizing the dyes in the fabric surface, leading to a colour-fading effect. In the case of the enzymatic colour-fading process, the enzyme in the aqueous medium effectively penetrated the interior part of the fabric. More fibres in the fabric were hydrolyzed by an enzyme action, and the weakened fibres were further removed by abrasion of the fabric and mechanical agitation [37]. The same research group reported low-pressure argon (Ar) plasma impingement (power <800 V and 57 mA) on denim fabric for 2.5, 5, 10 and 15 min [31]. Unlike the effect of stone or enzyme, in low-temperature plasma processing, the dyestuff is removed electrically by the impact of the plasma active species such as ions, radicals, electrons and UV light with the fibre, leaving a smooth and varnished surface on the fabrics. Power consumption of about 45 W (<800 V) was much lower compared with a few hundred watts reported in RF techniques. There were no significant changes in the breaking load of the samples owing to plasma bombardment. The colour strength *K/S* value decreased from 7.8 in the untreated sample to 6.1 and 5.3 in the 5-min and 15-min plasma-treated samples, respectively. The X-ray diffraction peaks of untreated as well as 15-min $O_2$ and Ar plasma-treated samples did not reveal noticeable changes in terms of crystal size (full width at half maximum), distance between them (*d*) and percentage of crystallinity (Inet). This indicates the effectiveness in colour removal without causing a detrimental effect on fibre strength.

Radetić et al. reported the discolouration of denim fabric by low-pressure capacitively coupled plasma (CCP) generated at an RF frequency of 13.56 MHz in the presence of argon and air gases; the results were compared with atmospheric pressure industrial corona plasma [49]. In another experiment, the efficacy of low-temperature plasma in the presence of two nonpolymerizing gases, argon and oxygen, was studied for the discolouration of denim [57]. The low-pressure plasma (LPP) apparatus consisted of two coaxial cylinders in which a low DC glow discharge (below 1200 V and 100 mA) could be produced radially between the anode and the cathode, and plasma treatment was carried out for 5, 10 and 15 min. In sputtering treatment, attacking active particles such as ions, radicals and electrons, to the surface of the fabric placed on the cathode would sputter the surface dyes from the fabric. Argon is not a reactive gas; hence no reaction occurred between the fabrics and the plasma gas. On the contrary, in oxygen-induced sputtering, the active radicals could create the functional groups on the denim fabric surface in addition to sputtering of the surface dye molecules, thus oxidizing some of the dye molecules. After washing, the varnish of the oxygen plasma-treated sample looked better than the argon plasma–treated samples. The argon plasma more profoundly discoloured the dye molecules in the denim fabric compared with oxygen plasma, although it is a nonreactive gas. Nevertheless, the results were reversed when the samples were subjected to washing after plasma treatment [57]. The *K/S* value of 12.6 in the untreated sample was found to reduce significantly to 9.9 and 7.2 in the 5-min $O_2$ and Ar plasma-treated samples, respectively, and values were 5.3 and 4.7 after 15 min of plasma exposure. The lowest *K/S* value in the washed samples was 3.5 and 2.7 in the 15-min Ar and $O_2$ plasma-treated samples, respectively. In the case of the LPP gas, the power applied to the CCP reactor of 100, 200 and 300 W and the treatment times of 5, 10 and 15 min were the variables, while maintaining the constant pressure of the reactor at 0.27 mbar. In APP, the samples were subjected to corona treatment at a speed of 4 mm/min with a variable power of 600, 800 and 1000 W and the number of passages was kept at 15, 30 and 45 [49]. The changes in colour attributes were assessed by the CIE LAB colourimetric system and the morphological changes were analyzed in a scanning electron microscope (SEM). SEM images before and after plasma treatment at a suitable plasma regime exhibited decolourization of denim, and the specific fibre surface changes were always observed in the form of submicrometre-sized striations, pits and cracks. However, prolonged plasma treatment, i.e., an increase in the number of passages, caused a considerable increase in colour difference of the untreated and corona-treated samples irrespective of the applied power. The effect of power was much more pronounced in the case of samples with 30 and 45 passages. When the experiment was conducted in the LPP in argon gas at a power of 100 W, it led to marginal changes in colour attributes, which was highly noticeable because of the increase in plasma power. The prolongation of plasma treatment was also not prominent. The breaking strength in the weft direction increased marginally after both the corona and LPP treatments, whereas the breaking strength in the warp direction increased only a little bit after corona treatment. Kan et al. reported the performance of APP treatment in the desizing of grey cotton denim followed by colour fading using the enzyme process [58]. Four important plasma process parameters were considered: treatment time, ignition power, oxygen gas concentration and plasma jet distance. Based on the experimental design, the treatment time of 5 s/mm, ignition power of 160 W, oxygen concentration of 1% and 3-mm jet distance were found to be optimal. When these parameters were evaluated based on their efficacy of performance, they followed the trend: jet distance > ignition power > oxygen concentration > treatment time. The colour fading of denim fabric occurred more in the plasma-treated sample compared with the conventional enzyme-treated sample. All of the treated samples possessed lower *K*/*S* sum values than their respective untreated samples. APP-treated denim turned into a paler shade than the enzyme-desized sample, possibly owing to oxidative fading of indigo dyes in the presence of plasma. On the contrary, enzyme reacted only with the sizing material of the fibre surface in the enzyme desizing process and no breaking of the indigo dye molecule was realized, resulting no significant change in shade. Atmospheric pressure cold plasma was also generated in the presence of helium (He), He-air and He-oxygen gas mixtures in a study in the presence of an indigo-dyed denim cotton textile [32]. The control and treated materials were characterized using SEM, a computerized colour matching system and a Fourier transform infrared technique. After plasma treatment, the water absorbency time decreased significantly to <2 s in the treated denim textile. The *K*/*S* value of the grey denim fabric was 19.2 and reduced to 18.4 after plasma treatment; it went down to 18.1 when the sample was desized before plasma treatment. Dow Corning also introduced a denim processing solution [59]. Their patented granulation platform technology could reduce the number of steps required to process the fabric that ultimately minimizes the amount of water requirement and the generation of effluent.

### 12.5.4 Denim processing using water-free laser technology

#### 12.5.4.1 UV excimer and laser

Similar to the plasma treatment of textiles, UV treatment has also been used for the surface modification of natural and man-made fibres to improve hydrophilic and other functionalities. A large number of lasers capable of operating at different wavelengths and different modes are available for the surface modification of various substrates, including textiles [60]. Surface modification by laser can be carried out in a variety of ways, depending on the purpose of the surface modification, such as etching, ablation, deposition, evaporation and surface functionalization [61]. The purpose of surface modification decides the type of laser to be used. Excimer lamps are used for the surface modification of polymers and textile materials for improving wettability, antistatic, antifelting, antimicrobial, antisoiling, adhesion strength and crosslinking properties of the sample. Lasers, which use a noble gas for generation, are generally referred to as the excimer lasers [4]. They operate in the UV to near UV region, i.e., from 193 to 351 nm. The $CO_2$ and Nd:YAG lasers are the most powerful; however, they are not suitable for modification of polymeric substrates owing to their longer wavelength and lower photon energies that are insufficient to break down the molecular bonds of a polymeric material. On the other hand, excimer lasers are capable of operating in the UV region and are strongly absorbed by the polymers for photochemical reactions. However, it is difficult to expose large areas of the polymers in an efficient manner owing to the requirement of a bigger facility that escalates the cost of the production [62].

The term 'excimer' stands for 'excited dimmer'. Excimers, which exist only in the excited state, do not possess a stable ground state under a normal condition. Excimer molecules do not remain stable, and once formed, they give up their excitation energy in the form of UV photons within a few nanoseconds. UV lamps work on the dielectric barrier discharge principle and emit intense monochromatic light in the UV region of the spectrum. Organic materials, which have nonbonding electron pairs, absorb light strongly when the wavelength is shorter than 200 nm [4]. The penetration depth of a photon is approximately 10 nm. The bond dissociation power of the photon is different for the various excimer systems; e.g., if xenon gas is used, it emits 172-nm light with a photon energy of 7.2 eV that is suitable for breaking polymeric bonds such as C—H, —C=O and —OH.

A laser is a device that can emit light through a process of optical amplification based on the stimulated emission of electromagnetic radiation. The term 'laser' originated as an acronym for 'light amplification by stimulated emission of radiation' [27]. It works by creating an extensive heat, in which the material within a small region is subjected to an intense heat that may also lead to melting of the material owing to a phase transformation from solid to liquid. The optimal characteristics of the laser light have increased its application in many fields including medical, industrial and military [63]. Some popular uses of lasers are laser marking (e.g., only the surface of the fabric is processed for fading), laser engraving (i.e., controlled cutting to a depth), laser welding (i.e., a molten material that joins or welds two or more layers of fabric together), and laser cutting (i.e., the cutting through a fabric). The laser used in the textile industry is of great importance for discolouration of denim, cutting of fabrics and wool-shrinkage controlling purposes [3]. Among the different laser processing techniques, only laser engraving or the embossing process allows short-time surface designing of patterns with precision and desired variety, size and intensity for a wide range of textile surfaces, including knitted and woven fabrics inflicting minimum damage to materials at the selected wavelength and laser power density [3]. Notably, while cutting synthetic fabrics with a laser beam, often the heat effect could melt fibres, causing the fabric plies to stick together or pilling the fabric. Laser engraving has the possibility of being applied to textile materials to create a worn effect as well as a faded look. Laser finishing is a completely dry process and with careful control of laser processing parameters, it can provide fast and accurate production with good reproducibility and repeatability [33]. The major advantages of laser-based processing over other conventional technologies are: (1) low chemical and water consumption, (2) environmental friendliness, (3) process flexibility ensuring replication of existing stonewash designs or the creation of a new 'finish styles', (4) the possibility of finishing being applied on parts, assembled garments and uncut material, (5) application feasibility of automated control of laser-faded technology so that manufacturers can change the style of the finish from one to another without retooling, (6) micrographic creation on the garment, (7) development potential of special logos or characters on the denim fabric, (8) shorter duration, (9) high precision and (10) negligible damage to the denim fabric [64–67].

Several commercial lasers are available for industrial application, including Nd:YAG, $CO_2$ and excimer. The production of faded jeans using conventional technologies generates a large quantum of polluted water as by-products. Similar to water-free plasma processing, $CO_2$ laser treatment is considered a dry treatment and can be applied to denim materials as an alternative to conventional wet processing, such as stonewashing, sand washing and bleaching to achieve the faded look and worn effects. The laser action on coloured denim is mainly an interaction of the fabric with indigo colour by thermal means [64]. Such an interaction could lead to uniform fading or a text or fading of a graphical image that is easy to accomplish, because the mirrors are controlled by a computer. Different levels of colour removal with minimal or no damage to the denim textile have been achieved by altering the different laser parameters. The process has three steps: (1) generation of a computer map of the area that will be irradiated on the fabric at different frequency levels with the desired pattern; (2) control of the laser system, scanning system and conveyor system; and (3) the laser-fading process. Of these steps, the third is important. The interaction between the laser beam and the textile or other material depends largely on the laser beam parameters, such as energy per pulse, repetition rate, pulse duration and wavelength. Similarly, it depends on the material parameters such as the thermal diffusivity, reflection coefficient and heat of vaporization. As far as the removal of indigo dye from the denim textile substrate is concerned, four physical phenomena happen: vaporization, photoablation, plasma-induced ablation, and the creation of shock waves. In the process, as the laser pulse energy increases, the material reaches the vaporization state more rapidly, providing less time for thermal conduction. Simultaneously, the vaporized material can simply be diffused into the surrounding atmosphere without further interaction with the beam. Nevertheless, the vaporization process is not convenient for use in denim fading because of the longer interaction time. In addition, the laser energy has enough time to be absorbed into the textile and initiate burning of the textile. The photoablation phenomenon indicates that the photon energy is absorbed by the bond within an organic polymer (i.e., indigo dye with a covalent bond). The lowest dissociation energy of indigo bonds is more than 2.7 eV. On the other hand, the photon energy for an Nd:YAG laser is 1.2 and 2.4 eV at the fundamental and second harmonic, respectively. Only the UV photons are capable of producing photodissociation. The use of excimer lasers in denim fading is possible; however, for practical reasons such as maintenance, reliability and cost, only the $CO_2$ (10.6-μm wavelength) laser and an Nd:YAG laser working at 1.064-μm wavelength and its second harmonic are normally considered during denim processing.

### 12.5.4.2 Laser processing of denim fabric

Although conventional technologies are much more beneficial from the point of mass customization, the production of discoloured jeans using those processes involves large quantities of water, which gets discharged when it is contaminated with a chemical admixture [26,67]. On the other hand, the laser is a source of energy that can be directed onto the desired object, whose power as well as intensity can be controlled. This technology is computer-driven and replaces the use of water, chemicals and stones, yet it can produce precise, repeatable and intricate patterns. In addition, it can create excellent localized effects on denim fabric. However, major disadvantages of the process are that it is more costly, each garment is required to be positioned individually for treatment and there is a possibility of the garment receiving treatment only on one side. Furthermore, the laser wavelengths and other process parameters were altered and their efficacy on colour removal was also examined. A laser beam's strength can be adjusted to an impact width in a range of 0–1000 μs, preferably within 0–200 μs, and this method enables the user to make 2–1024, preferably 256, grey toning applications [26,29]. The laser beam impact frequency is used within 0–100 kHz range, preferably within 0–5 kHz. To achieve a constant initial laser power, a period of 1–50 pixel, preferably 4–10, is considered at the beginning of the scanning process, and this difference is later compensated for. It was possible to create the design as well as to draw geometrical technical models of the texture, in which the design will be applied. The laser systems are also used in fashion designing, pleating, cutting and modification of the fabric surface to impart some special finish. The nominal power of the laser equipment is 250 W and the wavelength of the laser beam is 10.6 μm [26]. It is also possible to transfer certain designs into the surface of textile materials by changing the dye molecules in the fabric and creating alternations in its colour quality by directing the laser to the material at a reduced intensity. Although the interaction of the $CO_2$ laser with denim fabric is the strongest, it was observed that the $CO_2$ laser beam affects the structure of the textile fibres, because of the heating and the vaporization phenomena. The $CO_2$ laser treatment has been used in different areas of textiles for several years because of its advantages of surface patterning within a short period, with good precision, desirable effects, sizes and intensity, without causing much damage to the inherent bulk properties of the textiles [27].

Dascalu et al. reported the application of an Nd:YAG laser (1.064-μm energy wavelength and its second harmonic at 532 nm) with an energy per pulse as high as 40 mJ and a repetition rate of 60 Hz for denim fading [64]. The pulse duration varied from 28 to 45 ns at both wavelengths. Similarly, a pulsed $CO_2$ laser working at the rate of 5-kHz pulse repetition and 100-μs laser pulse length was also used. The denim fabric was washed after irradiation with water with no detergents and dried before measurement of colour attributes. During further experimentation, it was observed that the Nd:YAG laser working at 532 nm could produce a better result. Hung et al. described the use of an artificial neural network (ANN) and linear regression (LR) model to predict the colour properties of laser-treated denim fabrics [65]. The fabrics were treated by changing the pixel time (μs), resolution [dots per inch (dpi)] and grayscale (lightness percentage) as the process inputs, and colour properties such as the *K/S* sum value, CIE L*, a* and b* values, and yellowness index were predicted as the outputs. Statistical analysis showed that the ANN approach could provide a more accurate prediction than the LR approach, especially for the L value. The efficacy of the $CO_2$ laser on denim fading and colour evenness based on the laser parameters and irradiation direction was examined [66]. A commercially available indigo-dyed 100% cotton denim fabric with a fabric weight of 378 g/m$^2$ (25 ends/cm and 20 picks/cm) was used for treatment; the laser machine parameters were selected as follows: excitation frequency 81 MHz; laser medium $CO_2$; wavelength 10.6 μm; wave mode pulsed; power (W) / Energy (mJ/p) 100 W (60–230 W); pulse energy 5–230 mJ; pulse activation time <45 μs; polarization linear polarization perpendicular to the bottom of the laser head. The resolution of the laser beam was kept at 30, 60 and 80 dpi with a pixel time of 110, 160, 220 and 300 μs, respectively. The efficacy of laser treatment was also compared with neutral cellulase enzyme-treated samples with an enzyme activity of 30 EGU/g in powder form while treating the denim fabric. One EGU/g (EG activity) is defined as the amount of enzyme required to reduce the viscosity of a solution of carboxymethylcellulose to one-half. The colour fading percentage (CF%) was used to express the degree of laser-induced colour fading. The CF% was found to increase gradually with the increase in laser power, possibly owing to the removal of more of dye molecules [66]. Thereafter, CF% reaches an equilibrium position when the optimum laser power reaches about 13 W/cm$^2$. The CF% was higher in the case of laser-treated sample compared with the conventional cellulase-treated sample. The colour evenness of the laser-treated denim fabrics was about 0.22, which was rated as good. The colour fading effect as well as the colour evenness of the laser-treated denim fabric was found to be independent of the direction of laser irradiation.

The same research institute also explored the possibility of $CO_2$ laser engraving on an indigo-dyed denim fabric (2/1 twill, 376 g/m$^2$), keeping nearly similar laser parameters as stated previously but with different resolutions of 35, 45, 55, 65 and 75 dpi and pixel times of 120, 130, 140 and 150 μs [3]. More degradation of fibre and engraving was noticed under the SEM at a magnification of × 100 when the laser resolution and pixel time (power density) were increased from 35 dpi and 120 μs to 75 dpi and 150 μs, respectively. During the process, the laser-thermal energy becomes absorbed by the fibre, resulting in the production of water vapour and $CO_2$ that ultimately causes swelling and expansion of the fibre with a sponge-like structure. The *K/S* value of the untreated sample decreased significantly in the different laser-treated samples with various resolutions and pixel times [3]. The researchers also explored the ANN technique to predict the colour properties of the laser-treated 100% cotton fabric [33]. Pure cotton fabric samples were laser engraved with a $CO_2$ source laser (10.6 μm) with an excitation frequency of 81 MHz. The laser power varied from 60 to 230 W with a pulse energy ranging from 5 to 230 mJ. The experiments revealed that the colour properties could be predicted accurately with the help of an ANN model. The study demonstrated the usefulness of laser treatment before colouring and may find applications in the textile industry in the near future [33]. In an another study, the $CO_2$ laser (10.6 μm) was used to fade the denim fabric to study the effect of laser treatment on the fabric's tensile strength, tear strength, abrasion resistance, and static and kinematic friction coefficients [26]. It was found that 100- and 150-μs pulse times were suitable for the laser fading of denim materials. The *K/S* value of the laser-faded samples decreased significantly until 150-μs pulse time. Subsequently, the *K/S* values were less in 200 μs and with higher pulse time. Because of the removal of carbon particles, the yellowness index changed significantly after washing. The fabric tear strength in the warp direction did not decrease between 50 and 150 μs in the laser pulse times but a noticeable decrease in strength was observed between 200- and 250-μs pulse time and the samples degraded profoundly when it was in the range of 500–1000 μs. A more or less similar trend was noted in the tensile strength of the fabrics. It was also noticed that the laser fading process can cause an increase in both the static and the dynamic friction coefficients up to 100 μs; thereafter they leveled off. From the results of colour parameters and mechanical properties, it was inferred that 100- to 150-μs pulse times were the most suitable for the laser fading of denim fabric [26].

Ortiz-Morales et al. explored three laser sources: beams from Nd:YAG laser (1.064 μm and its second harmonic, 532 nm), CTH:YAG laser (2.09 μm) and $CO_2$ laser (10.6 μm) along with various laser beam parameters [67]. They used each laser process to get the optimum flounce and power density to obtain a similar fading appearance, such as those one can obtain by conventional processes. In the experiment, the $CO_2$ laser-based processes were the most suitable because of their higher efficiency compared with other lasers. Following a similar line of work, [29] put forward the laser machine and its design aspects to transfer pictures, figures and graphics of the desired variety, size and intensity on all kinds of surfaces, such as knitted and woven fabrics, and on leather with the desired precision, but without damaging the texture of the material. Computer-controlled laser beams were used to change the colour of the dye material on textile surfaces by directing the laser beams at a desired wavelength and intensity. All of these technical modifications could give a rich look to the product by wrinkling the textile surfaces to create a three-dimensional effect by deformation in addition to setting pictures and patterns on leather and synthetic fabrics by means of heat. It was possible to create designs on different surfaces of the fabrics, such as synthetic drapery, cotton denim, and even on leather by controlling the power of the laser beam. Here, the design is not limited to two-dimensional figures; a three-dimensional surface design is also feasible.

In a study, a $CO_2$ laser was used for the surface modification of cotton fabrics for the development of antimicrobial textiles using silver nanoparticles [63]. From the inductively coupled plasma optical emission spectrometry analysis, it was observed that the untreated cotton fabric contained 0.024% of Ag ion and all laser-treated samples exhibited a higher silver content. The highest amount of silver ions of 0.047% was observed for a higher energy level of the laser (100 W, 1000 cm/s). A fabric sample containing more silver is expected to ensure a higher antibacterial activity of the denim garment.

## 12.6 Some important finishing of denim fabrics

### 12.6.1 Dry mechanical/physical finishing

Different mechanical and physical processes are also available in the industry to provide an aesthetic and attractive look to the denim fabric. One such type of finishing is done on the pocket edges and bottom hems of denim fabric by rubbing it against the abrasive surface or stone. The distinctive effect of the denim fabric is achieved by using a swift tag machine with a plastic tag attached to the fabric to ensure a high colour contrast effect on the denim fabric. Control warp removal and creating holes on both the warp and the weft of the denim fabric also provide a vintage worn look. The tie effect of the denim fabric has been achieved by transferring the irregularly dyed garment to the denim garment by a process similar to stonewashing. It gives an attractive abraded effect to certain areas of the fabric. Permanent creases in the specific area, such as the pocket and hems of the denim fabric, also create worn and attractive looks [39].

### 12.6.2 Leather and wax finishing

Generally, leather and wax finish form a thin layer on the denim surface and thus change the surface attributes of the treated fabric [38]. Leather finishing of a denim garment is performed by using 660 g of anionic acrylic solution, 20 g of a cross-linking agent, 283 g of water and 37 g of a concentrated black pigment. This finish formulation is added to the denim fabric with 55.5% pickup. Subsequently, the treated fabric is dried and cured at 150°C for 5 min [38]. Such a finish is simple to apply but provides a fast surface treatment for specific effects, in which short and dense fibres give a 'peach skin' effect [38]. The leather finish can be applied either on a small area or on the whole denim surface. On the other hand, the wax finish provides a smooth, oily and greasy touch to the treated denim. The application of a special binder and chemicals provides an elastic, soft and silicone-like film on the treated area. The wax finishing of denim fabric has been carried out using 200 g of cationic polysiloxane emulsion, 700 g of polyurethane binder, 20 g of a cross-linking agent and 80 g of black pigment with an appropriate quantity of water. The denim fabric was treated with this mentioned formulation with 49.9% pickup, followed by drying and curing at an adequate temperature for a sufficient time.

### 12.6.3 Soft resin finishing

A special resin finishing is used to impart an authentic three-dimensional permanent crease that is resistant to wear and washing [38]. Resin can also be applied by spraying it directly onto the denim fabric surface, and creases can be formed using binder clips or aluminium pipes, before or after spraying the resin. In addition, denim fabric has been treated with a mixed formulation of 200 g/L glyoxal resin, 100 g/L acrylate polymer and 20 g/L catalyst for the synthetic resin. The wrinkles were formed on the denim fabric by a flexible aluminium table, followed by drying and curing.

### 12.6.4 Spider finishing

This particular treatment of denim fabric was carried out using an anionic aqueous solution of polysaccharide. It was applied to the denim fabric with 20.5%, followed by drying at 80°C. Then, the fabric was cracked and blown using compressed air. The broken places of the denim fabric were treated with a permanganate (5%) solution and left for 10 min. Thereafter, the sample was neutralized and treated with a nonionic softener, centrifuged and air-dried [38,39].

### 12.6.5 Strato finishing

In this finishing process, denim fabric was treated with a mixed formulation of 100 g/L nonionic silicon emulsion and 5 g/L permanganate solution for 30 min at 30°C, maintaining a material to liquor ratio of 1:15. Thereafter, the sample was washed for 20 min at 60°C with a reducing agent and a softened, followed by centrifuging and air-drying [38].

### 12.6.6 Dirty effect finishing

A few pigment-based dyes, such as Diresal Orange RDT or Diresal Brown RDT have been used to achieve a dirty effect on the denim fabric. Black pigment dyes have been used to impart a deep shade to the denim fabric. A pigment dye solution can be applied to the fabric by spraying it in a controlled manner. The treated fabric will look careless and fashionably dirty after the treatment [39].

### 12.6.7 Ice wash effect finishing

In this particular finishing, the stones act as vehicles to deposit chemicals on the garments so as to strip the indigo colour from the surface of the denim fabric. The surface chemical deposition will remove the colour from the outer surface of the garment and deliver an attractive elegant, frosted appearance. Indigo and sulphur-dyed denim fabrics are normally applied for the ice wash effect.

### 12.6.8 Wrinkle finish

Resin is used by the denim industry to create a unique and vintage look, e.g., the crinkle effect that can add value to denim garments and improve its value in the market [34]. In 1980s, resin was used to provide nonpress, iron-free cotton trousers; now it is used on denims to create wrinkles and creases to impart a natural vintage look. The fabric gets a natural colour that stays after multiple home washing. A wrinkle, also known as a rhytide, is a fold, ridge or crease in the cloth/garments that could induce a vintage and aged look in the fabric.

### 12.6.9 Antimicrobial and UV protective finishing

Vijayalakshmi and Ramachandran reported the functional finishing of denim garments using two natural oils: sweet citrus and rose oils [68]. The 2/1 right hand twill was made from 100% cotton, cotton/polyester, 100% tencel and tencel/polyester blended fibres. These oils were applied by the exhaust method and the samples were tested for functional properties such as antimicrobial, antifungal and UV protection. From the antimicrobial test, it was inferred that the cotton/polyester and tencel/polyester denim fabrics inhibited the growth of the bacteria greatly more than was done by the pure cotton or tencel fabrics. The finished cotton and tencel/polyester denim fabrics completely inhibited the growth of fungi. As far as UV protection is concerned, the percentage of UV protection factor is excellent in the case of cotton (97.5%) and tencel/polyester (97.9%), followed by cotton/polyester (96.5%) and tencel (94.7%) denim fabrics [68].

## 12.7 Conclusion

Denim finishing is an important textile operation for adding value to denim/jeans fabrics and making them attractive to younger customers, particularly by equipping them with a faded or worn fashion look. Numerous operations exist in the denim industry to accomplish such a need, including sanding, sandblasting, sand washing, snow washing, stonewashing with enzymes, prewashing, brushing, rinsing and bleaching. During denim processing, a part of the surface colour is removed by pumice stones. This process can also be done in the presence of a bleaching agent varying in concentration and added to a fixed amount of pumice stone. Emerging waterless or less water–based technologies, such as laser, plasma, ozone and sandblasting, have also been explored for denim washing from the point of sustainability, and have produced excellent fabric handle. Laser treatment has been found to be a useful tool for good colour fading of denim fabric with proper evenness, and for faster and accurate reproduction of the fabric look by controlling the laser process parameters. The laser action on coloured denim is mainly an interaction with indigo dyes by the thermal effect. The process has three steps: (1) the generation of a computerized map of the fabric area that will be irradiated with different frequencies; (2) control of the laser system, scanning system and conveyor system; and (3) a laser-fading process. The colour fading effect is not only affected by the laser power density but depends considerably on the laser resolution and pixel time. Similar to the laser treatment, the plasma technology can be used to replace conventional stonewashing or enzymatic bio-stoning processes that are used to impart the popular 'worn look' to the denim. Mostly, oxygen, argon and water vapour plasma generated at an atmospheric to a low-pressure condition has been used for colour removal of the denim fabric. Plasma can etch the surface of the indigo-dyed warp yarn besides oxidizing the dye molecules. The extent of etching was found to be more in the oxygen plasma–treated denim fabric than the water vapour plasma-treated sample. On the other hand, although the ozone finishing of denim does not eliminate the use of water completely, it could noticeably reduce the consumption of water, energy, chemicals, enzymes and stones. The rate of fading of indigo dye depends on the portion of the dye in the fibre and its solubility, and diffusion of ozone within the fibre. Sandblasting is a modern technique applied to the specific portions of the indigo-dyed denim apparel using special types of sands that do not cause problems associated with stonewashing. The sandblast fading process is popular in the industry for achieving a worn vintage effect in denim fabric.

# References

[1] Samanta KK, Basak S, Chattopadhyay SK, Gayatri TN. Water-free plasma processing and finishing of apparel textiles. In: Muthu SS, editor. Hand book of sustainable apparel production. CRC Press (Taylor and Francis); 2015. p. 3–37.
[2] Agarwal S. World denim production and consumption report: 2012. November 21, 2016. http://www.denimsandjeans.com/latest-denim-reports/global-denim-market-2012/5968.
[3] Kan CW, Yuen CWM, Cheng CW. Technical study of the effect of $CO_2$ laser surface engraving on the colour properties of denim fabric. Color Technol 2010;126:365–71.
[4] Samanta KK, Basak S, Chattopadhyay SK. Environment friendly textile processing using plasma and UV treatment. In: Muthu SS, editor. Roadmap to sustainable textiles and clothing: eco-friendly raw materials, technologies and processing methods. Springer; 2014. p. 161–201.
[5] Samanta KK, Gayatri TN, Saxena S, Basak S, Chattopadhyay SK, Arputharaj A, Prasad V. Hydrophobic functionalization of cellulosic substrates using atmospheric pressure plasma. Cellul Chem Technol 2016;50(7–8):745–54.
[6] Samanta KK, Jassal M, Agrawal AK. Atmospheric pressure glow discharge plasma and its applications in textile. Indian J Fibre Text Res 2006;31(1):83–98.
[7] Samanta KK, Basak S, Chattopadhyay SK. Potential of ligno-cellulosic and protein fibres in sustainable fashion. In: Muthu SS, Gardetti MA, editors. Sustainable fibres for fashion industry, vol. 2. Berlin: Springer; 2016. p. 61–110.
[8] http://www.fashionmegreen.com/what-are-sustainable-fibres, November 21, 2015.
[9] http://www.naturalfibres2009.org/en/iynf/sustainable.html, November 21, 2015.
[10] Sunder AE, Nalankilli G. Polyfunctional finishes on cotton textiles. Indian J Fibre Text Res 2012;37:364–71.
[11] Basak S, Samanta KK, Chattopadhyay SK, Das S, Narkar R, Desouza C, Saikh AH. Flame retardant and antimicrobial ligno-cellulosic fabric using sodium metasilicate nonahydrate. J Sci Ind Res 2014;73:601–6.
[12] Basak S, Samanta KK, Chattopadhyay SK, Das S, Narkar R, Desouza C, Saikh AH. Flame retardant and antimicrobial jute textile using sodium metasilicate nonahydrate. Pol J Chem Technol 2014;16(2):106–13.
[13] Samanta KK, Basak S, Chattopadhyay SK. Sustainable dyeing and finishing of textiles using natural ingredients and water-free technologies. In: Muthu SS, editor. Handbook of textiles and clothing sustainability, vol. 3. Springer; 2017. p. 99–131.
[14] Samanta KK, Gayatri TN, Saxena S, Basak S, Chattopadhyay SK, Arputharaj A. Effect of plasma treatment on physico-chemical properties of cotton. Int J Eng Res Technol 2014;3(3):2467–77.
[15] Samanta KK, Basak S, Chattopadhyay SK. Speciality chemical finishes for sustainable luxurious textiles. In: Muthu SS, Gardetti MA, editors. Handbook of sustainable luxury textiles and fashion. 2015. p. 145–84.
[16] Samanta KK, Basak S, Chattopadhyay SK. Sustainable UV protective apparel textile. In: Muthu SS, editor. Handbook of sustainable apparel production. CRC Press (Taylor and Francis); 2015. p. 113–37.
[17] Basak S, Samanta KK, Saxena S, Chattopadhyay SK, Narkar R, Mahangade R. Flame retardant cellulosic textile using banana pseudostem sap. Int J Cloth Sci Technol 2015;27(2):247–61.
[18] Basak S, Samanta KK, Saxena S, Chattopadhyay SK, Narkar R, Mahangade R, Hadge GB. Flame resistant cellulosic substrates using banana pseudostem sap. Pol J Chem Technol 2015;17(1):123–33.

[19] Basak S, Samanta KK, Chattopadhyay SK, Narkar R. Self-extinguishable lingo-cellulosic fabric using banana pseudostem sap. Curr Sci 2015;108(3):372–83.
[20] Basak S, Samanta KK, Chattopadhyay SK, Narkar R. Thermally stable cellulosic paper made using banana pseudostem sap, a wasted by-product. Cellulose 2015;22(4):2767–76.
[21] Basak S, Samanta KK, Chattopadhyay SK, Pandit P, Maiti S. Green fire retardant finishing and combined dyeing of proteinous wool fabric. Color Technol 2016;132(2):135–43.
[22] Basak S, Patil PG, Shaikh AJ, Samanta KK. Green coconut shell extract and boric acid: new formulation for making thermal stable cellulosic paper. J Chem Technol Biotechnol 2016;91(11):2871–81.
[23] Basak S, Samanta KK, Chattopadhyay SK. Fire retardant property of cotton fabric treated with herbal extract. J Text Inst 2014;106(12):1338–47.
[24] https://www.acs.org/content/acs/en/pressroom/newsreleases/2012/june/sustainable-denim-manufacturing-process-creates-green-jeans.html November 11, 2016.
[25] http://www.denimsandjeans.com/denim-fabric-developments/denim-processing-finishing-in-environmentally-friendly-context/1865 November 11, 2016.
[26] Ozguney AT, Ozcelik G, Ozkaya K. A study on specifying the effect of laser fading process on the colour and mechanical properties of the denim fabrics. Tekstil ve Konfeksiyon 2009;2:133–8.
[27] Khalil E. Sustainable and ecological finishing technology for denim jeans. AASCIT Commun 2015;2(5):159–63.
[28] Sumithra M, Raja NV. Mosquito repellency finishes in blended denim fabrics. Int J Pharm Life Sci 2012;3(4):1614–6.
[29] Ondogan Z, Pamuk O, Ondogan EN, Ozguney A. Improving the appearance of all textile products from clothing to home textile using laser technology. Opt Laser Technol 2005;37:631–7.
[30] Khan MdMR, Mondal MdIH. Physico-mechanical properties of finished denim garment by stone-bleach treatment. J Chem Eng 2013;28(1):36–40.
[31] Ghoranneviss M, Bahareh M, Shahidi S, Anvari A, Rashidi A. Decolourization of denim fabrics with cold plasmas in the presence of magnetic field. Plasma Process Polym 2006;3:316–21.
[32] Arputharaj A, Saxena S, Raja ASM, Kawlekar SR, Patil PG, Samanta KK. Development of atmospheric pressure plasma assisted process for denim materials. In: Proceedings of international conference on industrial textiles – products, applications and prospects, PSG college of technology, Coimbatore, India. August 26–27, 2016. p. 48.
[33] Hung ON, Song LJ, Chan CK, Kan CW, Yuen CWM. Using artificial neural network to predict colour properties of laser-treated 100% cotton fabric. Fiber Polym 2011;12(8):1069–76.
[34] Khalil E, Islam M. Wrinkle finish on denim by resin treatment: a review. AASCIT Commun 2015;2(3):82–7.
[35] Sumithra M. Nanoencapsulaed finishes for physical and comfort properties for durable 100% cotton denim fabrics. Int J Arts Humanit 2015;3(3):055–62.
[36] Gabr BG, Morgan NN. Efficiency of plasma onto denim faded garments. Int J Sci Res Eng Technol 2016;5(2):84–90.
[37] Cheung HF, Kan CW, Yuen CWM, Yip J, Law MC. Colour fading of textile fabric by plasma treatment. J Text 2013;2013:1–4. http://dx.doi.org/10.1155/2013/214706.
[38] Dekanic T, Punic T, Soljacic I. Influence of special finishes on denim properties. Indian J Fibre Text Res 2015;40:170–4.
[39] Denim book, cotton to fashion, ARCHROMA life enhanced, English, global. 2014. http://www.advanceddenim.archroma.com.

[40] Maryan AS, Montazer M. One pot denim washing and finishing using organo-montmorillonite: introducing nano mineral washing and finishing. Text Res J 2015;85(1):91–100.
[41] Aly AS, Moustafa AB, Hebeish A. Bio-technological treatment of cellulosic textiles. J Clean Prod 2004;12:697–705.
[42] Cavaco-Paulo A. Mechanism of cellulase action in textile processes. Carbohydr Polym 1998;37:273–7.
[43] Schindler WD, Hauser PJ, editors. Finishing with enzymes: bio-finishes for cellulose in chemical finishing of textiles. Cambridge (England): CRC Press; 2004. p. 181–7.
[44] Mazumder S. Effects of sand blasting with industrial enzyme silicon wash on denim apparel characteristics. Daffodil Int Univ J Sci Technol 2010;5(1):6–9.
[45] Montazer M, Maryan AS. Application of laccases with cellulases on denim for clean effluent and repeatable biowashing. J Appl Polym Sci 2008;110:3121–9.
[46] Pazarlioğlu NK, Sariisik M, Telefoncu A. Treating denim fabrics with immobilized commercial cellulases. Process Biochem 2015;40(2):767–71.
[47] Md.M. Hasan, Daffodil International University, http://www.slideshare.net/mhmudul/sand-blasting-washing-process-for-denim-garments-55638522 November 21, 2016.
[48] Khan H, Ahmad N, Yasar A, Shahid R. Advanced oxidative decolourization of red Cl-5B: effects of dye concentration, process optimization and reaction kinetics. Pol J Environ Stud 2010;19(1):83–92.
[49] Radetić M, Jovančić P, Puač N, Petrović ZL, Šaponjić Z. Plasma-induced decolourization of indigo-dyed denim fabrics related to mechanical properties and fiber surface morphology. Text Res J 2009;79(6):558–65.
[50] Samanta KK, Jassal M, Agrawal AK. Atmospheric pressure plasma polymerization of 1,3-butadiene for hydrophobic finishing of textile substrates. J Phys Conf Ser 2010;208:012098.
[51] Samanta KK, Jassal M, Agrawal AK. Improvement in water and oil absorbency of textile substrate by atmospheric pressure cold plasma treatment. Surf Coat Technol 2009;203:1336–42.
[52] Samanta KK, Jassal M, Agrawal AK. Antistatic effect of atmospheric pressure glow discharge cold plasma treatment on textile substrates. Fibers Polym 2010;11(3):431–7.
[53] Samanta KK, Joshi AG, Jassal M, Agrawal AK. Study of hydrophobic finishing of cellulosic substrate using He/1, 3-butadiene plasma at atmospheric pressure. Surf Coat Technol 2012;213:65–76.
[54] Teli MD, Samanta KK, Pandit P, Basak S, Chattopadhyay SK. Low-temperature dyeing of silk fabric using atmospheric pressure helium/nitrogen plasma. Fibres Polym 2015;16(11):2375–83.
[55] Samanta KK, Jassal M, Agrawal AK. Formation of nano-sized channels on polymeric substrates using atmospheric pressure glow discharge cold plasma. Nanotrends A J Nanotechnol Appl 2008;4(1):71–5.
[56] Garment washing techniques for cotton apparel, [Technical Bulletin, North Carolina, Cotton incorporated].
[57] Ghoranneviss M, Shahidi S, Moazzenchi B, Anvari A, Rashidi A, Hosseini H. Comparison between decolourization of denim fabrics with oxygen and argon glow discharge. Surf Coat Tech 2007;201:4926–30.
[58] Kan CW, Yuen CWM. Effect of atmospheric pressure plasma treatment on the desizing and subsequent colour fading process of cotton denim fabric. Color Technol 2012;128:356–63.
[59] https://www.dowcorning.com/content/publishedlit/26-1715-01, October 25, 2016.
[60] Basak S, Samanta KK, Chattopadhyay SK, Narkar R. Development of dual hydrophilic/hydrophobic wool fabric by Q172 NM VUV irradiation. J Sci Ind Res 2016;75:439–43.

[61] Basak S. Process development of wool fabric by 172 nm VUV excimer lamp [M.Tech. thesis]. India: Indian Institute of Technology (IIT)-Delhi; 2009.
[62] Sparavigna A. Laser treatment of textiles. J Mater Sci Technol 1993;68:242–7.
[63] Nourbakhsh S, Ashjaran A. Laser treatment of cotton fabric for durable antibacterial properties of silver nanoparticles. Materials 2012;5:1247–57.
[64] Dascalu T, Acosta-Ortiz SE, Ortiz-Morales M, Compean I. Removal of the indigo colour by laser beam-denim interaction. Opt Lasers Eng 2000;34:179–89.
[65] Hung ON, Chan CK, Kan CW, Yuen CWM, Song LJ. Artificial neural network approach for predicting colour properties of laser-treated denim fabrics. Fiber Polym 2014;15(6):1330–6.
[66] Kan CW. Colour fading effect of indigo-dyed cotton denim fabric by $CO_2$ laser. Fiber Polym 2014;15(2):426–9.
[67] Ortiz-Morales M, Poterasu M, Acosta-Ortiz SE, Compean I, Hernandez-Alvarado MR. A comparison between characteristics of various laser-based denim fading processes. Opt Lasers Eng 2003;39:15–24.
[68] Vijayalakshmi D, Ramachandran T. Application of natural oil on light weight denim garment and analysis of its multi-functional performance. Indian J Fibre Text Res 2013;38:309–12.

# Index

'*Note*: Page numbers followed by "f" indicate figures and "t" indicate tables.'

## G



## H



## I

**Z**

FSC
www.fsc.org
MIX
Papier | Fördert
gute Waldnutzung
FSC® C083411

Kolibri 360 GmbH
Ferdinand-Jühlke-Straße 7
99095 Erfurt, Deutschland
produktsicherheit@kolibri360.de